TRENDS IN THEORETICAL PHYSICS

Volume I

TRENDS IN THEORETICAL PHYSICS

Volume I

Based on the 1988–89 Distinguished-Speaker Colloquium Series of the Theoretical Physics Institute at the University of Minnesota.

EDITED BY

P.J. Ellis
and
Y.C. Tang

Addison-Wesley Publishing Company
The Advanced Book Program
Redwood City, California • Menlo Park, California • Reading, Massachusetts
New York • Don Mills, Ontario • Wokingham, United Kingdom • Amsterdam
Bonn • Sydney • Singapore • Tokyo • Madrid • San Juan

Publisher: *Allan M. Wylde*
Production Manager: *Jan V. Benes*
Electronic Production Consultant: *Lori Pickert*
Promotions Manager: *Laura Likely*
Cover Design: *Iva Frank*

Library of Congress Cataloging-in-Publication Data

Trends in theoretical physics / [edited by] Paul J. Ellis. Y. C. Tang.
p. cm.
Papers presented at a colloquium series started in the fall of 1988 at the School of Physics and Astronomy of the University of Minnesota.
Includes bibliographical references (p.).
1. Mathematical physics. 2. Astrophysics. 3. Nuclear physics.
I. Ellis, Paul J. II. Tang, Y. C.
QC20.T69 1990 530.1—dc20 90-28373
ISBN 0-201-50393-X

ABCDEFGHIJ-MA-89

Dedicated to William I. Fine
in appreciation of his central role
in the establishment of the
Theoretical Physics Institute
at the University of Minnesota.

Foreword

The formal announcement, on January 28, 1987, of the creation of a new Theoretical Physics Institute at the University of Minnesota, revealed very little of how the Institute came into being. The story really began more than four decades ago, when William I. Fine first got excited about physics when reading the *World Book Encyclopedias.* Although his career took him through legal work as assistant district attorney in Dallas, Texas, judge advocate trial lawyer for the U.S.A.F. and a private attorney to real estate development, his interest in physics continued, and his reading in the field kept him well abreast of the advances of the last two decades. We first met at a party in the early 80's, and soon got past the usual chit-chat about quarks and the "large picture" in the unification of all forces. I found Bill Fine extremely well-informed about what was going on in a variety of areas in physics, so it was a pleasure to get together from time to time and talk about the field. On one occasion Bill asked me how a committed layman could contribute to physics. After some conversations we came up with the idea of starting a Theoretical Physics Institute. From that moment on, Bill's devotion to the project never flagged, even when the response of potential corporate donors was very discouraging. I recall consoling ourselves with the thought that things could be worse, as described by an anecdote told by a friend who visited a major university. At a party following his colloquium, he met a visiting Japanese scholar, who turned out to be a Chaucer specialist. When asked whether there was much interest in Chaucer in Japan, he answered, "none whatsoever, just like here".

In 1985 I introduced Bill to Gloria Lubkin, editor of *Physics Today*, who became a very involved and essential advisor and resource for the project, and is now on the steering board. It was she who led us to broaden our concept from a narrow concentration on particle physics, and it was she who suggested that Leo Kadanoff be brought in as advisor and consultant.

This group of people, together with Allen Goldman, Charles Campbell, then Head of the School of Physics and Astronomy at the University of Minnesota, and Marvin Marshak, current Head of the School began working on involving the university in a more massive way than was originally contemplated. President Kenneth Keller of the University of Minnesota agreed to match an outside fund both for the creation of two chairs in Theoretical Physics, and with a significant budget line permanent fund. It was Bill Fine who provided the outside funding necessary to trigger the creation of the TPI, and it is his remarkable generosity that is responsible for the existence of the TPI.

Several years ago a friend lent me a book of essays and reviews written by Ludwig Boltzmann. One of the essays entitled "Journey to Eldorado" described his visit to the University of California at Berkeley in 1905. One of the experiences that he remarked on was visiting Stanford University, which had been endowed by Leland Stanford's widow. Boltzmann writes how remarkable that seemed to him; at home, he said, he would have expected the founding of a hospice for abandoned kittens. The tradition of private funds being used to create centers of learning, be they universities, libraries, telescopes or institutes is an old one in the United States, but it is not always that it is rooted in the passionate interest of an individual in a field so far removed from practical pay-offs. Bill Fine continues to be deeply involved in all aspects of the activities of TPI. I hope that he is having fun getting to talk with a remarkable group of visitors, among whom are the authors of the articles in this volume. He and they share a curiosity and a devotion to understanding how the universe ticks.

Stephen Gasiorowicz
University of Minnesota
August, 1989

Preface

The Theoretical Physics Institute was established at the School of Physics and Astronomy of the University of Minnesota in January of 1987. As part of the program of the Institute, a colloquium series on Trends in Theoretical Physics was started in the Fall of 1988. Distinguished speakers were invited here to give colloquia on a wide range of topics in theoretical physics. These colloquia were very well received by the audience. Consequently, it was decided to collect manuscripts from the speakers and publish a book for wide circulation. In making this decision, we were also influenced by the fact that there seems to be no journal which deals with the whole field of theoretical physics at a level accessible to the non-specialist.

The topics are entirely the choice of the speakers. Our only request to them was that the talk should represent an overview of the topic, supplemented by their personal viewpoints and physical insights. As will be evident to the reader, a number of the speakers have been able to provide a more detailed written account than was possible in a one-hour colloquium. It is our hope that the book will be useful to all members of the physics community, experimentalists and theorists alike.

Since it is our feeling that physics should not be overly compartmentalized into subfields, we have not grouped the manuscripts in this way. We hope that the random order will assist in the cross-fertilization of ideas among subfields, which is so beneficial in understanding how nature works. Of course, many topics in theoretical physics could not be covered in this volume, but we plan to fill in some of the gaps with a second volume next year.

We would like to express our gratitude to many people; first, of course, to the speakers themselves and second, to our colleagues in the School of Physics and Astronomy. Among the latter, we especially wish to mention S. Gasiorowicz and L. McLerran for their encouragement to proceed with the idea of publishing the book, and J. Broadhurst for his technical assistance with the text editing system LaTeX used to produce this volume. We also thank Lori McWilliam Pickert (Archetype Publishing Inc.) for her invaluable help with the Addison-Wesley sei style. Several of the manuscripts were typed by Sandy Smith and we thank her for cheerfully carrying out this task. Last, but certainly not least, we acknowledge our gratitude to Allan Wylde for agreeing to publish this book and for the assistance that he and others at Addison-Wesley have provided in this endeavor.

P. J. Ellis
Y. C. Tang

University of Minnesota
August 1989

CONTENTS

1

La Grande Illusion: Aspects of the Non-Relativistic Description of the Three–Nucleon System

F. S. Levin
Physics Department
Brown University
Providence, Rhode Island 02912

Although a complete understanding of the three-nucleon system has not been attained, much has been accomplished: this system is rich in experimental and theoretical achievements. This chapter reviews a selected set of experimental data and the successes and failures of the non-relativistic, nucleons-only, quantal description of them. Instances are noted where relativistic effects must be included. The review is intended to be accessible by the non-expert, so that the treatment of material is qualitative with the stress on gaining a physical understanding of the phenomena. Many surprising theoretical results have been discovered over the years by researchers studying the three- nucleon system and a number of them are described. Included among these are: the need for a theoretical framework such as the Faddeev equations; the Efimov effect; the influence of the non-measurable, deuteron D-state probability; the non-existence of the conventionally-defined, proton-deuteron scattering length; the phenomenon of scaling; the failure, in the proton-deuteron case, of a standard approximation used in analyses of charged-projectile collisions with nuclei; the apparent failure of the method of stationary phase and nonetheless its apparent utility in describing breakup collisions; and the successful use of two-body dynamics to explain phenomena that previously only the very complicated, Faddeev-type of three-body dynamics could account for.

1.1 Introduction

This review is written mainly for the non–expert in the area of few–nucleon physics. Its stress is on qualitative aspects of both the few nucleon systems and the theoretical frameworks used to describe them. An important feature in this regard is a discussion of how well data are fitted (and sometimes how well they are measured); this includes comments on the nature of the assumptions that go into the theoretical calculations. An equally important feature is a recounting of some of the theoretical surprises that have been encountered in the almost three-decades of analyses based on Faddeev–type equations. Indeed, the failure of textbook scattering theory and the introduction of the Faddeev equations themselves can be considered the first surprise in the area of three–particle theory. These surprises as well as the current set of large and small failures to fit data combine to make this corner of physics a fascinating, frustrating and piquant area of research to report on.

Not all topics in few nucleon physics are treated herein, hence the eclectic nature of this review: it is a biased sampling, with the particular topics chosen to illustrate some of what was understood in early 1989 about few–nucleon observables and why the theory has been so delightful a theater in which to work. The framework chosen for this review has two components: non–relativistic quantum mechanics and the assumption that a description in terms of neutrons and protons is adequate. One aim of the review will be to assess not only the appropriateness of these two components but whether such a description is illusory and why. Another will be to convey to the non–specialist a sense of why research in this area is as exciting and alive as in any other.

Although neutrons and protons are assumed to be the basic constituents, it is often useful to regard them as two different charge states of a more fundamental constituent, the nucleon (spin 1/2, mass equal to an atomic mass unit, etc.). There are then two well studied versions of the three–*nucleon* system: one consists of two neutrons and a proton, denoted nnp; the other contains two protons and a neutron, denoted ppn [1]. Of the pairs that can be formed from the preceding triplets, only the neutron–proton system can exist in a particle–stable bound state, viz, the deuteron, $^2H \equiv d = (n+p)$. The bound tri-nucleons $^3H = (n+n+p)$ and $^3He = (p+p+n)$ are then the least complex nuclei after the deuteron.

Since ^{3}H and ^{3}He have no bound excited states, then all the tri-nucleon excited states are in the continuum. For nnp they are of two forms: either $n+d$ or $n+n+p$. Similarly, the continuum states of the ppn system are $p+d$ and $p+p+n$. Note that n+d and p+d always refer to a free nucleon and the ground state of the deuteron, since the latter nucleus also has no particle–stable excited states.

Three–nucleon continuum states occur in nuclear reactions, for example:

$$\begin{aligned} n + d &\rightarrow n + d \\ &\rightarrow n + n + p, \end{aligned}$$

and also in processes involving non–hadronic probes:

$$\begin{aligned} \gamma + {}^3He &\rightarrow p + d \\ &\rightarrow p + p + n, \end{aligned}$$

or

$$e + {}^3H \rightarrow e + n + n + p.$$

In contrast to the last process, *elastic* electron scattering is a means of investigating electromagnetic properties of the tri-nucleon *bound* states.

The three–nucleon system has long been studied, and a great deal is known about it, both experimentally and theoretically. A major theoretical breakthrough occurred in the 1960's, when Faddeev–type theories of three–particle collisions were successfully used to interpret experimental data [2], even though in order to circumvent then computational difficulties most of these early analyses were based on simplified and/or unrealistic forms of the nucleon–nucleon interaction or their corresponding transition operators (defined in Sec. 1.2 below). More recently the efforts of many workers have finally led to fully realistic interactions being employed in calculations. These latter calculations are highly non–trivial, and it was believed for some time that only through such complex three–body dynamics could the three nucleon system be understood. It has very recently become apparent, however, that at least for the lower energies and, to date, excluding three–particle (breakup) continuum states, one can use *two–body* dynamics to understand important features of this system.

This generally unexpected progression from complex to simpler dynamics is one of the most recent examples of a variety of surprising and stimulating theoretical developments in this field. A number of them will be chronicled in this review of selected aspects of the three–nucleon system. As noted in the preceding, a major emphasis will be on the successes and failures of the non–relativistic, nucleons–only description. Pions (and other mesons) are thus not considered to be active ingredients in the three–nucleon system, although they will enter the description indirectly, for example, in meson exchange current contributions to electromagnetic processes. Furthermore, they are fundamental in modern formulations of the nucleon–nucleon interaction. By suppressing the mesonic degrees of freedom, the nucleons–only description is thus one involving effective inter-

TABLE 1.1
Observed or Inferred Properties of the np, nnp and ppn Systems

Bound States	Continuum States
Binding Energies	Angular Distributions
Radii	Cross Sections
Asymptotic Normalization Constants	Scattering lengths
Electromagnetic Moments	Effective ranges
Form Factors/Densities	(Mixing parameters)
	Polarizations/Asymmetries
	Vector
	Tensor
	Spin transfer coefficients

actions, even when an interaction between all three nucleons is included in the description. Such three–body forces have been introduced in an effort to close the gap between some of the experimentally determined three–nucleon observables and the corresponding theoretical values obtained from calculations employing only nucleon–nucleon interactions, as will be discussed below.

The nucleons–only description is not the only possibility; an alternative is the one studied by Sauer and collaborators [3], employing nucleons, pions and deltas (the 1232 MeV excited state of the nucleon). Some results of this approach will be compared with those from the nucleons–only description in Subsec. 1.3.1 below. In addition to these two, a third choice of fundamental constituents on which to base a theoretical analysis is that of quarks. Unfortunately, a quark (and gluon) description of even the two–nucleon system, much less the three-nucleon system, is at present an intractable problem if quantum chromodynamics is chosen as the theoretical framework. The situation does not improve if an empirical bag–type model is used in place of QCD [4]. It therefore appears that a quark–based approach to the three–nucleon system which is capable of accounting for the wide range of existing experimental data lies in the realm of science fiction: a scenario that may be fulfilled sometime in the (perhaps distant) future. On the other hand, since so much has been achieved from the non–relativistic, nucleons–only approach, it is an approach worth reviewing. Only by delineating its accomplishments and limitations will one then be in a position to go beyond it.

The success of any description of the two–nucleon and/or the three–nucleon system is measured in part by how well it predicts the "measurable" properties of the system. Some of these properties can be observed directly, others are inferred. Table 1.1 lists the majority of them for the *np*, *nnp* and *ppn* systems. Relevant units are MeV $= 10^6$ eV for binding energies and fm $= 10^{-13}$ cm for lengths. In the past, many of these properties have been calculated using simplified non–relativistic forces, but, since the emphasis here is on calculations employing what are generally regarded as "realistic" (local) interactions, almost all of the other type of computation will be skipped over or referred to only briefly.

There is still no single form of the two–nucleon interaction which is generally accepted as either the best or the most accurate, nor is it yet evident where a description in terms of only pair–wise interactions becomes truly inadequate. These are some of the features which make the two– and three–nucleon systems so frustrating and so fascinating to deal with. They add both spice and an occasional testiness, examples of which shall be referred to in later sections.

Given the fundamental role played by the two–nucleon interaction, this review begins by recalling some aspects of the theoretical description of two–particle bound and continuum states, in particular those related to relevant measurable properties and those that require a reformulation of the description for three–particle systems. This is done in Sec. 1.2, subsection 1.2.3 of which reviews two–nucleon interactions. Readers familiar with the Lippmann-Schwinger equation, transition operators, scattering lengths, etc., could skip directly to this portion of Sec. 1.2. After this introduction to basic concepts, the main subject of this review, viz, the three-nucleon system, is explored in four parts in Sec. 1.3. Its first subsection contains a brief description of the Faddeev formulation of three-particle collision theory. Bound states and then continuum states are the topic, respectively, of each of the next two subsections, following which is a selected review of many of the simplifications that have been discovered in this field over the years. Sec. 1.4 considers illusions.

1.2 The Two–Nucleon System

In a non–relativistic description of a quantal system, the Hamiltonian consists of two types of terms: the kinetic energy operators and the interactions (which may contain operators as well as multiplicative functions). For the two–nucleon system, the only interaction is the potential between the two nucleons. This potential depends not only on which two nucleons form the system (i.e., *nn*, *np* or *pp*) but also on their spins. That is, the spin singlet and spin triplet potentials differ. Since the neutron and the proton can be considered as two different charge states of the nucleon, a two–component spinor representation can be introduced for the nucleon, in analogy to the Pauli matrices and spinors for spin 1/2. The two–component representation for the nucleon is referred to as the isobaric spin or isospin description. Just as with ordinary spin, two nucleons can be in states of total isospin 0 or 1, i.e., in a singlet or triplet isospin state [5].

In terms of this language, the nucleon–nucleon interaction is both spin and isospin dependent. It can be constructed of operators that depend on the isospins, spins, orbital angular momenta, momenta, and positions of the two nucleons, such that when acting on the two–nucleon state vectors the two-nucleon potential turns into the nn, np or pp interactions as appropriate.

The advantage of the isospin description is its universality: the same interaction is used for any pair of nucleons, their spin and isospin quantum numbers being the means used to specify whether the interaction takes the nn, np or pp form. Furthermore, since nucleons are identical fermions, an N–nucleon system will be governed by an N–nucleon Hamiltonian $H^{(N)} = H(1,2\ldots N)$ which is symmetric in the labels of the N nucleons and contains the same two–nucleon interaction as does $H^{(2)}$. It is due to this latter fact that the two–nucleon system plays such an important role in all attempts to understand the three–nucleon system.

A comparison of $H^{(2)}$ and $H^{(3)}$ may be helpful at this point. In the barycentric system, $H^{(2)}$ takes the form

$$H^{(2)} = H_0^{(2)} + V_{12}, \tag{1.1}$$

where the free Hamiltonian $H_0^{(2)}$ is given by

$$H_0^{(2)} = \frac{p_{12}^2}{2\mu_{12}}, \tag{1.2}$$

with $\vec{p}_{12}$ the relative motion momentum operator for the pair 12 and μ_{12} the reduced mass of the pair; V_{12} is the two–nucleon interaction.

If only pairwise interactions occur in the three–nucleon system, the corresponding (barycentric) $H^{(3)}$ is given by

$$H^{(3)} = H_0^{(3)} + V_{12} + V_{13} + V_{23}, \tag{1.3}$$

where the free Hamiltonian $H_0^{(3)}$ is a sum of two terms, e.g.,

$$H_0^{(3)} = \frac{p_{ij}^2}{2\mu_{ij}} + \frac{p_k^2}{2\nu_k}. \tag{1.4}$$

Here, $p_{ij}^2/2\mu_{ij}$ is the kinetic energy operator for the pair ij (recall Eq. (1.2)), while p_k is the momentum operator for the third nucleon (k) relative to the CM of the pair ij, with ν_k being the relevant reduced mass for nucleon

TABLE 1.2
Bound State Properties

	J^π	B_d(MeV)	$\langle r^2\rangle^{1/2}$(fm)	$\mu(\mu_N)$	Q(fm^2)	A_S	η
p	$1/2^+$		0.8	2.79			
n	$1/2^+$		0.36	-1.91			
d	1^+	2.2246	2.09	0.857	0.286	.8846	.0271

k [6]. V_{ij} is the two–nucleon interaction between the pair of nucleons i and j, equal to the V_{12} of Eq. (1.1) when $i = 1$ and $j = 2$.

For any reasonably–behaved V_{ij} the Schrödinger equations involving $H^{(2)}$ and $H^{(3)}$ can each be solved to arbitrary numerical accuracy for both bound and continuum states. Thus, given a V_{12} that reproduces the two–nucleon observables, it is only a technical matter to calculate the three–nucleon observables. In this context, V_{ij} plays a passive role. However, there is no universally–agreed–upon form for V_{12}, nor do *any* of the various "realistic" V_{12}'s fit *all* of the two–nucleon observables. The quest for the nuclear force is thus far from over; this search, such as it is, represents an active role played by V_{12} in Eq. (1.1). How well $H^{(3)}$ and thus different V_{ij}'s do in reproducing three–nucleon observables will be discussed in Sec. 1.3. The present section, however, deals only with aspects of the two–particle system, e.g. : various V_{12}'s; how well measurables are fitted by the different potentials and how they compare with one another; the various formalisms, but only to the extent needed either to clarify/illustrate aspects of the two–particle tapestry or to provide background for the main three–particle developments in Sec. 1.3; and the observables that can be measured, either directly or else deduced.

1.2.1 Bound States

Table 1.2 lists most of the bound state properties of the deuteron, along with the corresponding ones, where they exist, for the neutron and proton [7]. The most significant of these is the existence of the non–zero deuteron quadrupole moment Q, since it implies that the deuteron ground state contains in addition to the expected $L = 0$ or S–state, an admixture of an $L = 2$ or D–state [8]. That is, if there were only an $L = 0$ component, then Q would be zero. On the other hand, components with $L > 2$ are ruled out by angular momentum conservation. The combination of $J = 1$ and $L = 0$ means that the deuteron is a spin triplet, i.e., $S_d = 1$. Because the magnetic moment $\mu_d \cong \mu_n + \mu_p$, it can be shown that the S state is predominant in the deuteron [8]. This combination of S and D orbital angular momentum components in the deuteron ground state, i.e., its being a $^3S_1 +^3 D_1$ mixture, means that in the triplet spin state the nucleon–nucleon interaction contains a non–central portion. This non–central or tensor force plays an important role in determining the binding energy of both the two– and the three–nucleon systems. Its contribution varies from one two–nucleon potential to another.

The values of A_s, the asymptotic S-state norm, and η, the ratio of asymptotic D- to S-state normalization constants, in Table 1.2 can be understood on inspection of the deuteron wave function, denoted ψ_d. It obeys

$$(B_d + H_0^{(2)} + V_{12})\psi_d = 0, \tag{1.5}$$

and can be decomposed as

$$\psi_d = \psi_{L=0} + \psi_{L=2}, \tag{1.6}$$

in an obvious notation. The dependence of these wave functions on space, spin, and isospin degrees of freedom can be expressed as follows:

$$\psi_{L=0} \equiv \frac{u(r)}{r}\mathcal{Y}_{L=0}\,(\text{angles, spin, isospin}) \tag{1.7}$$

and

$$\psi_{L=2} \equiv \frac{w(r)}{r}\mathcal{Y}_{L=2}\,(\text{angles, spin, isospin})\,. \tag{1.8}$$

If α denotes the deuteron bound–state wave number ($B_d \propto \alpha^2$), then it can be shown that asymptotically,

$$u(r) \underset{r\to\infty}{\sim} A_S e^{-\alpha r} \tag{1.9}$$

and

$$w(r) \sim \eta A_S e^{-\alpha r}, \tag{1.10}$$

where $A_D = \eta A_S$ is the deuteron's asymptotic D–state normalization constant. Both η and A_S can be determined experimentally (methods for doing this are referred to, e.g., by Ericson [9]). In contrast to A_S and η, the S–state and D–state probabilities P_S and P_D, given by

$$P_S = \int_0^\infty dr\; u^2(r) \tag{1.11}$$

and

$$P_D = \int_0^\infty dr\; w^2(r)\,, \tag{1.12}$$

with $P_S + P_D = 1$, are not measureable [10]. Hence, though different V_{12}'s tend to predict different P_D's, P_D cannot be used to fix any parameters in V_{12}.

1.2.2 Continuum States

Although the two–nucleon interaction is spin–dependent, the spin (and isospin) degrees of freedom will be suppressed in most of this section, thus allowing for a concentration on fundamentals. The result is simplicity without loss of any of the relevant physics.

In the continuum case, the boundary condition on the Schrödinger wave function $\psi_{\vec{k}}(\vec{r})$ is that it be asymptotic to a plane wave plus an outgoing spherical wave:

$$(2\pi)^{3/2}\psi_{\vec{k}}(\vec{r}) \sim e^{i\vec{k}\cdot\vec{r}} + f_k(\theta)e^{ikr}/r, \tag{1.13}$$

where $\vec{k}$ is the incident wave vector, θ is the scattering angle (a spherically symmetric potential is assumed), and $f_k(\theta)$ is the scattering amplitude. The scattering amplitude provides the link between theory and experiment. Let $N(\theta)$ be the number of particles that are scattered from the incident beam into $d\Omega$ at a solid angle Ω. $N(\theta)$ is proportional to the differential cross section or angular distribution $d\sigma/d\Omega$, which itself is related to $|f_k(\theta)|$ [11]:

$$N(\theta) \propto \frac{d\sigma}{d\Omega} \equiv |f_k(\theta)|^2.$$

The continuum solution to the Schrödinger equation can be obtained by solving the following Lippmann–Schwinger (LS) integral equation [12], which incorporates the outgoing wave boundary condition of Eq. (1.13):

$$\begin{aligned} \psi_{\vec{k}}(\vec{r}) &= \sqrt{\frac{1}{8\pi^3}}\left[e^{i\vec{k}\cdot\vec{r}} + \int d^3r' \left\{-\frac{m}{2\pi\hbar^2}\frac{e^{ik|\vec{r}-\vec{r}\,'|}}{|\vec{r}-\vec{r}\,'|}V_{12}(r')\right\}\psi_{\vec{k}}(\vec{r}\,')\right] \\ &\equiv \sqrt{\frac{1}{8\pi^3}}\left[e^{i\vec{k}\cdot\vec{r}} + \int d^3r' \left\{K_k(\vec{r},\vec{r}\,')\right\}\psi_{\vec{k}}(\vec{r}\,')\right]. \end{aligned} \tag{1.14}$$

As long as its kernel $K_k(\vec{r},\vec{r}\,')$ is sufficiently well–behaved, the LS equation can be solved via standard numerical techniques. This is ensured if K_k is traceclass, i.e., if $Tr[K_k^2] < \infty$. The traceclass condition is sufficient but not necessary, and it turns out that the kernel of Eq. (1.14) is not traceclass. This lack of a traceclass kernel for Eq. (1.14) is generally not a problem: assuming that V_{12} obeys not unusual mathematical restrictions, then a simple modification of K_k guarantees that it *can* be made traceclass [12]. Thus, in the two–particle case, the LS equation is both a formal and a practical tool. In the three–particle case, however, the kernel of the corresponding LS equation contains a non-removable delta-function singularity and can never be made traceclass. One result of this is that the solutions are not unique [14]. To obtain unique solutions to the three–particle Schrödinger equation requires reformulation of collision theory. The best known as well as the first of the various reformulations is that of Faddeev, which will be discussed in Sec. 1.3. The three–particle Faddeev equations can be expressed in terms of the two–particle transition operators, and before going on to examine some properties of $f_k(\theta)$, it is useful to define the two–particle transition operator $t_{12}(E)$, one of whose momentum–space matrix elements is related to $f_k(\theta)$.

A comparison between Eqs. (1.13) and (1.14) evaluated for $r \to \infty$ yields [11]

$$\begin{aligned} f_k(\theta) &= -\frac{(2\pi)^2 m}{\hbar^2} \int d^3r' \, \frac{e^{-i\vec{k}'\cdot\vec{r}'}}{(2\pi)^{3/2}} \, V_{12}\psi_{\vec{k}}(\vec{r}'), \\ &\equiv -\frac{(2\pi)^2 m}{\hbar^2} < \vec{k}'|V_{12}|\psi_{\vec{k}} >, \end{aligned} \tag{1.15}$$

where $\vec{k}' = k\vec{r}/r$, $k' = k \propto \sqrt{E}$, and $\vec{k}'\cdot\vec{k} = k^2\cos\theta$. The operator $t_{12}(E)$ may be defined via

$$t_{12}(E)|\vec{k} >= V_{12}|\psi_{\vec{k}} >; \tag{1.16}$$

hence, the plane wave matrix elements of $t_{12}(E)$ are proportional to $f_k(\theta)$:

$$f_k(\theta) = -\frac{(2\pi)^2 m}{\hbar^2} < \vec{k}'|t_{12}(E)|\vec{k} > . \tag{1.17}$$

Since $k'^2 = k^2 = \frac{2mE}{\hbar^2}$, $< \vec{k}'|t_{12}(E)|\vec{k} >$ is known as an on–shell (or on–energy–shell) matrix element. Although t_{12} has been introduced via Eq. (1.16), it can also be shown to be the solution of an operator form of LS integral equation [15]. It is then natural to define $< \vec{p}\,|t_{12}(E)|\vec{q} >$ as an off–shell matrix element of t_{12} as long as at least one of p and q is not equal to $\sqrt{2mE/\hbar^2}$. Such matrix elements play an important role in one of the integral forms of the Faddeev equations.

The scattering amplitude $f_k(\theta)$ (and thus $< \vec{k}'|t_{12}(E)|\vec{k} >$) is not an observable. However in the extreme low energy limit its magnitude can be measured, and its phase inferred. The relevant low–energy quantity is the scattering length a, related to the $\ell = 0$ phase shift. The phase shift expansion of $f_k(\theta)$ is [11]

$$f_k(\theta) = \sum_{\ell=0}^{\infty}(2\ell+1)f_\ell(k)P_\ell(\cos\theta), \tag{1.18}$$

where P_ℓ is the ℓth order Legendre polynomial and $f_\ell(k)$ is given by [11]

$$f_\ell(k) = e^{i\delta_\ell}\sin\delta_\ell/k = [kcot\delta_\ell - ik]^{-1}, \tag{1.19}$$

TABLE 1.3
Effective Range Parameters (in fm)

	Spin Triplet		
	a_t	r_t	
	5.424	1.759	
	Spin Singlet		
a_s^{np}	a_s^{nn}	a_s^{pp}(no e.m.)	a_s^{pp}(exp)
-23.748	-18.7	-17.5	-7.8
r_s^{np}	r_s^{nn}	r_s^{pp}(no e.m.)	r_s^{pp}(exp)
2.75	2.80	2.84	2.79

with $\delta_\ell(k)$ being the ℓth–order phase shift.

Of interest is the low energy (small k) limit. It suffices to consider S–waves, viz, $\ell = 0$. As long as $V_{12}(r)$ is short–ranged, then [16]

$$k \cot \delta_0 = -\frac{1}{a} + \frac{1}{2} r_0 k^2 + O(k^4), \tag{1.20}$$

where a is the scattering length and r_0 is the effective range parameter. Note that even for an interaction as "short–ranged" as a polarization potential, proportional to r^{-4}, Eq. (1.20) fails [17]. Its replacement contains a $k \ln k$ term, so that for $V_{12}(r) \propto r^{-4}$, $\lim_{k\to 0}\ k \cot \delta_0$ is infinite. In contrast, the $k \to 0$ limit of Eq. (1.20) is simply $-a^{-1}$ when $V_{12}(r)$ is truly short–ranged. One of the amusing recent developments is that the usual, two-proton, coulomb–modified, strong–interaction scattering length [18], long thought to be a well–determined quantity both theoretically and experimentally for *all* low energy scattering systems, does not exist for the case of proton-deuteron scattering. It too is infinite because of polarization potential effects that had been overlooked for many years. This situation, its consequences, and its resolution is discussed in Sec. 1.3.3.

In the case of the two–nucleon system, a and r_0 are replaced by the spin–triplet and spin–singlet parameters a_t, r_t and a_s, r_s. Only the $n - p$ system can exist in the triplet state, so that the singlet state encompasses the $n - n, n - p$ and $p - p$ systems. Values of these quantities are given in Table 1.3. Note that a_s^{pp} and r_s^{pp} refer to the experimentally determined values while a_s^{pp} (no e.m.) and r_s^{pp} (no e.m.) refer to theoretically deduced values with coulomb effects removed [19]. Were isospin an exact quantum number, i.e., if the interactions were charge independent, then the three

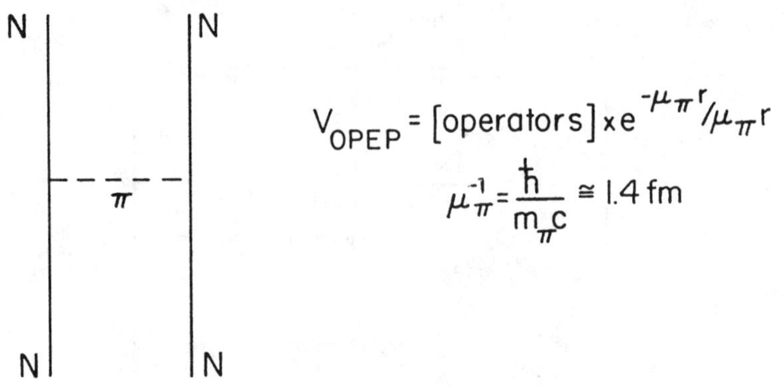

N N

π

N N

$$V_{OPEP} = [\text{operators}] \times e^{-\mu_\pi r} / \mu_\pi r$$

$$\mu_\pi^{-1} = \frac{\hbar}{m_\pi c} \cong 1.4 \text{ fm}$$

FIGURE 1.1
The one-pion exchange diagram and the corresponding (schematic) one-pion exchange potential. Here, N denotes a nucleon.

a_s's would be equal, while if nuclear forces were charge symmetric, then a_s^{nn} and a_s^{pp} would be equal. Discussions of these points can be found, e.g., in Ref. [5]. The negative values of a_s indicate that in the singlet state, V_{12} is attractive but not strong enough to bind, while the positive value of a_t means that when $S = 1$ the np system can and, of course, does bind.

As soon as incident energies are in the range 10–15 MeV and greater, $\ell > 0$ partial waves become important, and there is then a wide variety of measurements that can be and have been carried out, including not only of angular distributions but also of polarizations, asymmetries and polarization transfers. The normal arrangement is to express these quantities in terms of phase shifts and then compare phase shifts determined from a least squares fit to data with those from one or more realistic potentials. In addition to phase shifts, there are mixing parameters to be determined for those cases where the tensor force couples different partial waves. An example is the ${}^3S_1 - {}^3D_1$ pair for the $n - p$ system, for which the mixing parameter is denoted ϵ_1. The value of ϵ_1 is not known to high accuracy and it has been proposed to remeasure it in the hope of better assessing the contentious Bonn potential (more on this in the next subsection and in Sec. 1.3).

1.2.3 Potentials and Fits to Data

Among their other features, "realistic" nucleon–nucleon interactions are local in coordinate space and have the one–pion exchange potential (V_{OPEP}) as their longest range portion. This part of V_{12} has the form V_{OPEP} = [operators]$e^{-\mu_\pi r}/\mu_\pi r$, where the pion Compton wave length $\mu_\pi^{-1} = \hbar/m_\pi c \cong 1.4$ fm (m_π = pion mass); V_{OPEP} corresponds, schematically, to the generic one–pion exchange diagram of Fig. 1.1. This diagram is a representation of the original Yukawa hypothesis that the forces between nucleons arise from the exchange of pions. Since heavier mesons can also be exchanged, e.g., rho's, omega's, etc., they will evidently give rise to shorter range components of the interaction, some of which are attractive and others repulsive (in the same spin–isospin states). Because the $^1S_0$ and $^3S_1$ phase shifts change from positive to negative at a lab energy of around 250 MeV, V_{12} is conventionally taken to be strongly repulsive at very short ranges, so that the probability for finding two nucleons at very small separations is very small or zero, depending on whether the repulsive core is "soft" or "hard". A hard core, i.e., $V = \infty$ for $r \leq r_c$, was an early feature but is now no longer in vogue.

It is, of course, a convenient fiction that V_{12} is local in coordinate space. One source of non–locality is the possibility of creating and annihilating nucleon–antinucleon pairs (vacuum polarization effects) during the interaction of two nucleons. Because the annihilated nucleon could be one of the two initially interacting, the resulting contribution to the potential from this process becomes non–local. Nevertheless, apart from purely separable potentials, which will be considered later, all nucleon–nucleon potentials that are being or have been used are local. However, some of them can be said to be quasi–non–local due to their containing a few momentum–dependent terms. (The equivalence between a non–locality and an exponential dependence on momentum is discussed, e.g., in Refs. [5]. "Quasi–non–local" as used here refers to the occurrence of a quadratic momentum dependence.)

This nomenclature is even more apt than suggested above, since a $\vec{p}^{\,2}$ momentum dependence not only can be regarded as approximating a non–locality, but it directly takes some account of relativity. That is, the lowest–order relativistic correction to a static potential is of order $(v/c)^2 = (p/mc)^2$. Now, on the average, relativistic corrections are expected to be small, since the average relative momentum $\bar{p}$ is in the range 100–200 MeV/c so that $(\bar{p}/mc)^2$ for a nuclear system is only a few percent. However, this need not mean that all relativistic effects are small, for a strong repulsion at small r implies large momenta : such "hidden" relativistic effects may not be negligible.

In addition to locality and the one–pion tail, all realistic nucleon–

nucleon interactions contain a central and a tensor potential, spin–spin and spin–orbit terms, and a quadratic spin–orbit potential of some type [20], as well as an isospin dependence (the $n - p$ force in the deuteron is stronger than the $n - n$ or nuclear part of the $p - p$ force). Only three realistic potentials have been used in tri–nucleon calculations. They are (in alphabetical order) the new Bonn potential [21], the Nijmegen potential [22] and the Paris potential [23], the first and last of which come in several guises. The Bonn potential is the newcomer among them; it is also the somewhat provocative one, having had the effect of re-injecting into the field of three–nucleon physics a special sense of excitement that had been slipping away. Details of this development will be taken up in Sec. 1.3. Analyses of two–nucleon data based on the Bonn and Paris potentials are presented later in this section. In order to understand some of these analyses, it is necessary to consider a particular kind of non–realistic interaction.

There are two types of non–realistic potentials: those that are local but do not contain the full complexity outlined above, and those of separable form. Both types have been used in analyses of two–nucleon and three–nucleon data. Of immediate interest here is the separable model, which is a finite expansion in terms of projectors:

$$V_{sep} = \sum_{\alpha} \lambda_\alpha |g_\alpha >< g_\alpha|, \tag{1.21}$$

where λ_α is the strength of the αth term and g_α is the αth form factor. Although a potential such as given by Eq. (1.21) is unrealistic, it has been of great value. First, it is a simplifier, reducing the two–body LS equations to quadratures and the complicated, two–dimensional Faddeev–type equations that arise in the three–particle problem to coupled, one–dimensional integral equations. The resulting computations are much less tedious. Using relatively few parameters and simple functional forms for $g_\alpha(\vec{k})$ or $g_\alpha(\vec{r})$, separable potentials have been employed in fitting a great deal of three–nucleon data, starting from the pioneering calculations of Mitra and collaborators and of Amado and collaborators [2]. These were crucial, because they clearly established that three–particle collision theories could explain the data, including breakup cross sections. More recently, separable potentials have been used to establish the importance of the small but nevertheless non–negligible P–wave portions of the two–nucleon interaction [24] in fitting polarization–type data.

This simplifying aspect is especially important because a few–term separable expansion often yields an accurate approximation to a local potential [25] or its concomitant t operator. Such separable expansions have been studied and employed for many years. The efforts of the Graz group in approximating the Paris potential is pertinent, since this work [25] pro-

TABLE 1.4
Low Energy Predictions of Some Two-Nucleon Potentials

	Expt.	Bonn	Paris	RSC	V14
B_d (MeV)	2.2246	2.2246	2.2249	2.2246	2.2250
P_D(%)		4.38	5.77	6.47	6.08
Q(fm^2)	0.286	0.274	0.279	0.280	0.286
A_S	0.8846	0.8862	0.8869	0.8776	0.8911
η	0.0271	0.0262	0.0261	0.0262	0.0266
a_t(fm)	5.424	5.424	5.427	5.39	5.45
r_t(fm)	1.759	1.760	1.766	1.720	1.80

vided not only some assessment of the ability of the Paris potential to fit three–nucleon scattering data but also of how well it fitted some recent two nucleon data.

Once the parameters of a potential have been fitted to data, its off– and on–shell momentum–space behavior is determined by fourier transform. Based on the assumption that three–particle forces are negligible, the three–nucleon system had long been thought to be a means for discriminating between different two–nucleon potentials, that is, for fixing the off–shell behavior. For each potential, it was thus (and still is) necessary to fit existing two–nucleon data as accurately as possible and then to predict, again with adequate accuracy, additional data. Many less realistic potentials, in addition to the three realistic ones, have been so employed. Two of these are the older Reid soft core potential (RSC) [26] and the newer Argonne $V14$ potential [27]. In a wryly amusing and somewhat testy conference review, de Swart in 1984 [28] compared the Bonn, Nijmegen, Paris, and $V14$ potentials, remarking not only on the number of free parameters in each but more importantly pointing out that in all of them the long and intermediate range parts of the isospin tensor and spin–orbit interactions are incorrect. While these discrepancies were not large, they were definite and occurred in spite of the fact that the bound–state data and some of the effective range parameters were fitted quite well.

A recent comparison of various low–energy predictions with each other and experiment is given in Table 1.4, the values of which are taken from the article of Brandenburg et al., Ref. [7]. Overall, the agreement with experiment is very good, with a few obvious discrepancies. For the present purposes, the comparison between the different value of P_D, a non–measurable, is significant: the value of P_D for the Bonn potential is noticeably smaller than any of the others, indicating that the Bonn tensor force is weaker than for any of the other interactions. In order to produce the proper binding energy B_d, it is thus necessary to have a stronger triplet central force. The consequence of this for the three–nucleon system is dramatic, as will be seen in Sec. 1.3.

The goodness of fit seen in Table 1.4, at least for the Bonn and Paris potentials, is also evident in Fig. 1.2, wherein various phase shifts and the ϵ_1 parameter are displayed. This figure is taken from the conference report of Plessas and Haidenbauer [29]. The research was not only undertaken to compare the Bonn and Paris potentials, but also to test the Ernst, Shakin and Thaler separable potential expansion [30] to the Paris potential, the results of which are labelled PESTN, where N is the rank or the number of separable terms used in expanding a particular angular momentum portion of the two–nucleon interaction. The two potentials yield very similar phase shifts whose agreement with the empirically determined (the "experimental") phase parameters is good, except for $\delta(^1P_1)$ and ϵ_1. Chulick *et al.* have examined the experimental values of ϵ_1 (and other spin–dependent parameters) at energies up to 325 MeV and have concluded that sufficient uncertainties exist to warrant remeasurements, since the existing data is not sufficiently incompatible with the results of various potentials to discriminate between them [31].

1.3 Three Nucleon Systems

Although the three-nucleon systems are more complex and thus display a much richer set of phenomena than the two-nucleon systems, there are some fundamental similarities. For example, the angular momentum states in ^{3}H and ^{3}He are preponderantly S and D, with the former being dominant. Measured values of $\eta = C_2/C_0$, the ratio of the D to S state asymptotic normalizations, substantiate this (see Table 1.5).

The magnetic moments of ^{3}H and ^{3}He are approximately equal to those of the unlike nucleon, not only suggesting S-state dominance in the $J^\pi = 1/2^+$ ground states, but also that in the isospin zero state, a pair of bound nucleons each having the same orbital angular momentum tends to couple to total angular momentum zero, a feature found in all nuclei. On the other hand, no such qualitative arguments suffice either to understand the binding energies of ^{3}H and ^{3}He or their difference, which is not attributable simply to the coulomb energy due to the two protons in ^{3}He. Detailed calculations, involving the full spin dependence of the nuclear interactions, are required in order to determine $B_{^3\mathrm{H}}$ and $B_{^3\mathrm{He}}$.

The situation for the very low energy continuum states is analogous. In this case, the nucleon-deuteron relative orbital angular momentum is zero,

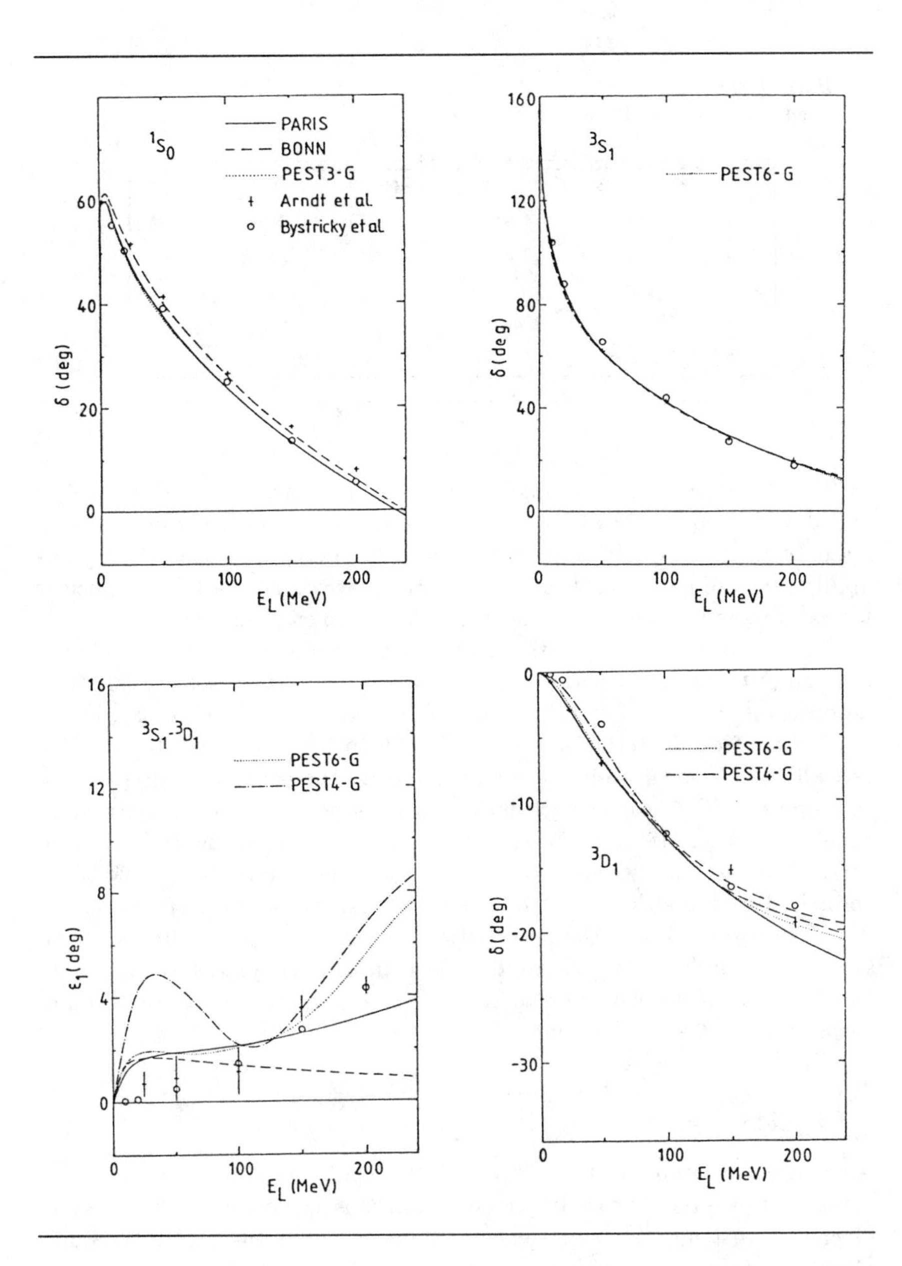

FIGURE 1.2
Various phase shifts and the ε_1 mixing parameter for the Bonn and Paris potentials, calculated using the separable expansion noted in the main text (from Ref. [29]).

TABLE 1.5
Selected Low Energy Three-Nucleon Data

	B_3(MeV)	$\langle r^2\rangle^{1/2}$(fm)	$\mu(\mu_N)$	C_0	η
^{3}H	8.48	1.54	+2.979	1.84	-.045
^{3}He	7.72	1.69	-2.128	1.905	-.048
	$^4a_{nd}$	$^2a_{nd}$	$^4a_{pd}$	$^2a_{pd}$	
	6.35 fm	0.65 fm	11.9 fm?	2.73 fm?	

so that there are two total angular momentum states: $J = 1/2$ (doublet) and $J = 3/2$ (quartet). The latter corresponds to a fully symmetric spin state, so that the spatial state is fully antisymmetric and hence, the probability for finding two nucleons at small separations is small. Since nuclear forces are short-ranged, this means that the quartet scattering length is insensitive to fine details of the nuclear interaction, an implication borne out computationally. On the other hand, the doublet scattering length is an observable that is far more problematic: the small value of $^2a_{nd}$ in the case of neutron-deuteron scattering ($\sim$ 0.6 fm) is very hard to fit theoretically, while for proton-deuteron scattering, the conventionally-defined, coulomb-modified scattering length $^2a_{pd}$ does not even exist, contrary to a belief held for decades. Furthermore, no $p - d$ experiments have been carried out at low enough energies to yield a reliable enough value to be compared with the theoretical one: hence, the question marks in Table 1.5.

These are a few of the pecularities associated with the three-nucleon systems, a number of which are described in this section. Since they tend to arise in a theoretical context, Subsec. 1.3.1 is devoted to a discussion of theory, relevant and irrelevant.

1.3.1 Theory

Although the bound states of ^{3}H and ^{3}He could be determined by solving the appropriate Schrödinger equation, this is no longer the method of choice: instead, the homogeneous Faddeev equations (or their equivalent) are solved. Collision amplitudes have been determined by solving both the differential form of the Faddeev equations supplemented by the appropriate boundary conditions and the corresponding integral equation form. Thus, one general formalism suffices for both bound and continuum state problems. Until recently, most of the computations were carried out in momentum space. In the past few years, a greater emphasis has been placed on determining the Faddeev components of the wave function in a coordinate representation. Some intriguing problems which have arisen in this connection will be commented on later.

Since calculations are based on the Faddeev approach, a brief discussion of it follows. In order to proceed in the simplest manner, a few inessential assumptions are introduced, viz, that the particles (labelled 1,2,3) are distinguishable, spinless, and interact only via attractive two-body potentials, each strong enough to support at least one bound state per pair. Extensions of the formalism to include particle identity, spin, and three-particle forces are discussed, e.g., in Refs. [2,6,12].

The Schrödinger equation for the system is

$$(E - H^{(3)})\Psi = 0, \tag{1.22}$$

with $H^{(3)}$ given by Eqs. (1.3) and (1.4). Eq. (1.22) describes both bound and scattering states. In the latter situation there are two-body and three-body (breakup) configurations. The two-body asymptotic or non-interacting configurations of the system are of the form $i + (j,k)$, where (j,k) denotes a bound pair and the restrictions $i \neq j \neq k \neq i$ hold. It is a notational convenience to use the symbol i to label the two-body configurations since the label of the unbound particle uniquely defines the configuration. This same labelling is also used in the Faddeev decomposition.

In the Faddeev treatment, the desired Schrödinger solution Ψ is decomposed into a sum of the three Faddeev components ψ_i^F, $i = 1, 2, 3$:

$$\Psi = \psi_1^F + \psi_2^F + \psi_3^F, \tag{1.23}$$

where the labelling is that noted in the preceding paragraph. The ψ_i^F obey a triad of coupled equations, one form of which is [32]

$$\begin{aligned} (E - H_0^{(3)} - V_{23})\psi_1^F &= V_{23}(\psi_2^F + \psi_3^F) \\ (E - H_0^{(3)} - V_{13})\psi_2^F &= V_{13}(\psi_1^F + \psi_3^F) \\ (E - H_0^{(3)} - V_{12})\psi_3^F &= V_{12}(\psi_1^F + \psi_2^F)\,. \end{aligned} \tag{1.24}$$

Equations (1.24) are to be supplemented by the appropriate boundary conditions. In the case of bound states, there is a unique solution to Eq. (1.24) such that the sum on the RHS of Eq. (1.23) yields Ψ, the solution to Eq. (1.22), while the eigenvalues of Eq. (1.22) and of Eq. (1.24) coincide [32].

For continuum states, two paths to the unique solution of Eq. (1.24) exist. In the one, boundary conditions are imposed so as to yield those ψ_i^F which contain the proper collision amplitudes, including those for breakup. These will be discussed in a later subsection. In the other, the boundary conditions are imposed by converting Eq. (1.24) to a triad of integral equations, analogous to the LS equation in the one-particle problem. One of the integral equation forms convenient for the present discussion is [32]

$$\begin{pmatrix} \psi_1^F \\ \psi_2^F \\ \psi_3^F \end{pmatrix} = \begin{pmatrix} \Phi_1 \\ 0 \\ 0 \end{pmatrix} + G_0^{(+)}(E) \begin{pmatrix} 0 & t_{23} & t_{23} \\ t_{13} & 0 & t_{13} \\ t_{12} & t_{12} & 0 \end{pmatrix} \begin{pmatrix} \psi_1^F \\ \psi_2^F \\ \psi_3^F \end{pmatrix} . \tag{1.25}$$

Here, Φ_1 is the incident state, equal to the product of a plane wave state of momentum $\vec{k}_1$ describing the motion of particle 1 relative to the CM of the pair (2,3), times φ_{23}, the bound state of the pair (2,3). In Eq. (1.25), t_{ij} is still the transition operator for the pair (i,j) but now embedded in the three-particle Hilbert space, and $G_0^{(+)}(E) = \lim_{\epsilon \to 0} (E + i\epsilon - H_0^{(3)})^{-1}$ is the outgoing-wave, free-particle resolvent operator [33].

In an obvious matrix/vector notation, Eq. (1.25) can be re-expressed as

$$\psi^F = \Phi + \mathbf{K}^F \psi^F ; \tag{1.26}$$

comparison of Eq. (1.26) with Eq. (1.25) serves to define the (matrix) Faddeev kernel $\mathbf{K}^F$. Under relatively mild conditions on the potentials V_{ij} and by implication on the t_{ij}, Faddeev [32] has proved that the fifth power of the kernel $\mathbf{K}^F$ is compact (i.e., is traceclass). It is this that guarantees the uniqueness of the continuum solutions to Eq. (1.26), in contrast to those of the corresponding three-particle LS equation. In addition, it is usually assumed (for well behaved V_{ij} and t_{ij}) that *any* set of three-particle integral equations with a kernel whose structure is similar to $\mathbf{K}^F$ also yields unique and well-behaved solutions, even though proofs analogous to Faddeev's have not been carried out for any of these. The consequence of compactness is that standard methods may be used to obtain accurate numerical solutions to Eq. (1.25).

Not only does Eq. (1.25) define the continuum Faddeev components, but it also leads to the bound state part of the spectrum via solution of its homogenous portion, just as with the two-particle LS equation. This is, of course, a trivial similarity. In contrast there is an important, very non-trivial, and beautiful difference between the LS and the Faddeev integral equations when it comes to describing bound states. It is well known that a short-range, finite-depth, attractive potential in three-dimensions can support only a finite number N_2 of bound states [34]. Solving the homogenous part of the two-particle LS equation for the bound states does not violate this result. This, however, is *not* the case for a three-particle system interacting via attractive, short-range forces of finite depth: Efimov [35] discovered that under certain conditions, the number of three-particle bound states N_3 can become infinite, even though N_2 for each pair remains finite! A detailed mathematical analysis of this "Efimov effect" was first carried out by Amado and Noble [35], who showed that an infinite number of (S-wave) bound states occurs when, e.g., the momentum space Faddeev kernel $\mathbf{K}^F$ is itself infinite. The structure of $\mathbf{K}^F$, i.e., of the Faddeev integral equation, allows one to understand this in a simple way.

The key to this understanding is Eq. (1.20) and the fact, roughly speaking, that $\mathbf{K}^F$ can, under certain conditions, become divergent for "momentum" zero [36]. First, note that Eqs. (1.17) and (1.18) imply that

$$\langle \vec{k}'|t_{12}(E)|\vec{k}\rangle = -\frac{\hbar^2}{(2\pi)^2 m}\sum_{\ell=0}^{\infty}(2\ell+1)[k\cot\delta_\ell - ik]^{-1}P_\ell(\cos\theta), \quad (1.27)$$

where $k' = k$ and $\vec{k}'\cdot\vec{k} = k^2\cos\theta$. Second, from the angular momentum expansion of the LHS of Eq. (1.27), *viz*,

$$\langle \vec{k}'|t_{12}(E)|\vec{k}\rangle = (\frac{1}{2\pi})^3\sum_{\ell=0}(2\ell+1)\langle k|t_{12}(E)|k\rangle_\ell P_\ell(\cos\theta), \quad (1.28)$$

it follows from Eq. (1.20) that

$$\lim_{k\to 0}\langle k|t_{12}(E)|k\rangle_{\ell=0} = \frac{2\pi\hbar^2}{m}a, \quad (1.29)$$

where a is the S-wave scattering length for the pair (1,2). It should be evident now that the $\ell = 0$ portion of the low-momentum matrix element of K^F_{12} is proportional to a. It is this proportionality that underlies the Efimov effect: since a becomes infinite when there is a zero-energy bound-state or resonance in the two-body system (here (1,2)), then by implication K^F_{12} and thus $\mathbf{K}^F$ also become infinite.

An alternate means of expressing this exceptional result is as follows. Let $\lambda V_{12}(r)$ be the potential of range r_0, and let λ_0 be that strength which just supports the first zero-energy bound state. Then the relevant relation is [35] $\lim_{\lambda\to\lambda_0} N_3 \sim \frac{1}{\pi}\ln(|a|/r_0) \to \infty$. For $\lambda > \lambda_0$ or $\lambda < \lambda_0$, the infinite number of Efimov states becomes finite (or zero) as they are overtaken by the lowest two-body threshold in the complex energy plane [36]. The existence of the Efimov effect depends crucially on the dimensionality of space and on particle number: there is no Efimov effect in two-dimensions for three particles nor for systems containing more than three particles, i.e., in these cases the analog of $\mathbf{K}^F$ is non-singular [37].

The preceding explanation of the existence of the Efimov effect is based on a momentum space analysis. Because of the $N_3 = \infty$ limit, an equivalent description, using a coordinate-space, effective potential approach should be possible. In analogy to the long-ranged coulomb potential with its infinite number of bound states, one can expect that when $\lambda = \lambda_0$ there is an effective potential in the Efimov case which is also of long range. This expectation is indeed borne out: Efimov has argued [35] that at $\lambda = \lambda_0$, the

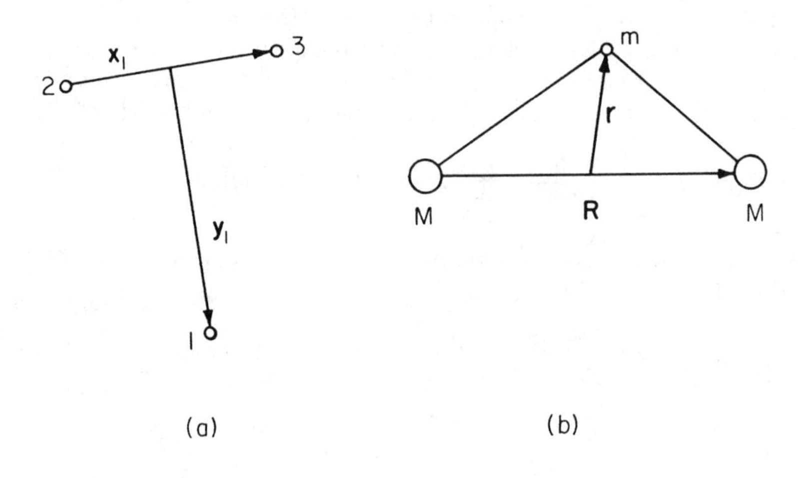

FIGURE 1.3
a: Coordinates for an equal mass three-particle system. b: Coordinates for the Born-Oppenheimer model of the Efimov effect.

three-particle system experiences an effective potential which behaves like ρ^{-2}, where ρ is the hyperspherical radius. For an equal-mass three-particle system, this radius is given by $\rho = (x_1^2 + 4y_1^2/3)^{1/2}$, where the Jacobi coordinates x_1 and y_1 are illustrated in Fig. 1.3a. Furthermore, Fonseca, Redish and Shanley [38] have used a Born-Oppenheimer model (two heavy masses and a light one) to demonstrate that a similar long-ranged potential exists at $\lambda = \lambda_0$, the potential in this case behaving as R^{-2}, where R is the separation between the two heavy particles, as in Fig. 1.3b.

Efimov states have been found to exist for the case of the ^{4}He trimer [39], but not for the three-nucleon system since the nucleon-nucleon interaction does not give rise to a zero-energy bound state or resonance. The Efimov effect has been discussed here not only because it is one of a number of unorthodox features characterizing three-particle systems, but also because the effective potential description can be used to provide a simple account of low-energy tri-nucleon properties. Like the Efimov effect itself, such a simplicity of explanation was a previously unexpected aspect of three-particle dynamics.

TABLE 1.6
Some Calculated and Measured ^{3}H Bound State Properties

	$B_{^3\mathrm{H}}$(MeV)	$\langle r^2\rangle_H^{1/2}$(fm)	E_c(keV)	$\eta = C_2/C_0$
Paris	7.64	1.66	626	-0.039
RSC	7.35	1.67	620	-0.040
V14	7.67	1.67	623	-0.041
Expt	8.48	1.54	638	-0.045

1.3.2 ^{3}H and ^{3}He Bound States

Since 1969 there has been an ongoing series of international conferences on few-body problems. In the first few such meetings, the low-energy three–nucleon system was the centerpiece. As time passed, other systems and higher energies began to push the three-nucleon system to the side. The number of researchers working on this system decreased as the difficulties, either in carrying out computations or in fitting data, increased. By the 1983 meeting in Karlruhe, research on the three-nucleon problem was so clearly regarded as old-fashioned that the system, except in a few of its more esoteric aspects, could almost have been characterized as dull, a by-way too often trampled to produce any new blooms. And then, at the Tokyo/Sendai meeting in 1986, results were presented that utterly shattered this image! The developments leading up to this event, its nature, and the subsequent state of affairs to date dealing with bound state properties, are the subject of this subsection.

Many $N = 3$, Faddeev-equation-based calculations have employed separable potentials, starting with the pioneering work of Amado, Mitra and their collaborators [2]. Despite impressive agreements with data, the unrealistic character of such potentials makes most of their results irrelevant to the purposes of this review, although a few such results and their consequences are reported on in Subsecs. 1.3.3 and 1.3.4 below. The change from separable potentials was at first to semi-realistic potentials. The early calculations using them concentrated on the ^{3}H binding energy and the neutron-deuteron doublet scattering length (discussed in the next subsection). With the measurement, e.g., of the root mean square charge radii $\langle r^2\rangle^{1/2}$, the asymptotic normalization constants C_0 and C_2, and the charge form factors and their inferred charge densities — which allow the ^{3}He-^{3}H coulomb energy (E_C) to be determined [40], more stringent conditions had to be met by the semi-realistic and realistic interactions used in later calculations. Exemplifying the results of such calculations is the comparison of $B_{^3\mathrm{H}}$, $\langle r^2\rangle_H^{1/2}$, E_C and $\eta = C_2/C_0$ from the RSC, Paris and V14 potentials with experiment given in Table 1.6, taken from Ref. [41]. Most of the theoretical numbers are from fully converged 34-channel calculations, i.e., sufficiently many partial waves were employed that use of additional partial waves would yield negligible changes. It is obvious that the data are not well fitted.

Two general explanations were proposed for these failures:

1. A nucleons-only, non-relativistic quantal description is the wrong theoretical framework.

2. The assumption that only two-nucleon forces are important in the three-nucleon system is inadequate and thus effects of three-nucleon forces must be taken into account.

At this point in time, only one analysis based on explanation **1** appears to be relatively well-founded theoretically, viz, the nucleons-plus-delta model of Sauer and collaborators [42]. It seems to be the only contemporary type-**1** description that has actually been applied to the three-nucleon system without being overburdened by too many ad-hoc parameters. Some of its results will be referred to shortly.

Investigations involving the other explanation, viz, inclusion of three-nucleon forces, goes back many decades [43]. Only in the past ten years, however, have the somewhat sporadic, but important, efforts of the past been translated into the activities of several groups simultaneously, and only in the past 3 years have computations of reliable binding energies which include some three-nucleon force effects been reported. As will be seen, the results are not unambiguous. They are, nevertheless, an essential ingredient in the attempt to push a nucleons-only, non-relativistic quantal description to its limits.

The simplest and longest-range type of three-nucleon potential involves the exchange of two pions. An example of this involves one of the three nucleons being excited to the Δ resonance at 1232 MeV. A schematic diagram for this is given in Fig. (1.4c), which is to be compared with Fig. (1.4b). The latter actually represents either a sum of two-nucleon interactions or the iteration of a single two-nucleon interaction and thus does not correspond to a true three-nucleon potential: to produce one requires a change of state. Heavier meson exchanges, such as $\rho\pi$ or $\rho\rho$, also yield three-nucleon forces, but none have been used to date in any calculations on three-nucleon systems.

The contemporary three-nucleon potentials (3NP's) of three different groups have been used in recent calculations of $B_{^3\mathrm{H}}$, the testing ground for such computations. The Tucson-Melbourne (TM) 3NP [44] has been

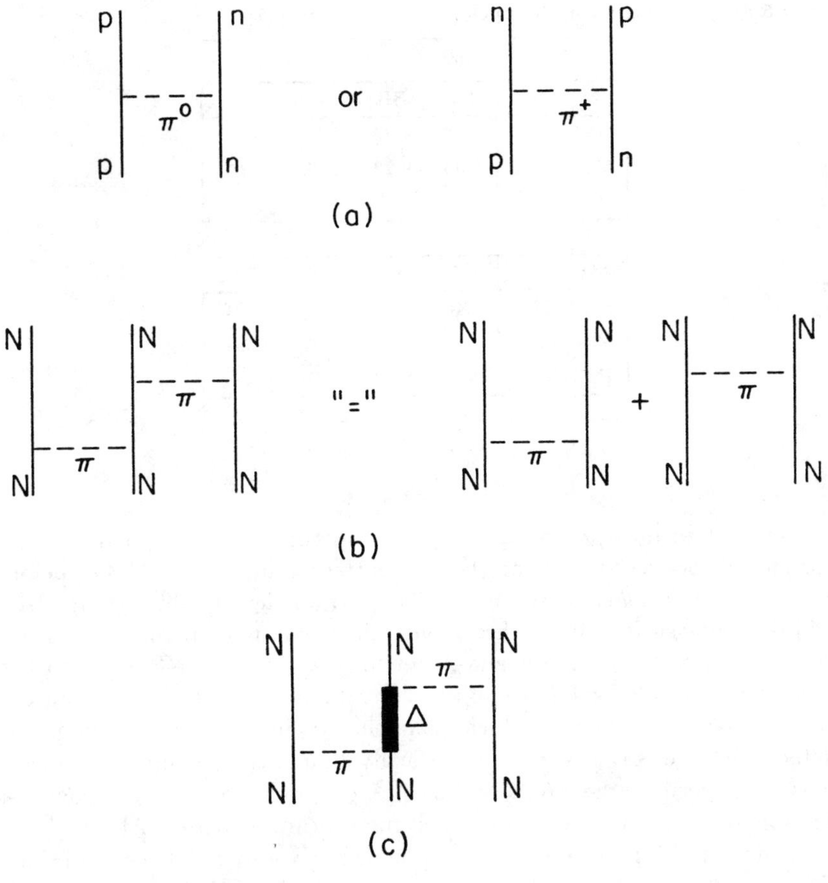

FIGURE 1.4
a: Two schematic examples of contributions to the one-pion exchange potential between a proton (p) and a neutron (n). b: A pseudo or "reducible" three-nucleon interaction which is actually the sum of two separate one-pion exchange, two-nucleon interactions. c: Contribution to a true, two-pion exchange three-nucleon interaction; the appearance of the Δ makes this an "irreducible" diagram.

TABLE 1.7
Variation in $B_{^3H}$ (MeV)with Different NN and 3NP Models ($\Lambda = 5.8 m_\pi c^2$) and with Changes in Λ (from Ref. [41])

(a)

	TM	BR	UA
RSC	8.86	8.89	8.70
V14	9.36	9.22	8.99
Paris	9.71		

(b)

$\Lambda(m_\pi c^2)$	4.1	5.0	5.8	7.1
RSC/TM	7.46		8.86	11.16
V14/TM		8.42		
Paris/TM		8.32		

constructed to incorporate as many a-priori constraints as possible (e.g., current algebra restrictions on the πN scattering amplitude [41,44]) before employing any *ad hoc* prescriptions. The 3NP of the Brazilian group (BR) [45] does not satisfy all of these constraints, but numerically this is not very significant [41]. The Urbana-Argonne (UA) group's 3NP [46] is rather more phenomenological. None of the three groups claim to understand the very short-range behavior, which, as in the two-nucleon case, remains undetermined and is thus treated empirically. In addition, a cutoff parameter Λ is introduced in the πNN, $\pi N\Delta$, etc., couplings in order to provide a suitable damping as the pion off-shell momentum increases. Although arguments have been given to fix Λ [47], none are wholly reliable and in some analyses various values of Λ have been used to determine how strongly the final binding energy depends on it.

Shown in Tables 1.7a and 1.7b are results of converged (34 channel) calculations of $B_{^3H}$ for various combinations of two- and three-nucleon interactions with varying choices for the cutoff parameter Λ (in multiples of the pion rest mass $m_\pi c^2$). The V14 NN potential is softer than the RSC interaction, so results for these two potentials are reasonably suggestive for the general effects of the 3NP. The value of $\Lambda = 5.8 m_\pi c^2$ has some theoretical pedigree [47], and it is thus used as a benchmark. It is quite clear that for this "reasonable" value of Λ, all calculations yield an overbound ^{3}H, the increase in binding being about 1.5 MeV in each case. Interesting ingredients in the various 3NP's are that they contribute much more strongly in odd parity states than do the different NN interactions and that they are strongly dependent on the spatial configuration of the three nucleons, a feature that will be considered later.

It is instructive at this point to compare these results with those of the nucleons plus delta model of the Hannover group [42]. Unlike the additional 1.5 MeV binding produced by inclusion of the 3NP (with $\Lambda = 5.8 m_\pi c^2$), the coupled channel approach of the latter group yields typically only some tenths of an MeV additional binding for the various forces employed, i.e., the desired 8.48 MeV is not achieved. Explanations beyond the scope of the review as to why this model provides less additional binding than any of the 3NP's are given by Coon [48] and by Gibson and McKellar [41].

Other calculated properties of ^{3}H also do not agree with their experimental counterparts when these realistic two-nucleon interactions (and no 3NP's) are used to compute them. From Table 1.6 it is seen that the rms radii are too large, while the coulomb energies and asymptotic D to S ratios are too small; furthermore, it is seen in Table 1.7 that the inclusion of the 3NP's (which failed to yield $B_{^3\mathrm{H}} = 8.48$ MeV) also does not improve the situation for these latter three quantities. Therefore, failure to yield the correct $B_{^3\mathrm{H}}$ generally means that other observables are not well fitted. In contrast, however, is the fact that if the correct value $B_{^3\mathrm{H}}$ *is* obtained in a calculation, then so are the values of the other bound state observables. This feature, denoted scaling [49], is one of the surprises in this field. It will be examined in some detail later, but its existence clearly explains why the 3NP calculations do not fit $\langle r^2 \rangle_H^{1/2}$, E_C or η.

It is worth noting that comparison of the calculated 764 keV difference $B_{^3\mathrm{H}} - B_{^3\mathrm{He}}$ with the experimental value of 638 keV for E_C shows that about 1/6 of the 764 keV arises not from the coulomb interaction between the two protons in ^{3}He but from charge symmetry breaking forces [50], i.e., from the difference between the nuclear part of the proton–proton interaction and the (purely nuclear) interaction between a pair of neutrons. These symmetry-breaking forces are thought to arise from short range meson-exchange processes but not directly from 3NP's, which have been estimated to add only a few keV of the missing 125 keV [51]. That only 638 keV of the 764 keV ^{3}H - ^{3}He binding energy difference is due to coulomb forces is determined from the hyperspherical formula approximation, developed independently by Friar and by Fabre de la Ripelle [40]. In it, E_C is written as an integral over the tri-nucleon charge form factors $F_c(q)$, which can be taken from experiment. (The charge form factor is essentially the Fourier transform of the charge distribution and q is the momentum transfer.) The hyperspherical formula approximation is 99% accurate, so that the value of 638 keV is reliable [40,51].

Electromagnetic properties of the tri-nucleon bound states such as

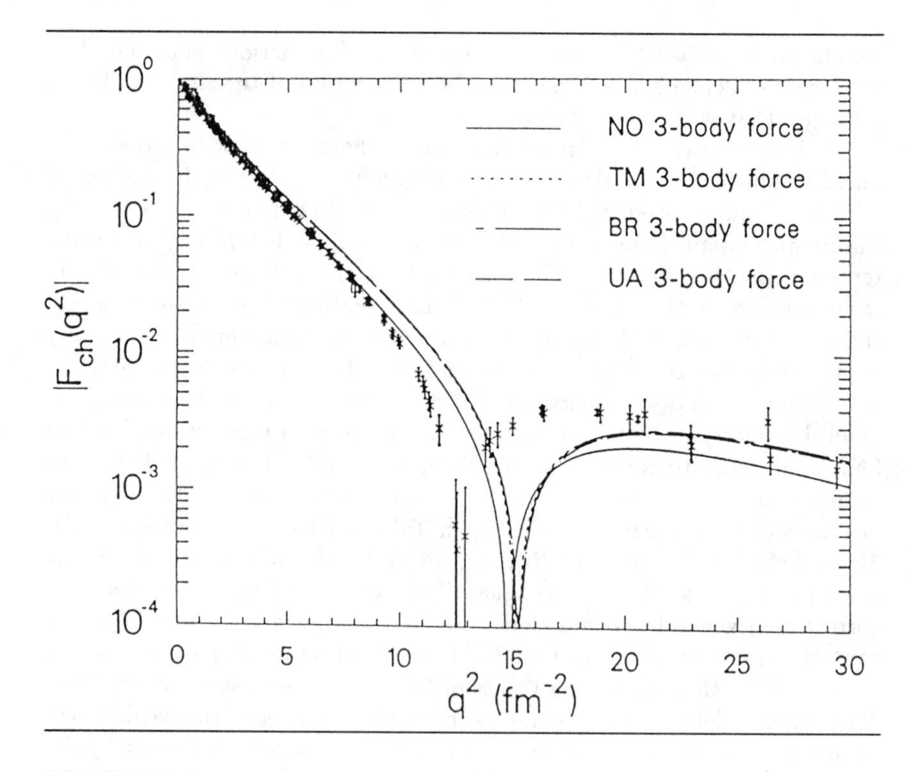

FIGURE 1.5
Effect on the ^{3}H form factor of various three nucleon potentials (from Ref. [41])

the charge form factors (determined in elastic electron scattering measurements) are also not well fitted by calculations employing only two-nucleon potentials. The predicted minimum occurs at a q^2 of about 15 fm^{-2} not 12.5 fm^{-2} as observed, and the theoretical maximum at around $q^2 = 20$ fm^{-2} is too low and occurs at too large a value of q^2. Inclusion of a 3NP does not improve the theoretical curves sufficiently to come close to fitting the data, as seen in Fig. 1.5, even though the secondary maximum is increased by about 50%.

It is now generally accepted that the form factor discrepancies arise from use of the wrong charge and current operators. That is, in addition to the ordinary operators for charge and current densities, one should also include the operators corresponding to meson exchange, since meson exchange mediates the nuclear force in the picture being considered here. The importance of meson exchange currents was established by Riska and Brown, who showed that their use eliminated the long-standing 10% dis-

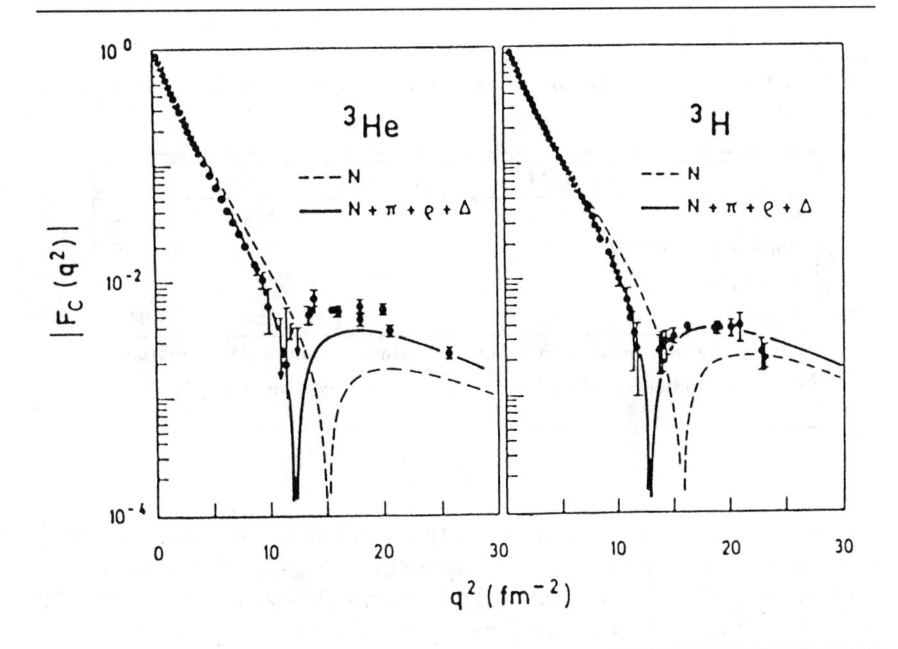

FIGURE 1.6
Improvement of the agreement between theory and experiment for the ^{3}He and ^{3}H form factors due to the inclusion of meson exchange currents (from Ref. [41]).

crepancy between theory and data for radiative neutron-proton capture [52]. There are two forms for these currents, one based on pseudoscalar (PS) and the other on pseudovector (PV) nucleon-meson (pion) couplings. Both correspond to relativistic corrections. The PV coupling is the more realistic, although most of the early calculations employed the PS form, which for the deuteron is not a problem, though it is for ^{3}H and ^{3}He [53]. Shown in Fig. 1.6 is the excellent fit to $F_c(q^2)$ for ^{3}H and the vastly improved fit for ^{3}He that results when PV meson exchange contributions to the charge density are included in the computations [54]. This improvement is the first unambiguous indication of the inadequacy of a purely non-relativistic description of the three-nucleon systems. Unfortunately, the calculations themselves are not unambiguous and thus must be regarded as *ad hoc*. The problems are those of completeness and consistency [50,53]. First, there are a number of momentum-dependent ($1/c^2$) correction terms which simply have not been computed for ^{3}H and ^{3}He. Second, the form of the relativistic corrections are not completely specified, but, whichever forms are

TABLE 1.8
^{3}H Bound State Properties for a Configuration Space Version of the Bonn Potential

	$B_{^3\mathrm{H}}$(MeV)	$r_{ch}(fm)^a$	$B_{^3\mathrm{H}}$ - $B_{^3\mathrm{He}}$	η
Bonn	8.33	1.73	697	-.0418
Bonn + TM 3NP ($\Lambda = 5m_\pi c^2$)	9.01	1.69	718	-.0446
Expt	8.48	1.68 - 1.81^b	760	-.045

^{a}) r_{ch} is the root mean square radius obtained for finite sized nucleons.
^{b}) The experimental values have been compiled by Kim et al [57].

chosen, then the two-nucleon interaction (and thus the wave functions) must contain the same relativistic corrections. None of the realistic two-nucleon interactions enjoy this degree of consistency. Despite almost 20 years of intensive effort on meson exchange/relativistic corrections, much further work remains to be done.

These results seem to imply a clear message: non-relativistic quantal calculations are unable to account accurately for the tri-nucleon bound state properties, thus suggesting the need for relativistic effects if not a fully relativistic formulation, a suggestion strongly supported by the meson-exchange current analyses of the charge form factor. Such a conclusion, however, is premature. The preceding, somewhat gloomy situation was electrifyingly transformed in 1986 by calculations of the Sendai group [55] and of Brandenburg et al. [56]. Using a form of the Bonn two-nucleon interaction and no 3NP's, these groups carried out Faddeev computations which yielded $B_{^3\mathrm{H}} \cong 8.33$ MeV, a result in error by only 200 keV (2%)! This is to be contrasted with errors in $B_{^3\mathrm{H}}$ of the order of 1000 keV arising from use of any of the other realistic two-nucleon interactions.

With these calculations, what was an increasingly problematic description, viz, that based on nucleons as the constituents and non-relativistic quantum mechanics as the theoretical framework, now appears quite reasonable. Inevitably, though, questions arise. For example, What is meant by the phrase "a form of the Bonn two-nucleon interaction"?, or, How is the Bonn potential able to yield $B_{^3\mathrm{H}} \cong 8.48$ MeV?, or, How well are other tri-nucleon properties predicted? This last question was addressed by the Sendai group, which also computed some of these properties, first using the same pair-wise Bonn potential between the three nucleons and then adding in the TM 3NP with $\Lambda = 5m_\pi c^2$ [55]. A selection of their results is presented in Table 1.8. Possibly the most interesting of the numbers is the value of 9.01 MeV for $B_{^3\mathrm{H}}$ obtained when the 3NP is included in the calculation. (In view of the comments made concerning the 3NP, its use in this computation is obviously exploratory: the consistency of using a 3NP with any NN potential requires investigation.) The overall very good agreement between η in lines one and three of Table 1.8 is encouraging, although in connection with r_{ch}, the ability of the Bonn potential to fit the tri-nucleon charge form factors is no better than for any other realistic potential [57]. This latter result is not surprising, since the charge form factor is sensitive (at large q^2) to meson-exchange current effects, which do not play an important role at small q^2 where r_{ch} is determined.

Given that $B_{^3\mathrm{H}}$ is so close to 8.48 MeV for the Bonn potential calculations, one wants to know how this is achieved, i.e., how the Bonn potential differs from other two-nucleon interactions. The answer is suggested by Table 1.4, *viz*, by the value of $P_D = 4.38\%$ for the Bonn potential. This is a much lower deuteron D state percentage than for the other two-nucleon potentials. The implication is that the tensor force for the Bonn potential is weaker and thus its attractive central force is stronger than for the other potentials. Since the tensor force is of longer range than the attractive central forces and the tri-nucleon systems are smaller than the deuteron, the central (tensor) force attractions make a greater (lesser) contribution to the binding energy for the Bonn potential than for the others. This idea and related ones are discussed, e.g., by Brandenburg et al. [58].

Two phrases used in the preceding paragraphs still require clarification: "*the* Bonn potential" and "a form of the Bonn two-nucleon potential." In the original formulation by the Bonn group (Machleidt, Holinde and Elster [21]), there are four versions of the potential. The most fundamental is the full Bonn potential; it is energy dependent. Three approximations, denoted "one-boson-exchange" (OBE) versions, have also been introduced, denoted the OBEPT, the OBEPQ and the OBEPR approximations. The first of these is energy-dependendent, while the latter two are energy-independent "static" versions in momentum space (OBEPQ) and in coordinate space (OBEPR). In the OBE approximation the parts of the full potential due to multi-meson exchange are replaced by a single boson (a σ-meson) with an effective mass. The report of the Sendai group describes the approximations in reasonable physical detail [55]; their computation used the OBEPR version. The results in Table 1.4 are from the OBEPQ version.

As Holinde has stressed [21], the full Bonn potential and the energy-dependent OBEPT version fit the NN data very well, including the ϵ_1 parameter (even at 325 MeV): it is the energy-independent OBE approximations that fail to fit ϵ_1. Differences between these latter two versions and the full Bonn potential are manifested also at low energy: $P_D = 4.25\%$, 4.38% and 4.81% for the full, OBEPQ and OBEPR interactions, respectively. OBEPR is not simply the fourier transform of OBEPQ, since the latter contains retardation effects in the propagator while a static approximation to this propagator is introduced prior to taking the fourier transform. In other words, there are three different OBE potentials, and while each is unique as compared to the full potential, the latter two (OBEPQ and OBEPR), are ad hoc. So far, the full potential, due to its complexity, has not been used in tri-nucleon calculations. The OBEPR version, like the Paris and Nijmegan potentials, contains $\vec{p}^2$ terms, and a coordinate-space set of tri-nucleon bound-state calculations taking the $\vec{p}^2$ dependences explicitly into account for these three potentials have been performed recently by Friar, Gibson and Payne [59]. No surprises were encountered among the results.

Even though the OBEPQ and OBEPR versions are ad hoc and fit data less well than the full Bonn potential, the overall situation is in fact a very positive one: a nucleons only, non-relativistic quantum mechanics description of at least the bound state properties is now no longer a completely unreasonable one. Much remains to be done but it is quite clearly now worth doing, using the nucleons only, non-relativistic quantal framework as a starting point.

1.3.3 Tri–Nucleon Continuum States

Many calculations of nucleon-deuteron elastic scattering and breakup cross sections and of polarizations as well have been made using separable potentials fitted to the two-nucleon data. Although the details of these computations are not relevant here, the results are: great success was obtained in fitting data. Not only did this establish the primacy of the Faddeev-type formulation for analyzing continuum states, it also suggested that collision data would in general be less sensitive to details of the two-nucleon interaction than bound state data. In addition, it pointed up the need to accurately fix the (small) P-wave parts of the two-nucleon interaction, to which vector polarizations and asymmetries are sensitive.

These separable potential calculations extended from very low energies to 50 MeV bombarding energy and greater. In contrast, relatively few computations of collision amplitudes have been carried out using realistic potentials, so they remain less well tested in this energy regime. Nevertheless, most of these latter calculations yield good agreement with the data, without the use of 3NP's. An example is the computation of Koike, Plessas and Zankel [60], who used an accurate separable expansion for the Paris potential in their momentum space Faddeev equations. The overall agreement with $n + d$ elastic scattering data at 8, 10.25 and 12 MeV and spin transfer coefficients from polarized nucleon deuteron elastic collisions is very good, as seen in Fig. 1.7.

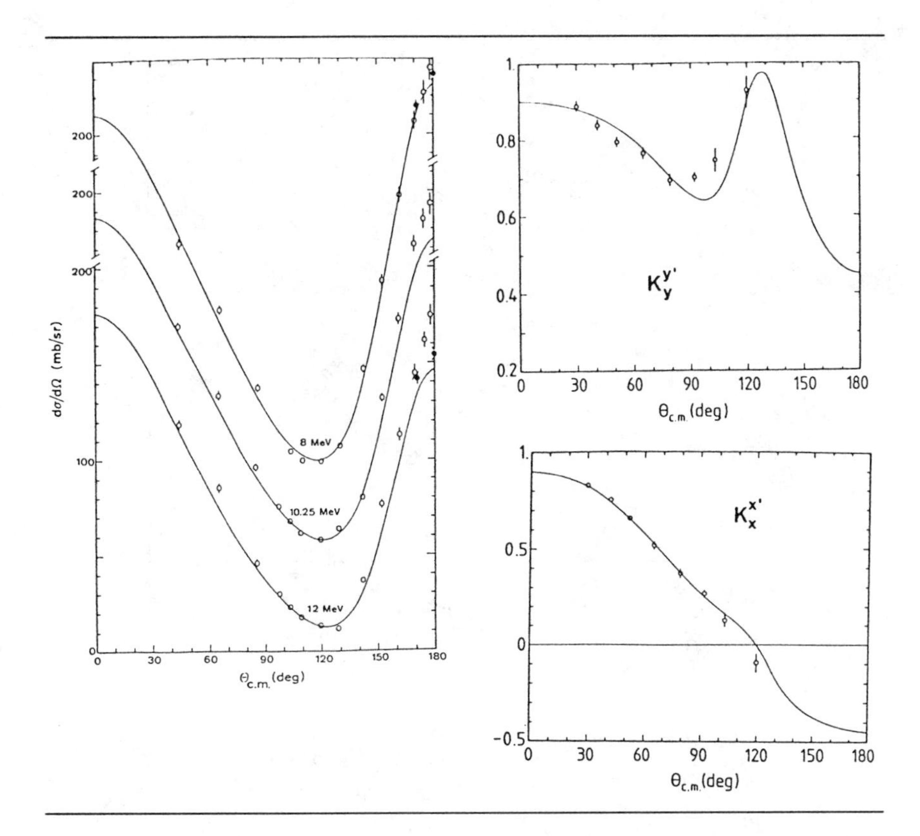

FIGURE 1.7
Comparison of data and Paris-potential-based computations of elastic scattering angular distributions and spin transfer coefficients (from Ref. [60]).

Because so many continuum calculations have been able to fit experiment, agreement with data is at present less interesting than possible disagreements, since these could be evidence for the effect of 3NP's. To date, no continuum calculations involving 3NP's have been carried out even though discrepancies have been found between experiment and the computations, which have used only (realistic) two-nucleon potentials. Two such calculations will be discussed in this subsection; each is of special interest because related theoretical developments are of a "surprising" nature. The first discrepancy involves the pair of doublet nucleon-deuteron scattering lengths ($^2a_{nd}$ and $^2a_{pd}$); the second concerns nucleon-deuteron breakup.

For short range forces, the two-particle effective range function $K(E) =$

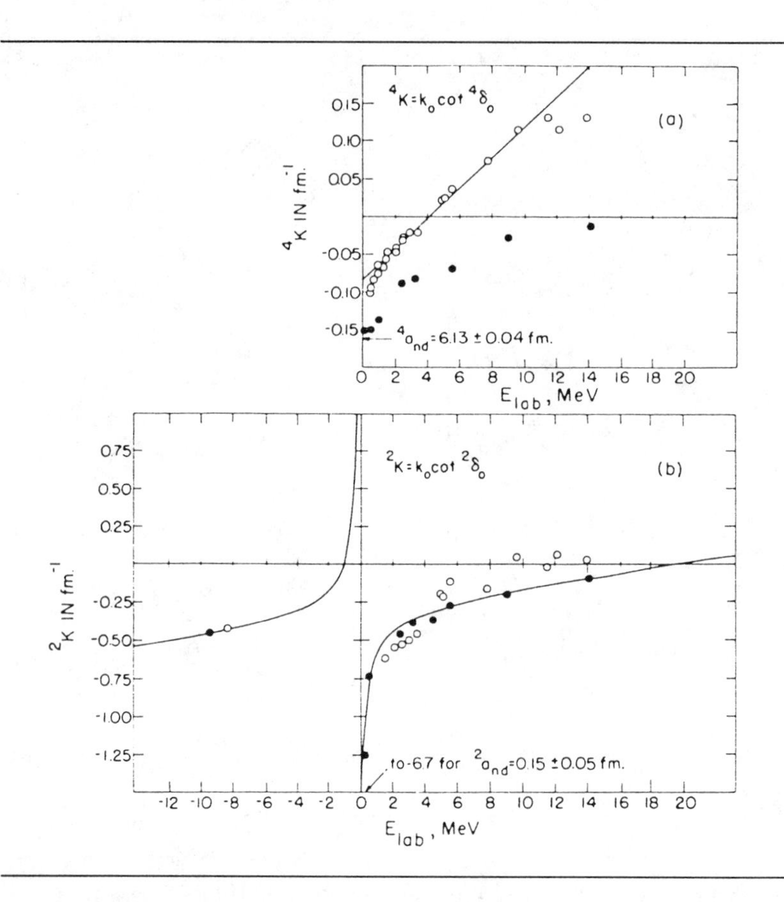

FIGURE 1.8
Neutron-deuteron effective range functions, from the review of J.D.Seagrave, in *Three Body Problems*, ed. J.S.C. McKee and D.M. Rolph (North Holland, Amsterdam, 1970).

$k \cot \delta_0$ takes the form of Eq. (1.20). It is at first natural to expect that $K(E)$ in the case of nucleon-deuteron scattering will have a similar form. However, the example of the Efimov effect suggests that this expectation may need revising. Revision is indeed required for $^2a_{nd}$, as was noted over 20 years ago: in this case, the doublet effective range function $^2K_{nd}(E)$ has a pole slightly below the threshold for elastic $n-d$ scattering; the effect of the pole is to change the straight line behavior of $^2K_{nd}(E)$ as a function of $k^2 \propto E$ for small k given by Eq. (1.20) to the more complex form [61]

$$^2K_{nd}(E) = -A + Bk^2 - C/(1 + Dk^2),$$

where the pole structure for $k^2 < 0$ is manifest. A naive explanation is that in the doublet state, the forces are almost strong enough to produce an excited bound state; between this virtual state and the triton bound state the scattering amplitude would have to have a pole, i.e., ${}^2K_{nd}(k^2)+ik = 0$. More detailed, sounder explanations were provided many years ago; a good source for references and a summary are the reviews of Noyes [62]. The data and results of analysis that established the pole behavior is shown in Fig. 1.8, although the value ${}^2a_{nd} \cong 0.15 fm$ extracted from this data was, via a later experiment, revised to ${}^2a_{nd} = 0.6 fm$ [63]. As discussed by Noyes [62], the extraction of ${}^4a_{nd}$ is straightforward, and the early values of ${}^4a_{nd} \approx 6 fm$ have remained unchanged.

In contrast to this long settled $n-d$ situation, the case of very low energy $p-d$ scattering has gone through various upheavals and, at least experimentally, not only remains uncertain but is likely to continue to. The first suggestion of an anomalous situation came in the 1983 paper of Friar, Gibson and Payne [64], who made a zero-energy, coordinate-space $p-d$ calculation and found ${}^4a_{pd} \cong 14$ fm and ${}^2a_{pd} \cong 0.15$ fm, in contrast to the then experimental values of about 11.9 fm and 2.73 fm, respectively. The theoretical results were obtained using a semi-realistic but simple local interaction, the Malfliet-Tjon, MT I-III potential [65], the results from which were believed to be sufficiently accurate. (For example, agreement with the ${}^4a_{np}$ value from a separable potential calculation of Alt [66] was obtained.) Friar et al. proposed that new experimental investigations be undertaken. The following years saw a flurry of new results and controversies.

One reason for the proposal in Ref. [64] to remeasure was the relatively high values of the lowest energies at which a_{pd} was extracted experimentally. In contrast to the long settled $n-d$ situation, very low energy $p-d$ scattering has recently been shown to be much more problematic than had been thought previously. Three new and surprising theoretical results have been obtained in the past few years; two of them point up strongly the inadequacy of the experimental situation. The most recent of these three findings is one that should have been inferred years ago, but of course was not. It is the existence of a below-threshold pole in ${}^2a_{pd}$, in precise analogy to the ${}^2a_{nd}$ case [62]. Verification of this behavior seems to have been contingent on the ability to carry out accurate configuration space $p-d$ calculations. Fig. 1.9, taken from the analysis of Chen, Payne, Friar and Gibson [67] not only quite clearly displays the phenomenon in question, but also shows that no really low energy $p-d$ data exists with which the calculations can be compared.

Just recently, however, it was believed — albeit briefly — that there wasn't even an experimental situation which could remain unsettled. This conclusion was based on the fact that a *conventionally defined*, coulomb-modified scattering length does not exist for the proton-deuteron system,

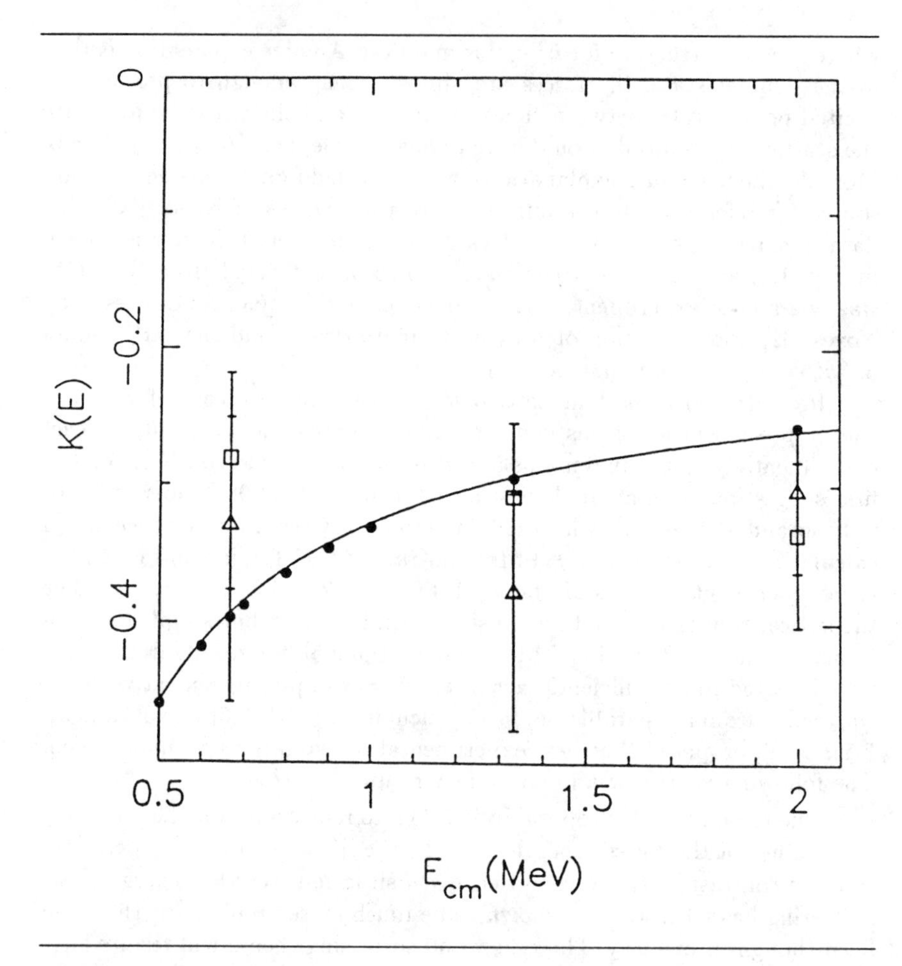

FIGURE 1.9
Comparison with data of the p-d effective range function calculated in Ref. [67].

thus implying that there is nothing to measure (much less to be unsettled). The non-existence of the *standard* coulomb-modified $p-d$ scattering length is a consequence of the need to describe the tri-nucleon system via Jacobi coordinates, say $\vec{x}_1 = \vec{r}_n - \vec{r}_{p_1}$ and $\vec{y}_1 = \vec{r}_{p_2} - \frac{1}{2}(\vec{r}_n + \vec{r}_{p_1})$ (in an obvious notation for a two-proton, one-neutron system), so that the interproton distance is $r_{p_1p_2} = |\vec{r}_{p_1} - \vec{r}_{p_2}| = |\vec{y}_1 - \frac{1}{2}\vec{x}_1|$. Given that the deuteron bound state keeps x_1 from growing very large, $r^{-1}_{p_1p_2}$ can be expanded in powers of x_1/y_1; this eventually leads to an asymptotic y_1^{-4} polarization potential contribution. As noted by Berger and Spruch [17], the effect of such a potential is to introduce a $\ln k$ dependence in the effective range function, whose $k \to 0$ limit is thus non-existent. Discussions and reviews of the then situation can be found in the papers of Kok and of Merkuriev in the Proceedings of the Ninth Conference on Few Body Problems in Physics held in 1984 in Tbilisi [68], which is the time when the $p-d$ scattering length problem was first aired.

Two subsequent papers demonstrated that the polarization potential anomaly was, in fact, not an unsurmountable problem. First, Berthold and Zankel found that when a pole is included in the doublet pd effective range function, the effect of the polarization potential was negligible down to 200 keV [69]. Second, Bencze et al. have introduced an altered definition of the coulomb modified scattering length which, in the presence of an additional polarization potential, leads to a new $K(E)$ having a finite limit when k (or E) $\to 0$ [70]. Furthermore, these latter authors demonstrate that ignoring the y_1^{-4} term is a good approximation in a zero-energy, coordinate-space calculation as long as the asymptotic boundary conditions are imposed at distances less than several hundred fm [70]. An experimentally determined $p-d$ scattering length is thus a meaningful quantity to seek if experiments could be carried out at low enough E.

The question implied by Friar, Gibson and Payne [64] thus arises again: are the values of E sufficiently low that reliable ${}^4a_{pd}$ and ${}^2a_{pd}$ have been extracted from experiment? In addition, are the theoretical values reliable? The answer to both questions could be NO if (i) the pole term noted by Berthold and Zankel exists and influences the low-E $p-d$ data in the same way as the pole term does in the $n-d$ case and (ii) if the $E=0$ results of Ref. [64] are not the $E \to 0$ limits of $E > 0$ results, as had been suggested. To settle this question, Chen, Payne, Friar and Gibson [67] solved the configuration space Faddeev equations at various E and numerically determined ${}^4K_{pd}(E)$ and ${}^2K_{pd}(E)$. The "surprise" was a pole behavior in the latter quantity, precisely analogous to the corresponding one in ${}^2K_{nd}(E)$. The behavior of ${}^2K_{pd}(E)$ is shown in Fig. 1.10, where the curvature demonstrates quite convincingly that a straight line extrapolation from data at $E \geq 400$ keV cannot possibly yield a correct result. (The $E=0$ limits (for the MT I-III potential) are precisely the same as found in 1983 [64].) Unfortunately, the coulomb repulsion in low energy $p-d$ scattering will very likely make experimental verification of the predicted ${}^2a_{pd}$ value close to impossible.

The last of the surprising developments arising in the $p-d$ context concerns the treatment of systems in which both short-range and coulomb (long-range) interactions are present. In the case of two-particle systems, the treatment is standard, as described in many texts and monographs, see, e.g., Taylor [71] or Newton [12]. For a three-particle system, one must

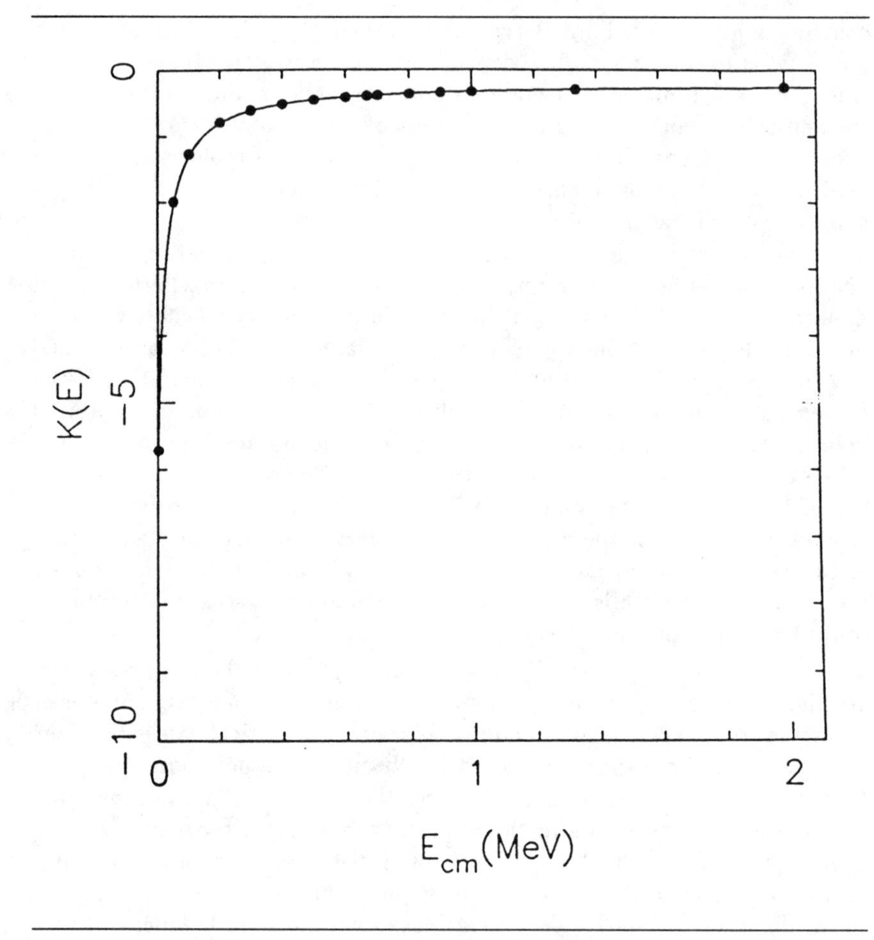

FIGURE 1.10
Extreme curvature at small E in the p-d doublet effective range function (from Ref. [67]).

include in a Faddeev-type of analysis a correct prescription for handling the long range force and its concomitant phase factors. There are two basic approaches to this: the configuration space description of Merkuriev [72] and the momentum space description of Alt, Sandhas and collaborators [73]. Comparisons of the results of early calculations based on each description is given by Kok [68]. Kok also briefly compares the two approaches, whose alleged validity has been a source of controversy, a point beyond the scope of the present review. Since the surprising development occurs in the momentum space approach, that is the one discussed herein.

In this latter treatment, the long-range interaction is first made short-range by means of a screening factor involving a screening radius R, beyond which the screened interaction is essentially zero. The full collision amplitude is then given by a sum of the screened coulomb amplitude plus a short-range but screened nuclear amplitude. Calculations are done for various values of increasing R until an R-independent result is obtained. A recent review is given in Ref. [74].

The momentum space method is apparently now accepted as a reliable one [68]. Alt, who never doubted the reliability of his method (!), has studied the accuracy of various approximations to it [75]. One such is well known in nuclear reaction analysis: it involves the replacement of the short range but screened nuclear part of the amplitude by the unscreened, pure nuclear amplitude (i.e., the amplitude for a neutral projectile to be scattered by the target). This is then added to the appropriate coulomb amplitude; the absolute square of the sum is proportional to the differential cross section. Alt has calculated $p{-}d$ differential cross sections at proton lab energies of 6.78 MeV and 35 MeV using this approximation (and a similar one), and found angle-dependent errors which can be as large as 10%. This is a surprisingly big discrepancy for a procedure long believed to be reasonably accurate. Larger errors may show up in polarization calculations, due to amplitude interferences not present in a differential cross section calculation. The implications are obvious for collision analyses where this approximation was used; whether this will mean reanalysis, and if so, how, is a matter for future investigation.

The remaining topic dealt with in this subsection is breakup. Breakup calculations, like the coulomb plus short-range force computations noted in the preceding, can in principle be performed in either coordinate or momentum space. However, no successful coordinate-space breakup calculations have yet been reported in the literature, although a number of such calculations have been attempted. Until very recently, all have failed. The reasons for the failures and also for the very recent success are not clear, despite careful mathematical analysis. This surprising coordinate-space situation will be discussed next.

It suffices to consider the case of S-waves in each of the coordinates $\vec{x}_1$ and $\vec{y}_1$. If the hyperspherical coordinates ρ and θ are introduced via $x_1 = \rho\cos\theta$ and $y_1 = \sqrt{\frac{3}{4}}\rho\sin\theta$, then Glöckle [76] has shown that in the breakup region the S-wave Faddeev component $u(\rho,\theta) = \psi(x_1 y_1)/x_1 y_1$ behaves asymptotically ($x_1 \to \infty$, $y_1 \to \infty \Rightarrow \rho \to \infty$) as

$$u(\rho \to \infty, \theta) \sim -\frac{K_1}{\pi i}\int_{-\pi/2}^{\pi/2} d\alpha \cos\alpha e^{iK\rho\cos(\alpha-\theta)}\bar{T}(K_1 \sin\alpha), \qquad (1.30)$$

where $K = \sqrt{mE/\hbar^2}$, $K_1 = K\sqrt{4/3}$, and $\bar{T}$ is one form for the S-wave breakup amplitude. In Eq. (1.30), the hyperspherical angle of observation θ must be kept smaller than some maximum value θ_0, since for $\theta \geq \theta_0$, $u(\rho \to \infty, \theta)$ will also contain contributions from elastic scattering (here it is assumed that V_{ij} supports only one bound state and that all three masses are equal to m).

Eq. (1.30) contains the desired breakup amplitude as part of an integral, in contrast, say, to the case of elastic or inelastic scattering, where the relevant amplitude simply multiplies the outgoing wave term $\exp(ikr)$. Eq. (1.30) as it stands, therefore, is not an appropriate boundary condition to be imposed in order to obtain both u and the breakup amplitude numerically. However, an appropriate form *is* obtained when the method of stationary phase [77] is used to approximate the integral. The result is [76]

$$u(\rho \to \infty, \theta) \approx u_{sp} = K_1\sqrt{\tfrac{2}{\pi}}e^{i\pi/4}e^{iK\rho}\cos\theta\bar{T}(K_1\sin\theta)/\sqrt{K\rho}. \quad (1.31)$$

The form (1.31) allows a breakup amplitude to be extracted from a numerical integration of Eq. (1.24), i.e., Eq. (1.31) is an appropriate boundary condition. The next order correction in the asymptotic series of which Eq. (1.31) is the leading term goes as ρ^{-1}. Thus, if $\rho = 100$ fm, Eq. (1.31) can be expected to be accurate to 10% (or better).

Breakup boundary conditions were first employed by Merkuriev, Gignoux, and Lavergne to extract elastic and breakup amplitudes from an $n-d$, coordinate-space Faddeev calculation [78]. Eq. (1.31) was later used by Payne in neutron-deuteron calculations [79] and an analog of it was employed by Kozack and Levin in model deuteron-nucleus calculations [80]; neither of these latter two sets of calculations were successful: in contrast to the work of Merkuriev et al., both sets of computations failed to yield stable amplitudes (elastic or breakup) when the bombarding energies were greater than the breakup threshold energy, even for $\rho \approx 120$ fm. This is much larger than the 30 fm employed by Gignoux et al. Initial efforts to understand the origin of the unstable results concerned the form of the asymptotic boundary conditions and were unsuccessful. What was needed instead was an investigation as to whether, at $\rho \lesssim 120$ fm, asymptotia is being achieved. For example, either Eq. (1.30) might not be correct at such ρ or the stationary phase approximation, Eq. (1.31) might not be valid.

That the method of stationary phase could be the problem was suggested by Glöckle [81]. Following this, Kuruoglu and Levin [82] and Glöckle [83] made detailed studies of the accuracy of the stationary phase approximation to Eq. (1.30). Kuruoglu and Levin constructed a variety of analytic models for $\bar{T}$ and found that for the simplest functional forms for $\bar{T}$, Eq. (1.31) could be 99% accurate for ρ as low as 50 fm, while for more realistic forms, Eq. (1.31) was found to be in error by 25% at $\rho \approx 8300 fm$. Glöckle used numerically-determined, Faddeev equation breakup amplitudes in his test, and found reasonable accuracy (a few percent) for the stationary phase approximation only when $\rho \gtrsim$ several thousand fm.

These unexpected results were most unwelcome, since they meant (again in contrast to the findings of Merkuriev et al.) that coordinate-space Faddeev calculations would be prohibitively time and storage intensive. But, as it turns out, this is not the end of the story: near the end of 1988, it was discovered for the MT I-III model [84] that by improving the numerical procedures (smaller mesh sizes, etc.), stable amplitudes *could* be obtained from a coordinate-space Faddeev calculation by imposing Eq. (1.31) as the breakup boundary condition! Two aspects about this development remain to be sorted out: why does a better numerical procedure appear to overcome an apparently intrinsic non-asymptotic form and, does stability imply accuracy? The answers to these question are essential to any future coordinate-space Faddeev computations and are under active investigation.

Coordinate space approaches to breakup calculations are a fairly recent innovation: the traditional means for performing breakup computations has been via momentum space integral equations. While there is a 20 year history of performing such calculations using separable potentials, it is only within the past few years that breakup computations whose input consists of modern realistic potentials have been undertaken. The work discussed herein is that of Glöckle and collaborators; it was motivated not only by a desire to search for any discrepancies which could require the inclusion of three-nucleon potentials in the computations, but also by the results of kinematically complete, proton-induced breakup experiments, $p+d \rightarrow p+p+n$ [85]. Certain spatial configurations have been proposed as especially significant in the search for 3NP effects, including the collinear and the equilateral triangle or space-star configurations: in the former, 3NP's tend to be repulsive, while in the latter they tend to be attractive, a configuration dependence found in other three particle systems [86]. In addition to these two configurations, additional emphasis has been placed on the so-called final state interaction (FSI) region, i.e., the kinematic configuration in which two of the three nucleons are in a relative motion state having almost zero energy, while the third nucleon carries away almost all of the final state kinetic energy. This is an important kinematic situation because the breakup cross sections are sensitive to the $^1S_0$ parts of nucleon-nucleon forces, in particular to any charge dependences, as manifested, e.g., in the difference between $n-n$ and $n-p$ $^1S_0$ forces.

Glöckle and collaborators have made a large number of calculations in their attempts to fit data, using both the Paris potential and the OBEPQ form of the Bonn potential. Reference [87], which analyzes the 14.1 MeV p–d breakup data, contains a summary of many of their other results as well as figures comparing theory and experiment, and it is used as the source of the following comments. The main finding is that significant discrepancies exist between the data and the calculations, which employ only two-nucleon forces. These discrepancies occur for both the FSI and collinear configurations in the 14.1 MeV case, for the equilateral triangle configuration at 13.0 MeV (although the agreement is very good in this configuration at 10.3 MeV), as well as for the neutron analyzing power in elastic $n-d$ scattering. Balanced against this are the facts that in the $p-d$ analyses, coulomb effects were not taken into account nor were nucleon-nucleon partial waves with j greater than 2. The former is not thought to be an important omission due to the coulomb barrier height being about 200 keV while the significant energies (those in the vicinity of the experimental peaks in the FSI case) are 6 MeV or greater [87]. Nevertheless, both points should, if possible, be investigated, as should the role, noted in Ref. [87], that 3NP's may play in these processes. These are computations whose undertaking is among the most important in this field.

1.3.4 Simplifications

Three-nucleon computations are not easy. Even separable-potential calculations of breakup are very non-trivial. Any procedures that simplify the complexity of three-particle dynamics are thus of interest. In this closing section on the three-nucleon system, some examples of such simplifications are discussed, including some additional surprises.

One category of simplification is that of analytically soluble models. It has been known for 25 years that the N-particle problem in one dimension is exactly soluble when the interactions are pairwise delta functions [88]. Gerjuoy and Adhikari [89] exploited this for the case N=3 to demonstrate analytically that a single Lippmann-Schwinger equation description of this system fails to yield a unique solution in the continuum, thereby helping to end a controversy on the adequacy of such a description. Despite the absence of three-particle breakup states in this system, the analysis of Gerjuoy and Adhikari is a tour-de-force that yielded results unobtainable by numerical means.

The preceding example demonstrates how useful a schematic interaction like the zero-range delta-function potential can be. Zero-range situations have long played an interesting role in three-particle physics, the earliest example being Thomas's demonstration that the three-particle binding energy is infinite when the interactions are of zero-range [90]; this was later followed by the zero-range treatment of the three-particle system by Skornyakov and Ter-Martirosyan [91]. More recently Brayshaw [92] and later Noyes and collaborators [93] have used such models to obtain interesting results. It thus came as somewhat of a surprise when Friar, Gibson and

Payne [94] obtained an analytic solution to the N-dimensional Faddeev equations for the ground state of three spinless particles in a particular, finite-range potential, viz, that for a harmonic oscillator. This allows one to examine the behavior of the individual Faddeev components analytically. The behavior of these components in configuration space is useful to know and to compare with their sum, the Schrödinger solution, especially as it is only the latter quantity that has physical significance. In particular, even in this simple model, it was found that the individual components display features not seen in the full solution (i.e., features which cancel in the sum). A similar result has been found in numerical calculations based on short-range pair interactions [95].

Mentioned in Subsec. 1.3.2 was the demonstration by Fonseca, Redish and Shanley of an R^{-2} Efimov-effect potential in a system of two heavy particles and one light one. Use of this model was an outgrowth of earlier work by Fonseca and Shanley which explored the limits of validity of the Born-Oppenheimer approximation when the light-heavy interaction was separable [96]. They found, contrary to previous belief, that the B-O approximation was extremely accurate even for heavy to light mass ratios as low as 5, a result attributed to the short range nature of the interaction. Extensions to the case of scattering have been made and are both relatively simple and amusing to study [97].

A major simplification occurs, of course, when separable interactions are used, since two-dimensional integral equations are reduced to coupled, one-dimensional integral equations. Because such interactions are unrealistic, computations involving them are not being reported in detail here. One exception to this is the calculation of Phillips, which established the linear relation between ${}^2a_{nd}$ and $B_{^3\mathrm{H}}$ [98]. Phillips constructed a variety of ${}^1S_0$ and ${}^3S_1 - {}^3D_1$ separable potentials, each of which gave correct values of the deuteron binding energy, triplet scattering length and triplet effective range but yielded different values of the deuteron D-state probability P_D and the singlet scattering length and phase shifts. A comparison of the resulting sets of ${}^2a_{nd}$ and $B_{^3\mathrm{H}}$ led to the linear correlation of Ref. [98]. An immediate consequence was the identification of the "correct" value of ${}^2a_{nd}$ from the then pair of competing experimental values!

This "Phillips line" was the first example of "scaling" in the tri-nucleon systems [49], i.e., of the fact that the values of a set of seemingly independent observables are actually fixed once one of them is determined. Following Phillips's discovery, Brayshaw showed, using a boundary condition model, that low energy breakup observables were correlated with ${}^2a_{nd}$ [99], while in a related development, Tjon found a correlation between $B_{^3\mathrm{H}}$ and $B_{^4He}$ [100], though subsequent analyses indicated rather more scatter in the pairs of points than in the A=3 case [101]. Girard and Fuda established that ${}^2a_{nd}$ was correlated linearly with C_0 [102] (the asymptotic S-state nor-

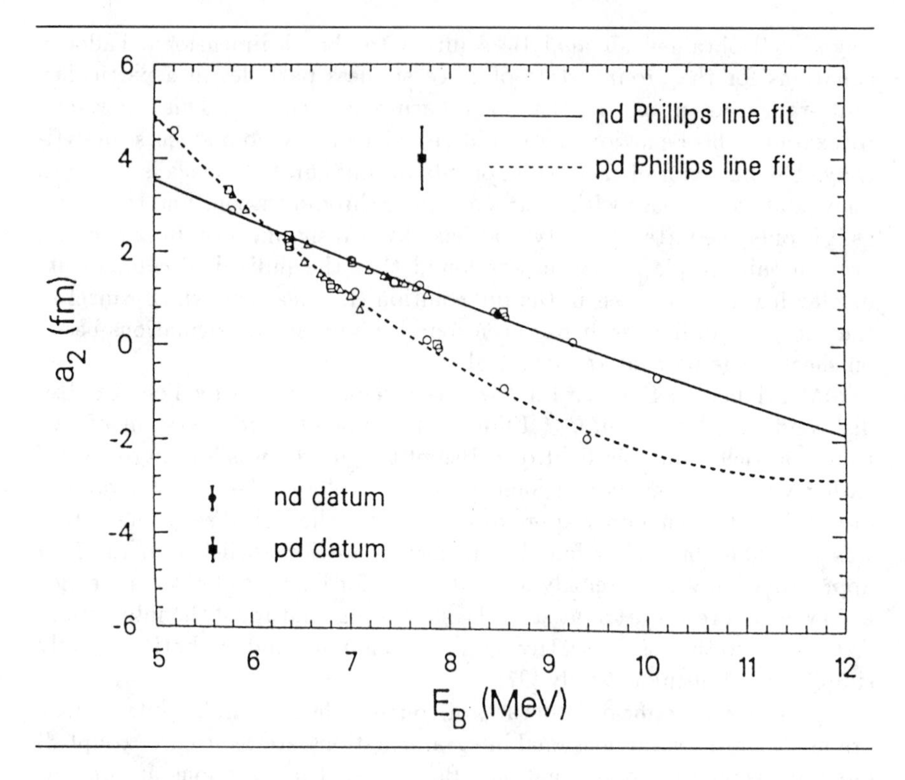

FIGURE 1.11
Scaling phenomena: dependence of the nucleon-deuteron doublet scattering lengths on the three-nucleon binding energy (from Ref. [41]).

malization constant) and with the pole in ${}^2K_{nd}(E)$ [103] (and thus with the virtual (excited) state of the triton). Finally, Friar, Gibson, Chen and Payne introduced the notion of scaling as used herein in a paper that extended the above correlations by showing [49] that the tri-nucleon charge radii and the triton S′-state probability each scale with $B_{^3\mathrm{H}}$, the observable now most commonly chosen as the independent one.

A recent compilation of this set of correlations has been made by Gibson and McKellar [41], from which Figs. 1.11–1.13 are taken. Of particular interest is the huge discrepancy between the Phillips curve for ${}^2a_{pd}$ vs $B_{^3\mathrm{He}}$ and the data of Fig. 1.11. This is the only example in which the datum does not fall on the Phillips curve at the correct value of B_3. This otherwise inexplicable aberration is easily explained by the p-d analysis of the preceding subsection as the result of measurements at insufficiently small values of E. That is, no measurements have yet been made at energies low enough

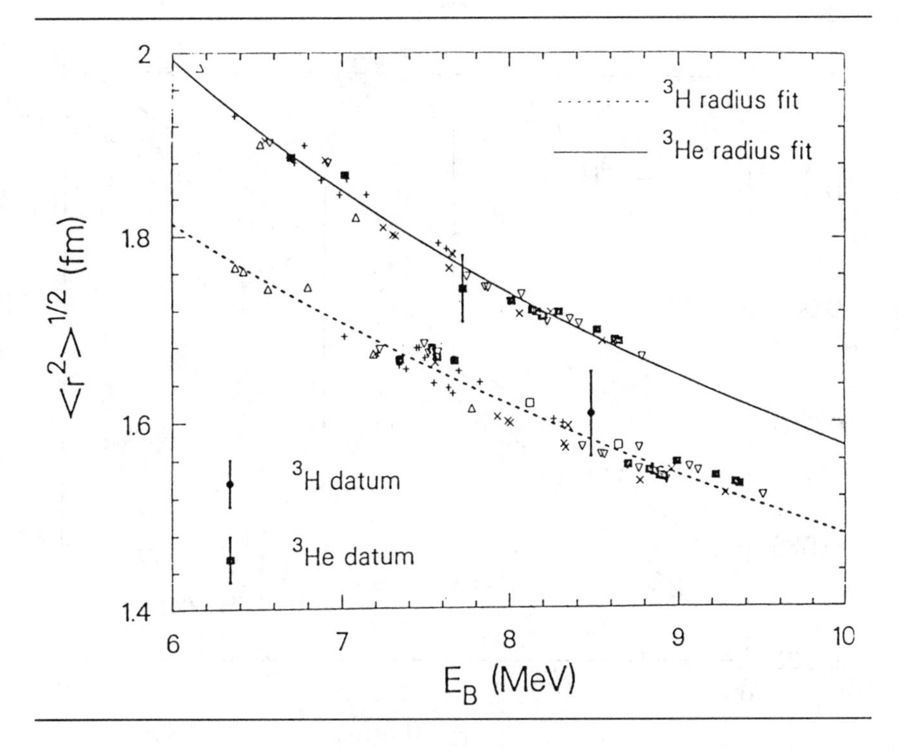

FIGURE 1.12
Scaling phenomena: dependence of the root mean square charge radii on the three-nucleon binding energy (from Ref. [41]).

to display the curvature of ${}^2K_{pd}(E)$ predicted theoretically, and hence the straight line extrapolation of the present data to E=0 gives an entirely false value of ${}^2a_{pd}$.

All the theoretical "data" points in these figures are the results of calculations employing full three-nucleon dynamics, i.e., they are based on Faddeev-type of analyses. The existence of these correlations strongly suggests that there must be simple, physical (i.e., qualitative) arguments which could be invoked to account for these results. A further implication is that a simpler, non-Faddeev dynamics may be able to describe aspects of the three-nucleon systems, at least at low energies. Both inferences are correct, and this section concludes with a brief discussion of each.

Just the nomenclature "scaling" itself is suggestive of a set of functional relations such as ${}^2a_{nd} = {}^2a_{nd}(B_{^3\mathrm{H}})$, $\langle r^2\rangle^{1/2} = \langle r^2\rangle^{1/2}(B_3)$, $C_{0,2} = C_{0,2}(B_3)$, etc. Efimov and Tkachenko [104,105] have noted that the Phillips

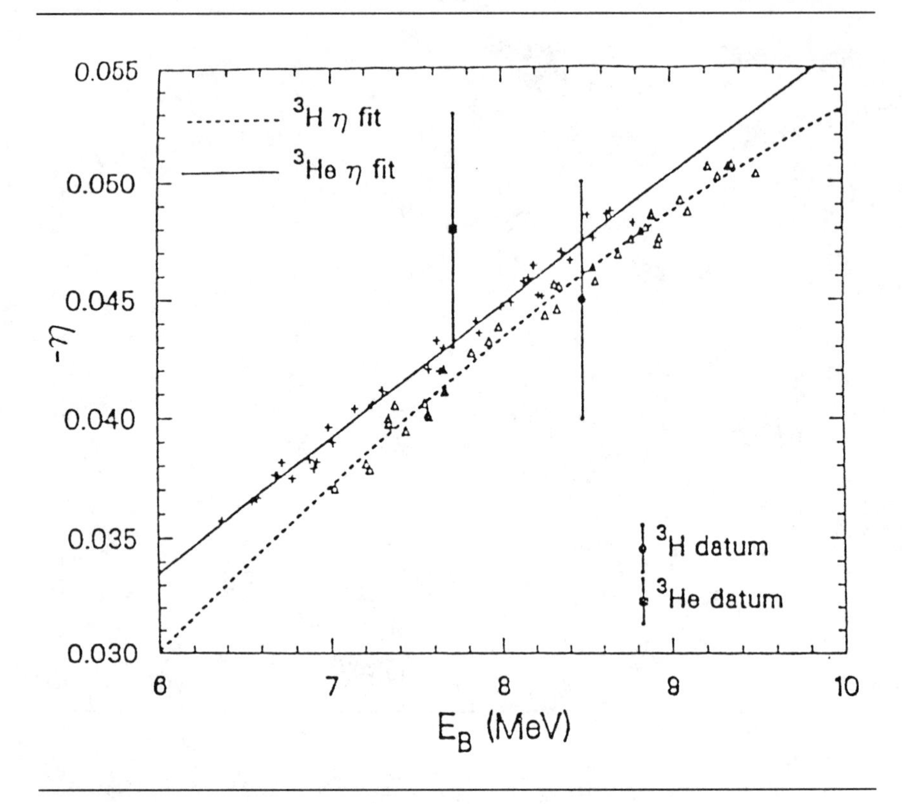

FIGURE 1.13
Scaling phenomena: dependence on the three-nucleon binding energy of the ratio of the tri-nucleon asymptotic D-state to S-state normalization constants (from Ref. [41]).

line relating ${}^2a_{nd}$ and $B_{^3\mathrm{H}}$ can be regarded as the N=3 analog of the N=2 relation between the deuteron binding-energy and the n-p triplet scattering length a_t, viz, $B_2 \cong \hbar^2/ma_t^2$ [8]. This latter relation is a consequence of B_2 being much smaller than the strength V_0 of the attractive part of the n-p, spin-triplet potential, e.g., $B_2/V_0 \cong .05$. Since $V_0 \gg B_3$, Efimov and Tkachenko argue that zero-range theory plus a simple linear correction can be used to derive an implicit relation between ${}^2a_{nd}$ and B_3; the comparison of the results of this relation with numerically determined values is shown in Fig. 1.14, taken from Ref. [105]. The overall agreement is impressive. Simple approximations, based on a dispersion-theoretic N/D formula, have also been used by Adhikari and Torreão [106] to provide a fit to the ${}^2a_{nd}$ vs $B_{^3\mathrm{H}}$ curve.

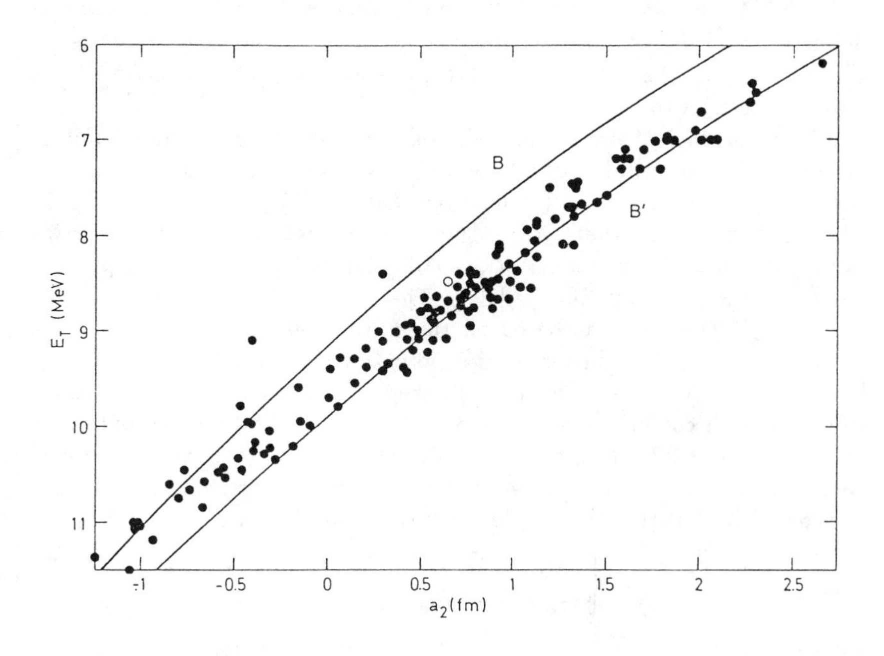

FIGURE 1.14
$^2a_{nd}$ vs $B_{^3\mathrm{H}}$ calculations of Efimov and Tkachenko [105]. Curve B: zero-range theory; curve B': zero-range theory plus linear correction. The open circle is the experimental result.

In their scaling analysis Friar *et al.* [49] approximated the triton bound state wave function everywhere by its asymptotic form $\exp(-K\rho)/\rho^{5/2}$, where ρ is the hyperspherical radius introduced near the end of Subsec. 1.3.1, and $K^2 = MB_3/\hbar^2$. They also used an isospin decomposition of $\langle r^2 \rangle$ into a isoscalar part $\langle r^2 \rangle_s$ and an isovector part given by two terms, one proportional to $\langle r^2 \rangle_s$ and another, denoted $\langle r^2 \rangle_V$, which is largely determined by the overlap of the S and the S′ states in the tri-nucleon. They found $\langle r^2 \rangle_s^{1/2} \cong 1/2K \cong 1/2(\hbar^2/MB_3)^{1/2}$. Fitting curves of the form $\lambda(\hbar^2/MB_3)^\beta$ to the theoretical data points for $\langle r^2 \rangle_s$ and $\langle r^2 \rangle_V^{1/2}$ yields $\lambda_s = 0.8$, $\beta_s = 0.47$, and $\lambda_V = 0.14$, $\beta_V = 0.89$; the former pair are in reasonable agreement with the qualitative values $\lambda_s = 0.50$, $\beta_s = 0.5$ deduced from use of the asymptotic wave function. Furthermore, within an oscillator shell model picture used to qualitatively represent the triton, Friar et al. noted that the S′ state is composed of $2\hbar\omega$ excitations, which ultimately leads to a $P_{S'} \propto B_{^3\mathrm{H}}^{-2}$. The analytic curve fitted to the theoretical points is $P_{S'} \propto B_{^3\mathrm{H}}^{-2.1}$, showing the validity of their qualitative arguments. The accuracy of the hyperspherical formula for the coulomb energy [40] is also discussed in this paper.

The existence of so many correlations led Tomio, Delfino and Adhikari to re-examine the validity of a two-body description of the low energy trinucleon systems in terms of a nucleon-deuteron effective potential [107]. These authors noted that typical short-range, effective potential descriptions, e.g., optical model descriptions [108], fail in the N-d case, as pointed out some time ago by Efimov [109]. The successful insight of Tomio et al. into this problem was based on the Efimov effect and its coordinate space characterization in terms of a potential behaving like ρ^{-2} (Efimov [35]) or R^{-2} (Fonseca et al. [38]). Tomio *et al.* constructed a potential which, though cut off at large nucleon-deuteron separations y, contains a y^{-2} dependence for y larger than a typical nuclear force range. Two different three-parameter, effective interactions were constructed. The example considered here is their interaction A, which has the form

$$V(y) = \begin{cases} -V_0 = -\widetilde{V}_0 \exp(-\mu y_0)/y_0^2 & , y \leq y_0 \\ -\widetilde{V}_0 \exp(-\mu y)/y^2 & , y \geq y_0 . \end{cases}$$

Added to $V(y)$ for the p-d case was the coulomb interaction.

The parameters V_0, y_0 and μ were obtained by fitting them to the experimental values of $^2a_{nd}$, $B_{^3\mathrm{H}}$ and $B_{^3\mathrm{He}}$. The validity of this two-body, effective potential description will be established if the subsequent calculation of $^2a_{pd}$, $^2K_{nd}(E)$ and $^2K_{pd}(E)$ yields reasonable results. Tomio et al. calculated a value of $0.2 \pm .05$ fm for $^2a_{pd}$ using interaction A, and then suggested the smaller value 0.15 ± 0.1, based on qualitative arguments, as the correct range to compare with any eventual, very low-E p-d experimental results. Both numbers are in quite reasonable agreement with the value of 0.17 fm obtained in the calculations of Chen et al. [67]. And, just as found in these latter calculations, interaction A yields a $^2K_{pd}$ which has a pole slightly below the elastic scattering threshold. Comparison of the p-d and n-d effective-range functions of Tomio et al. with data is only fair, although the general trend of the data is predicted. Finally, Phillips lines were determined and compared with the results of other calculations in Ref. [107]; the results are in reasonably good agreement.

In view of the previous failure of an effective two-body description, the success of the Tomio et al. calculation is another triumphant surprise. It very clearly demonstrates that important aspects of the three nucleon systems can be understood in terms of not only two-body dynamics, but two-body, *non-relativistic* dynamics. Even if this simplicity of description is limited to a total energy lying below the breakup threshold, it still demonstrates that complex three-particle dynamics is not required to describe all tri-nucleon properties. Though it seems unlikely that such a description could be extended to breakup reactions, it lends hope to any future searches for simple ways to understand breakup phenomena. Such searches cannot be undertaken until many more theoretical and experimental investigations of breakup have been carried out. Presumably, they will be, since the breakup regime is a key area in the future of tri-nucleon physics.

1.4 Illusion or Reality?

Not only is the phrase "nothing succeeds like success" an aphorism open to question, but its opposite, "nothing fails like failure", would be also. Bearing in mind that physical theories are never proved right but only wrong, it is appropriate to ask whether, due to its failures, the nucleons-only, non-relativistic quantal description of the three-nucleon system should be considered illusory. This review has tried to make clear that the answer to this query is NO: just as in the film whose title has been used here, the old order is neither passé nor powerless. The nucleons-only, non-relativistic quantal description is alive and well, thank you. It will certainly be used for years to account for data. And, it has all the elements one associates with a successful theory, viz, elegance, predictive power, accuracy, and the ability to create surprises. In other words, it works and works reasonably well. Naturally, there are limitations to this approach, some of which have been noted. But, as the limits of these limitations are not all precise, there are still opportunities for further successes: the failures to fit experimental data, apart from those corrections due to meson exchange effects (and possibly other relativistic effects), are not yet clearly assignable to the framework. No further assessment of this framework will be possible until three-nucleon potentials and breakup phenomena have been much more fully explored. The future of this field should be as exciting as the recent past.

Acknowledgements

This work was supported in part by the US Department of Energy.

Bibliography

[1] The other triplets, *viz*, *nnn* and *ppp*, do not exist in even a long-lived, weakly decaying bound state such as ^3H and are not considered in this review.

[2] These analyses were carried out by R. D. Amado and collaborators and A. N. Mitra and collaborators. Summaries of their work and relevant literature citations can be found in reviews authored by them: R.D. Amado, in *Elementary Particle Physics and Scattering Theory* (Gordon and Breach, New York, 1970) and A.N. Mitra, in *Advances in Nuclear Physics*, Vol. 3 (Plenum, New York, 1970).

[3] For a recent review see: P.U. Sauer, in *Few Body Methods: Principles and Applications*, ed. T.K. Lim, C.G. Bao, D.P. Hou and H.S. Huber (World Scientific, Singapore, 1986).

[4] A qualitative discussion of bag models is the article of G.E. Brown and M. Rho in Physics Today, Vol. 36, No. 2, February 1983. A more quantitative review, with references to both original articles and other reviews is that of F. Myhrer and J. Wroldsen, Rev. Mod. Phys. **60** (1988) 629.

[5] The concept of isospin as well as its implementation, is discussed in many texts on nuclear physics. Two of the many are: A. de Shalit and H. Feshbach, *Theoretical Nuclear Physics*, Vol. I, (Wiley, New York, 1974), and M.A. Preston and R.K. Bhaduri, *Structure of the Nucleus* (Addison-Wesley, Reading, 1975).

[6] The pairs $\vec{p}_{ij}$ and $\vec{p}_k$, and their canonical coordinate space partners introduced later in this review, are examples of Jacobi coordinates. Definitions, and relations between them, can be found in many places, e.g., in the article by I.R. Afnan and A.W. Thomas in *Modern Three Hadron Physics*, ed. A.W. Thomas (Springer-Verlag, New York, 1977).

[7] Nucleon values in Table 1.2 are from Preston and Bhaduri [5], while those for the deuteron are from R.A. Brandenburg, G.S. Clulick, R. Machleidt, A. Picklesimer and R.M. Thaler, Phys. Rev. **C37** (1988) 1245.

[8] Deuteron properties are discussed in Ref. [5].

[9] T.E.O. Ericson, in *Few Body Problems in Physics*, Vol. I, ed. B. Zeitnitz (North Holland, Amsterdam, 1984).

[10] R.D. Amado, Phys. Rev. **C19** (1979) 1473; J.L. Friar, Phys. Rev. **C20** (1979) 325.

[11] This relation can be found in any text on quantum mechanics that treats scattering theory, as well as in monographs on scattering theory.

[12] The LS equation is discussed at an elementary level in many quantum mechanics texts. For a treatment which deals with some of its mathematical properties, see, e.g., R.G. Newton, *Scattering Theory of Waves and Particles*, 2nd Ed. (Springer, New York, 1982).

[13] This is discussed by Newton [12]; see also K.M. Watson and J. Nuttall, *Topics in Several Particle Dynamics* (Holden-Day, San Francisco, 1967).

[14] This non-uniqueness was first pointed out by L.L. Foldy and W. Tobocman, Phys. Rev. **105** (1957) 1099. An elegant alternative demonstration has been given by W. Sandhas, in *Few Body Nuclear Physics* (IAEA, Vienna, 1978).

[15] See, e.g., Newton [12] or L.S. Rodberg and R.M. Thaler, *Introduction to the Quantum Theory of Scattering* (Academic, New York, 1967).

[16] The "effective range" expansion (1.20) is derived in many texts and monographs, see, e.g., Newton [12] or Rodberg and Thaler [15].

[17] R.O. Berger and L. Spruch, Phys. Rev. **138** (1965) B1106; see also Gy. Bencze and C. Chandler, Phys. Lett. **163B** (1985) 21.

[18] The coulomb-modified, strong-interaction scattering length is discussed in various nuclear physics texts. See J.D. Jackson and J.M. Blatt, Rev. Mod. Phys. **22** (1950) 77 for the original treatment.

[19] The numbers in Table 1.3 are from Brandenburg *et al.*, Ref. [7], who cite original sources.

[20] For discussions of symmetry imposed restrictions on the form of the nucleon-nucleon interaction, see Ref. [5] or P. Ring and P. Schuck, *The Nuclear Many Body Problem* (Springer, New York, 1980).

[21] R. Machleidt, K. Holinde and Ch. Elster, Phys. Rep. **149** (1987) 1; K. Holinde, Few Body Systems, Suppl. **2** (1987) 22.

[22] M.M. Nagels, T.A. Rijken and J.J. de Swart, Phys. Rev. **D17** (1978) 768; J.J. de Swart and M.M. Nagels, Forschr. Phys. **26** (1978) 215.

[23] M. Lacombe, B. Loiseau, J.M. Richard, R. Vinh Mau, P. Pires, and R. de Tourreil, Phys. Rev. **D12** (1975) 1495; M. Lacombe *et al.*, Phys. Rev. **C21** (1980) 861.

[24] See, e.g., W. Grüebler, Nucl. Phys. **A353** (1981) 31 and H.O. Klages, Nucl. Phys. **A463** (1987) 353.

[25] Recent studies, with citations to the earlier work, can be found in Y. Koike, J. Haidenbauer and W. Plessas, Phys. Rev. **C35** (1987) 396.

[26] R.V. Reid, Ann. Phys. (NY) **50** (1968) 411.

[27] R.B. Wiringa, R.A. Smith and T.L. Ainsworth, Phys. Rev. **C29** (1984) 1207.

[28] J.J. de Swart, W.A. van der Sanden and W. Derks, Nucl. Phys. **A416** (1984) 299.

[29] W. Plessas and J. Haidenbauer, Few Body Systems, Suppl. **2** (1987) 185.

[30] D.J. Ernst, C.M. Shakin and R.M. Thaler, Phys. Rev. **C8** (1973) 507.

[31] G.S. Chulick, Ch. Elster, R. Machleidt, A. Picklesimer, and R.M. Thaler, Phys. Rev. **C37** (1988) 1549.

[32] L.D. Faddeev, Zh. Eksp. Teor. Fiz. **39** (1960) 1459 (Sov. Phys. JETP **12** (1961) 1014) and Mathematical Aspects of the Three Body Problem in *The Quantum Scattering Theory* (Davey, New York, 1965).

[33] Resolvent operators, outgoing-wave Green's functions, etc., are discussed, e.g., in Refs. [12] and [15].

[34] See, e.g. B. Simon, in *Studies in Mathematical Physics*, ed. E.H. Lieb, B. Simon and A.S. Wightman (Princeton, 1976).

[35] V. Efimov, Phys. Lett. **33B** (1970) 563; Nucl. Phys. **A210** (1973) 157.

[36] R.D. Amado and J.V. Noble, Phys. Lett. **35B** (1971) 25; Phys. Rev. **D5** (1972) 1992. See also the discussion of this topic in the review by Amado: R.D. Amado, in *Few Particle Problems in the Nuclear Interaction*, ed. I. Slaus, S.A. Moszkowski, R.P. Haddock and W.T.H. van Oers (North Holland, Amsterdam, 1972) pp. 260–266.

[37] S.K. Adhikari, A. Delfino, T. Frederico, I.D. Goldman and L. Tomio, Phys. Rev. **A37** (1988) 3666.

[38] A.C. Fonseca, E.F. Redish and P.E. Shanley, Nucl. Phys. **A320** (1979) 273.

[39] T.K. Lim, K. Duffy and W.C. Damert, Phys. Rev. Lett. **38** (1977) 341; H.S. Huber and T.K. Lim, J. Chem. Phys. **68** (1978) 1006; H.S. Huber, Phys. Rev. **A31** (1985) 3981.

[40] J.L. Friar, Nucl. Phys. **A156** (1970) 43; M. Fabre de la Ripelle, Fisica **4** (1972) 1.

[41] B.F. Gibson and B.H.J. McKellar, Few Body Systems **3** (1988) 143.

[42] Reviews of this approach can be found in P.U. Sauer, Prog. Part. Nucl. Phys. **16** (1986) 35 and Nucl. Phys. **A463** (1987) 273.

[43] J.-I. Fujita and H. Miyazawa, Prog. Theor. Phys. **17** (1957) 360.

[44] S.A. Coon, M.D. Scadron, P.C. McNamee, B.R. Barrett, D.W.E. Blatt and B.H.J. McKellar, Nucl. Phys. **A317** (1979) 242.

[45] H.T. Coelho, T.K. Das and M.R. Robilotta, Phys. Rev. **C28** (1983) 1812.

[46] J. Carlson, V.R. Pandhiripande, and R.B. Wiringa, Nucl. Phys. **A401** (1983) 59.

[47] See, e.g., Gibson and McKellar, Ref. [41].

[48] S.A. Coon, Few Body Systems, Suppl. **1** (1987) 41.

[49] J.L. Friar, B.F. Gibson, C.R. Chen and G.L. Payne, Phys. Lett. **161B** (1985) 241.

[50] See, e.g., J.L. Friar, in *New Vistas in Electro-Nuclear Physics*, ed. E.L. Tomusiak, H.S. Caplan and E.T. Dressler (Plenum, New York, 1986).

[51] See, e.g., J.L. Friar, Nucl. Phys. **A463** (1987) 315.

[52] D.O. Riska and G.E. Brown, Phys. Lett. **38B** (1972) 193.

[53] See, e.g., J.L. Friar and B. Frois, in *The Three–Body Force in the Three–Nucleon System*, ed. B.L. Berman and B.F. Gibson (Springer–Verlag, Berlin, 1986).

[54] C. Hajduk, P.U. Sauer, and W. Strueve, Nucl. Phys. **A405** (1983) 581; see Refs. [50] and [53] for interpretive discussions.

[55] T. Sasakawa, Nucl. Phys. **A463** (1987) 327.

[56] R.A. Brandenburg, G.S. Chulick, R. Machleidt, A. Picklesimer and R.M. Thaler, Los Alamos National Laboratory preprint LA-UR-86-3700 (1986), unpublished.

[57] Kr.T. Kim, Y.E. Kim, D.J. Klepacki, R.A. Brandenburg, E.P. Harper, and R. Machleidt, Phys. Rev. **C38** (1988) 2866.

[58] R.A. Brandenburg, G.S. Chulick, R. Machleidt, A. Picklesimer and R.M. Thaler, Phys. Rev. **C37** (1988) 1245.

[59] J.L. Friar, B.F. Gibson and G.L. Payne, Phys. Rev. **C37** (1988) 2869.

[60] Y. Koike, W. Plessas and H. Zankel, Phys. Rev. **C32** (1985) 1796; additional results are given in Y. Koike, J. Haidenbauer and W. Plessas, Phys. Rev. **C35** (1987) 396.

[61] W.T.H. van Oers and J.D. Seagrave, Phys. Lett. **24B** (1967) 562.

[62] H.P. Noyes, in *Three Particle Scattering in Quantum Mechanics*, ed. J. Gillespie and J. Nuttall (Benjamin, New York, 1968) and in *Three Body Problem in Nuclear and Particle Physics*, ed. J.S.C. McKee and P.M. Rolph (North Holland, Amsterdam, 1970).

[63] W. Dilg, L. Koester and W. Nistler, Phys. Lett. **36B** (1971) 208.

[64] J.L. Friar, B.F. Gibson and G.L. Payne, Phys. Lett. **124B** (1983) 287.

[65] R.A. Malfliet and J.A. Tjon, Nucl. Phys. **A127** (1969) 161.

[66] E.O. Alt, in *Few Body Dynamics*, ed. A.N. Mitra, I. Slaus, V.S. Bhasin and V.K. Gupta (North Holland, Amsterdam, 1976).

[67] C.R. Chen, G.L. Payne, J.L. Friar and B.F. Gibson, Phys. Rev. **C39** (1989) 1261.

[68] See the articles by L.P. Kok and by S.P. Merkuriev in *Few Body Problems in Physics*, ed. L.D. Faddeev and T.I. Kopaleishvili (World Scientific, Singapore, 1985).

[69] G.H. Berthold and H. Zankel, Phys. Rev. **C34** (1986) 1203.

[70] Gy. Bencze, C. Chandler, J.L. Friar, A.G. Gibson and G.L. Payne, Phys. Rev. **C35** (1987) 1188.

[71] J.R. Taylor, *Scattering Theory* (Wiley, New York, 1972).

[72] S.P. Merkuriev, Sov. J. Nucl. Phys. **24** (1976) 150; Theor. Math. Phys. **32** (1977) 680; Lett. Math. Phys. **3** (1979) 141; Ann. Phys. (NY) **130** (1980) 395; Acta Phys. Austriaca, Suppl. **23** (1981) 65.

[73] E.O. Alt Ref. [66]; E.O. Alt, W. Sandhas, H. Zankel and H. Ziegelmann, Phys. Rev. Lett. **37** (1976) 1537; E.O. Alt, W. Sandhas and H. Ziegelmann, Phys. Rev. **C17** (1978) 1981; E.O. Alt and W. Sandhas, Phys. Rev. **C21** (1980) 1733.

[74] E.O. Alt, in *Few Body Methods: Principles and Applications*, ed. T.K. Lim, C.G. Bao, D.P. Hou and H.S. Huber (World Scientific, Singapore, 1986).

[75] E.O. Alt, invited papers presented at the *X European Symposium on the Dynamics of Few Body Systems*, 3–7 June 1985, Balatonfured, Hungary (Mainz University preprint MZ–TH/85-09 May 1985) and at the *International Conference on the Theory of Few Body and Quark–Hadronic Systems*, 16–20 June 1987, Dubna, USSR (Mainz University preprint MZ–TH/87-08, June 1987).

[76] W. Glöckle, Z. Phys. **271** (1974) 31.

[77] See, e.g., A. Erdelyi, *Asymptotic Expansions* (Dover, New York, 1956).

[78] S.P. Merkuriev, G. Gignoux, and A. Lavergne, Ann. Phys. (NY) **99** (1976) 30.

[79] G.L. Payne, priv. comm.

[80] R. Kozack and F.S. Levin, unpublished calculations, referred to in Phys. Rev. **36** (1987) 883.

[81] W. Glöckle, unpublished calculations (priv. comm. from J.L. Friar and W. Glöckle).

[82] Z.C. Kuruoglu and F.S. Levin, Phys. Rev. **C36** (1987) 49.

[83] W. Glöckle, Phys. Rev. **C37** (1988) 6.

[84] Priv. comm. from J.L Friar and G.L. Payne.

[85] M. Karas, M. Buballa, J. Helten, B. Laumann, R. Melzer, P. Neissen, H. Oswald, G. Rauprich, J. Schulte-Uebbing and H. Paetz gen. Scheck, Phys. Rev. **C31** (1985) 1112.

[86] See, e.g., Refs. [41], [50], [51] and [53].

[87] H. Witala, W. Glöckle and Th. Cornelius, Phys. Rev. **C39** (1989) 384.

[88] J.B. McGuire, J. Math. Phys. **5** (1964) 622.

[89] See E. Gerjuoy and S.K. Adhikari, Phys. Rev. **C34** (1986) 1, which contains additional references to their own work and to the controversy noted in the text.

[90] L.H. Thomas, Phys. Rev. **47** (1935) 903.

[91] G.V. Skornyakov and K.A. Ter-Martirosyan, Zh. Eksp. Teor. Fiz. **31** (1956) 975 (Sov. Phys. JETP **4** (1956) 648).

[92] See, e.g., D.D. Brayshaw, in *Modern Three Hadron Physics*, ed. A.W. Thomas (Springer–Verlag, Berlin, 1977).

[93] See, e.g., H.P. Noyes and J.V. Lindesay, Aust. J. Phys. **36** (1983) 601.

[94] J.L. Friar, B.F. Gibson and G.L. Payne, Phys. Rev. **C22** (1980) 284.

[95] See, e.g., J.L. Friar, B.F. Gibson and G.L. Payne, Z. Phys. **A301** (1981) 309.

[96] A.C. Fonseca and P.E. Shanley, Ann. Phys. (NY) **117** (1979) 268.

[97] A.C. Fonseca and P.E. Shanley, Nucl. Phys. **A382** (1982) 97.

[98] A.C. Phillips, Nucl. Phys. **A107** (1968) 209.

[99] D.D. Brayshaw, Phys. Rev. Lett. **32** (1974) 382.

[100] J.A. Tjon, Phys. Lett. **56B** (1975) 217.

[101] J.A. Tjon, in *Few Body Systems and Nuclear Forces*, Vol. II, ed. H. Zingl, M. Haftel and H. Zankel (Springer–Verlag, Berlin, 1978).

[102] B.A. Girard and M.G. Fuda, Phys. Rev. **C19** (1979) 583.

[103] B.A. Girard and M.G. Fuda, Phys. Rev. **C19** (1979) 579.

[104] V. Efimov and E.G. Tkachenko. Phys. Lett. **157B** (1985) 108.

[105] V. Efimov and E.G. Tkachenko, Few Body Systems **4** (1988) 71.

[106] S.K. Adhikari and J.R.A. Torreão, Phys. Lett. **132B** (1983) 257.

[107] L. Tomio, A. Delfino and S.K. Adhikari, Phys. Rev. **C35** (1987) 441.

[108] The optical model is discussed in many monographs and lecture notes on nuclear physics; see, e.g., G.R. Satchler, *Direct Nuclear Reactions* (Oxford, New York, 1983).

[109] V. Efimov, Nucl. Phys. **A362** (1981) 45.

2

Supernovae in Theory and Practice

G.E. Brown
Physics Department
State University of New York at Stony Brook
Stony Brook, New York 11794

Supernova 1987A (Shelton) had a large explosion energy. Such a large energy not only favors the prompt explosion mechanism, but also an equation of state at supranuclear densities which is substantially softer than obtained in conventional calculations. Inclusion of medium corrections, which are consistent with the implications of chiral symmetry restoration at higher densities, softens the high density equation of state, consistent with the supernova explosion requirements.

2.1 Introduction

Supernova 1987A (Shelton) has illustrated vividly that extremely energetic explosions resulting in supernovae do exist - at least one has been shown to exist - and I shall indicate that the explosion mechanism is a rather general one. As will be discussed later, one can deduce from the light curve that the explosion energy of SN 1987A is [1]

$$E_{expl} \cong 1 - 2 \text{ foe}. \qquad (2.1)$$

[1]The supernova unit of energy one foe equals ten to the fifty-one- ergs.

This requires the prompt shock mechanism of BCK [1] and probably a relatively soft equation of state for (neutron rich) nuclear matter in the region of densities $\rho \sim \rho_0$ to $4\rho_0$ where ρ_0 is nuclear matter density. Stiff equations of state or large iron cores lead to the Wilson delayed shock mechanism [2]. These result in an energy $\lesssim 0.5$ foe.

Nuclear matter calculations employing a nucleon-nucleon interaction which fits the phase shifts in nucleon-nucleon scattering and employing state-of-the-art many-body techniques [3] obtain a compression modulus $K_0 = 240$ MeV, which pertains to $\rho = \rho_0$, temperature $T = 0$, and an adiabatic index $\Gamma \cong 3.5$ for high densities $\rho \sim 3 - 4\rho_0$. This K_0 is nearly double that used by BCK [1] and Γ is substantially higher than the BCK $2 < \Gamma < 3$. (The Friedman-Pandharipande K_0 is also at the upper end of the $K_0 = 210 \pm 30$ MeV [4] obtained from the nuclear breathing modes. The BCK K_0 pertains, however, to a neutron-rich environment $\rho_n/\rho_p \sim 2$, since substantial electron capture has gone on, and this lowers $K_0 \sim 40$ MeV [5]. With $K_0(N = 2Z) \cong 140$ MeV, it was possible [1] to obtain prompt explosions.

Recently, the work of BCK [1] has been extended to include more complete neutrino transport, especially the downscattering of neutrinos by electrons, and to include the μ- and τ-neutrinos and antineutrinos. The result is that more energy is carried off by neutrinos, weakening the shock wave which carries out the explosion, with the result that the E_{expl} of Eq. (2.1) is substantially decreased. Depending on the initial model, the prompt explosion mechanism may not even work. However, it has also been shown that the compression modulus of infinite nuclear matter is much less than 180 MeV, indeed [6], $K_0 < 140$ MeV. Up to about twice nuclear matter density, the EOS is substantially softer than the BCK one. This should help the explosion considerably. Also, new burning rates may change initial models, as we shall discuss later. Decrease of the iron core in the initial model by $\sim 0.1\ M_\odot$ would greatly help the explosion. We shall therefore continue as if the BCK prompt mechanism [1] is a viable one, although Wilson's delayed mechanism is under active investigation at Brookhaven and Stony Brook. As noted earlier, it is difficult to see how the delayed mechanism will furnish the large explosion energy of Eq. (2.1).

As we shall outline later, medium dependent effects, not included in conventional nuclear matter calculations, would be expected to lower the adiabatic index substantially at high densities. These medium corrections can be interpreted as harbingers of the chiral restoration transition to quarks and gluons at high density. They are related, but not in a simple way, to the "swelling" of nucleons in the nucleus.

Highly relevant to our discussion is the investigation of the high-density equation of state at Bevlac energies $E/A \sim 1 - 2$ GeV. Early indications [7] were that it was extremely stiff, with $K_0 \cong 400$ MeV. It was qualitatively shown [8] that much of the apparent stiffness resulted from the strong momentum dependence in the nucleon-nucleon interaction. Quantitative calculations [9] showed that when soft EOS was supplemented by the empirically known momentum dependence, many of the features earlier ascribed to the stiff EOS could be reproduced although there were still some indications [9] of a stiff EOS in the flow angle. More recent work [6] shows the soft EOS plus known momentum dependence to give sideways flow essentially identical with that given by the stiff EOS. Note, however, that what the heavy-ion investigators call a soft EOS is close to the Friedman-Pandharipande [3] one, and not as soft as we now obtain.

In addition to the chiral restoration phase transition, which may come at rather high densities $\rho \sim 5 - 10\rho_0$, there are phase transitions which would be expected earlier on:

1. π^0 condensation at $\rho \sim 2\rho_0$ [11],
2. kaon condensation at $\rho \sim 2.7\rho_0$ [12],

so there is no paucity in the number of ways the conventionally calculated EOS, without inclusion of these phase transitions, can be softened.

Indeed, the graver problem may be in making the equation stiff enough to stabilize the masses of known neutron stars. Recent work by Prakash, Ainsworth and Lattimer [13] has shown that even with a compression modulus as low as $K_0 = 120$ MeV, neutron stars can be stable up to at least 1.47 $M_\odot$, sufficient to encompass the accurately measured $1.444 \pm .01$ $M_\odot$ star in PSR 1913-16. The central value of 4U0900-40 is larger, the value given as 1.85 ± 0.3 $M_\odot$, but the error is large. An interesting suggestion is that neutron stars of mass greater than ~ 1.5 $M_\odot$ cannot exist, because the EOS is soft.

2.2 Equation of State of Dense Nuclear Matter from Supernova 1987A (Shelton)

The first nearby supernova in 383 years, the first "close" one since the discovery of the telescope, is settling lots of questions and problems concerned with the supernova mechanism. It may be premature to draw definite conclusions, but analyses to date of the light curve [14] give an explosion energy of $\sim 1 - 2$ foe. This is the maximum energy that could be obtained from earlier analyses [1] and this large energy could be obtained in Ref. 1 only by assuming the equation of state of somewhat neutron-rich matter ($\sim$ twice as many neutrons as protons) to be quite soft at high densities $\rho \sim 2 - 4\rho_0$; specifically, with adiabatic index $\Gamma \sim 2.5$. With further improvements in the calculation, the EOS needs to be still softer.

We now sketch the scenario of stellar collapse, bounce and explosion. A primary role in this is played by the Chandrasekhar mass,

$$M_{ch} = 5.67 Y_e^2 \, M_\odot . \tag{2.2}$$

Here Y_e is the ratio of electrons to nucleons and $M_\odot$ denotes the mass of the sun, $M_\odot = 2 \times 10^{33}$ g or 2×10^{54} erg. The Chandrasekhar mass is the maximum mass which can be stabilized against gravitational collapse. Since the relativistic degenerate electrons provide essentially all of the pressure for densities $\rho < \rho_0$, it is natural that Y_e should play a dominant role here.

From the luminosity of the progenitor of SN1987A, Sanduleak 202-69, one can deduce that the progenitor mass was 15-20 $M_\odot$. Were the distance to the Large Magellanic Cloud, where the supernova explosion took place, more accurately known, we could pin down this mass more accurately. We shall take it to be 18 $M_\odot$.

As successive nuclear fuels are exhausted and the center of the star compresses with higher temperatures, finally silicon is burned into ^{56}Fe and the process stops there. ^{56}Fe has the largest binding energy per nucleon of any nucleus in the periodic table, so no energy can be obtained by burning it. Due to electron capture along the way, Y_e has dropped from an initial $Y_e = 0.5$ to $Y_e \cong 0.41$. Silicon burning will continue to add to the iron core until the latter reaches the Chandrasekhar mass relevant for $Y_e \cong 0.41$, $M_{ch} \gtrsim 1 \, M_\odot$. At this point the relativistic degenerate electrons can no longer stabilize the star, and the core collapses.

The entire collapse and bounce happen in seconds - indeed, the most important part, in milliseconds, as we shall detail - and this time is orders of magnitude less than the time in which the star exterior to the Fe core can respond. Therefore, we can consider the evolution of the Fe core independently of the rest of the star, but with the proviso that the mantle and envelope must be blown off, if a black hole is not to be formed. (For stars above a certain mass, the explosion is expected to result in a black hole. The dividing mass between those which give neutron stars and those which result in black holes is not known, but probably lies around 25 $M_\odot$.)

By burning to ^{56}Fe the star has established a great order before the onset of collapse, with entopy per nucleon, in units of Boltzmann's constant,

$$\frac{s}{k_B} \sim 1. \tag{2.3}$$

In fact, entropy is not increased by the strong interactions because there is always time for these to equilibrate, collapse times of $\sim$ 1 ms being many orders of magnitude longer than the strong-interaction times. Weak interactions equilibrate only for densities $\rho \gtrsim 10^{12}$ g/cm^3, but during the regime of densities in which they do not equilibrate the neutrinos from electron capture through

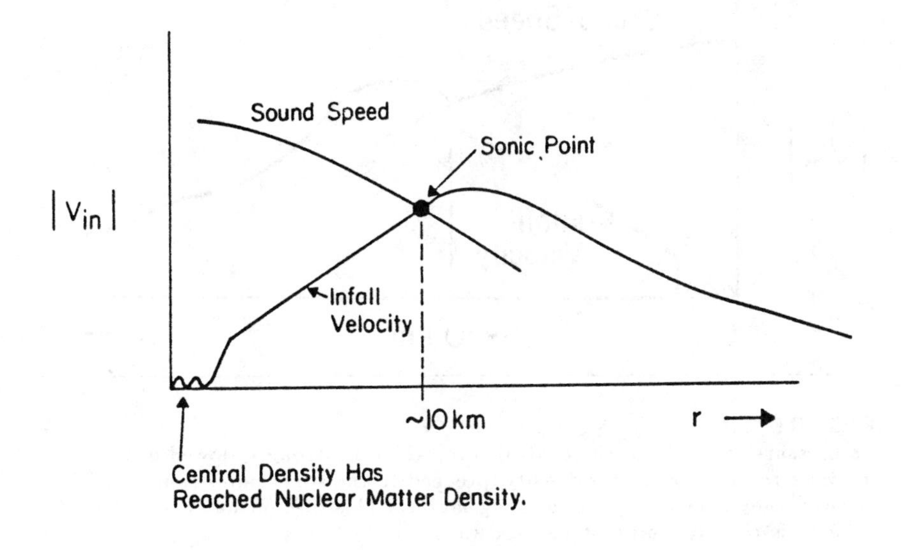

FIGURE 2.1
Profile of the infall velocity as function of r. The absolute value of the velocity rises linearly to the sonic point, where $s = -V_{infall}$. Here s is the speed of sound.

$$e^- + p \rightarrow n + \nu \tag{2.4}$$

mostly escape, carrying entropy off into space. Thus, one can consider $s/k_B \sim 1$ during the entire collapse.

As shown by BBAL [15] this is not a sufficient entropy for nuclei to break up, so nuclei stick together, and ultimately are squeezed into nuclear matter at a density of $\rho \sim 2 \times 10^{14}$ g/cm^3 before resistance is encountered to the collapse. By this time, the core has a considerable velocity, and overshoots up to several times nuclear matter density, the precise number depending upon the equation of state at these densities.

Some capture of electrons continues up to $\rho \sim 10^{12}$ g/cm^3, but after that density neutrinos are trapped and there is effectively equilibrium between electrons and neutrinos. For densities $\rho \gtrsim \rho_0 = 2 \times 10^{14}$ g/cm^3, $Y_e \sim 1/3$, meaning that there are about twice as many neutrons as protons.

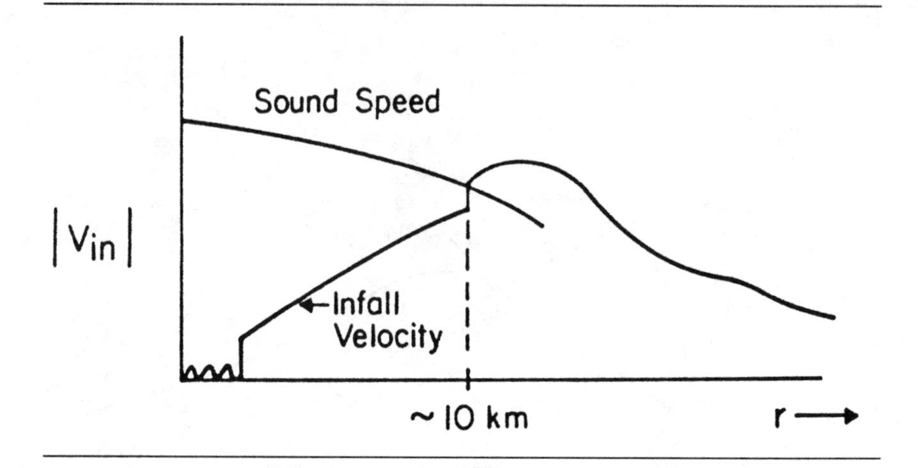

FIGURE 2.2
As pressure waves from the central core, which is strongly slowed down as it comes to nuclear matter density, proceed to the sonic point (and no further) they slow down the infalling matter. This discontinuity in the velocity is the beginning of a shock wave.

The real hot neutron star is formed only $\sim$ 0.5 s later, after the neutrinos can take away lepton number from the star.

During the collapse, the inner Chandrasekhar mass of the star collapses as a unit, i.e., homologously. We show the velocity profile in Fig. 2.1. The point at which the velocity of sound, which is greatest in the high density region in the interior, crosses the magnitude of the infall velocity is called the sonic point. Inside of this point, different parts of the infalling matter can communicate with each other by sound waves, so the core can collapse as a whole - therefore the homologous collapse. The behavior of the homologous core can be described analytically [16], generalizing the theory of polytropes to a time-dependent situation.

When the center of the core reaches nuclear matter density, suddenly pressure waves are emitted, because nuclear matter is hard to compress. These pressure waves proceed outwards with the velocity of sound. As pressure waves travelling with the velocity of sound - in the framework of the infalling matter - they cannot get beyond the sonic point; were they to venture beyond, they would be swept back by the supersonic infalling matter. Therefore, a discontinuity in the velocity builds up at the sonic point as the outgoing pressure waves try to slow down the infalling matter (see Fig. 2.2).

Because of the large infall velocity, the core overshoots and goes to high densities, up to $\rho \sim 4\rho_0$ in current calculations, before bouncing back. On the way back from the bounce, it does a lot of pdV work on the external matter, the energy from this work catching up with the shock wave, which has already taken off, feeding it power. (Sound, or pressure, waves behind a shock, catch up with it. The shock wave travels faster than the speed of sound in the matter ahead of it.) The bounce is highly damped, however, [17] by the infalling matter, and the core rapidly comes to rest.

In most calculations to date, the shock wave did not rip on out to the edge of the envelope of the star, as we believe it to have done in SN1987A, but slowed down and stopped at a radius of $\sim$ 100 km. It simply did not have the energy to get further. (The shock wave continuously loses energy by doing work on the infalling matter, especially by dissociating it, and by neutrino emission.)

During the frustrating time of these calculations, Wilson [2] found that over a long time period of $\sim$ 1 s, a small percentage of the neutrinos from the hot interior of the star were absorbed by the material behind the shock wave. This produced new pressure, which was sufficient to get the shock wave started again, and which produced shocks which blew off the mantle and envelope of the star. For stars in the mass region M $\sim$ 15 $M_\odot$ the explosion energies were only $\sim$ 0.5 foe, as noted earlier, but could be built up somewhat by convection.

In 1984, determined to see how the prompt mechanism worked, Cooperstein et al [18] manufactured an Fe core of mass 1.25 $M_\odot$ from a 15 $M_\odot$ star evolved by Woosley and Weaver [19] with original core of 1.35 $M_\odot$, and were able to obtain a marginally successful prompt explosion. These authors were still using a very stiff equation of state.

Why does decreasing the Fe core by 0.1 $M_\odot$ make such a difference? The shock wave from the bounce of the core following collapse begins, as we discussed above, at the edge of the homologous core, mass point 0.6 $M_\odot$ - 0.7 $M_\odot$. It then makes its way out through the rest of the core, raising the entropy per nucleon from $\sim$ 1 to $\sim$ 10 (in units of Boltzmann's constant), dissociating the nuclei in the process. Dissociation of 0.1 $M_\odot$ costs the shock $\sim$ 2 foe in energy, roughly the total energy of the final supernova explosion. Thus, decrease in core mass by $\sim$ 0.1 $M_\odot$ greatly facilitates the explosion.

Studies of the prompt explosion mechanism by Stephen Bruenn [20], who included all neutrino flavor and neutrino-electron downscattering in his transport (handled by multigroup diffusion) show that with the BCK EOS with $K_0 = 180$ MeV and $\Gamma = 2.5$, the iron core of the progenitor must be $\lesssim 1.10$ $M_\odot$ for the prompt explosion to operate. With $K_0 \sim 120$ MeV, the core might be somewhat larger.

Essentially no cores evolved to date for stars as large as 18 $M_\odot$ are this small, although their sizes have been coming down steadily with the

TABLE 2.1

Dependence of Supernova Explosion Energies on the Equation of State. Here the 0.33 in $K_0(0.33)$ means that 1/3 of the nucleons are protons; Γ is the adiabatic index. The column labelled GR indicates whether or not general relativity was used; ρ_c^{max} gives the maximum density reached in the collapse. The explosion energy E_{expl} is given in foe. The E_{lost}^{bounce} gives the energy lost in neutrinos before bounce, where E_{lost} gives the total energy lost in neutrinos.

Model	Mass ($M_\odot$)	$K_0(0.33)$ (MeV)	Γ	GR	ρ_c^{max} (g/cm^3)	E_{expl} (foe)	E_{lost}^{bounce} (foe)	E_{lost} (foe)
32	12	220	2	No	4.1(14)	-	0.9	2.6
33	12	170	2	No	4.9(14)	-	0.9	2.1
38	12	140	2	No	5.4(14)	0.1	0.9	3.2
40	12	140	2	Yes	2.9(15)	3.2	0.8	2.2
41	12	140	3	Yes	7.5(14)	0.8	0.8	3.3
29	15	220	2	No	4.1(14)	-	1.0	2.1
42	15	140	3	Yes	7.4(14)	-	0.9	2.5
43	15	140	2.5	Yes	9.9(14)	1.7	0.9	3.4
44	15	120	3	Yes	7.8(14)	-	0.9	3.0
45	15	90	3	Yes	9.4(14)	0.8	0.9	3.2

years. Recently the $^{12}C(\alpha,\gamma)^{16}O$ burning rate which enters in an essential way to the evolution of these cores has come down drastically [21] when measured in precision coincidence measurements. Even with these more accurate measurements, the precise value is not well determined. This lower rate is certain to decrease the size of the iron core somewhat, since during the longer carbon burning time there will be increased electron capture and neutrino losses, lowering both Y_e and the entropy. Both effects decrease the size of the Chandrasekhar mass and, hence, of the iron core. It looks difficult, even with the decreased $^{12}C(\alpha,\gamma)^{16}O$ to evolve a core as small as 1.10 $M_\odot$, and it is not clear that even the very soft EOS of Ref.[6] will facilitate the prompt explosion sufficiently. On the other hand, the distance to the progenitor is known only to $\sim \pm 10\%$, and with the lower limit of this distance, the mass of the star, as determined from the luminosity of the progenitor, might be substantially less than 18 $M_\odot$. In other words, by pushing parameters to their limits, the prompt explosion can be made to work. It does not, at the moment however, provide enough energy for the explosion unless all of the parameters discussed above are pushed to their limits.

Last year, Nomoto and Hashimoto [22] as part of the Brookhaven-Stony Brook collaboration, evolved a 13 $M_\odot$ star and found an Fe core mass of only 1.18 $M_\odot$. Comparison of this calculation with that of Woosley and Weaver [19] showed the discrepancy between the two calculations to have come chiefly from the neglect of Coulomb interaction by the latter authors. Reburning [23] the Woosley-Weaver 15 $M_\odot$ star with the inclusion of the Coulomb interaction brought the core mass down to 1.25 $M_\odot$, in line with that of the core "manufactured" by Cooperstein et al. [18]. So at this stage, it became clear that the prompt explosion mechanism would work.

Although there have by now been several studies of the energetics (see, e.g., K. van Riper [24]) and it is now clear that energetic explosions result from soft EOS's, a systematic quantification of precisely how this operates has not yet been carried out. The early BCK results [1] which showed the importance of the equation of state, are shown in Table 2.1.

Extensive calculations which include effects of neutrino-electron scattering and all three species of neutrinos have been carried out by Bruenn [20]. His results are shown in Fig. 2.3 for a soft BCK EOS with $K_0 = 180$ MeV and $\Gamma = 2$. The various models he used are indicated in the upper left hand corner of the figure and are described in detail in Table 2.2. The straight lines shown in Fig. 2.3 correspond to different definitions of explosion energy (see caption) and different prescriptions for the transport; as more complete transport is put in, the lines move towards the lower left hand corner.

As noted earlier, a small iron core is desirable, also a low initial entropy. If the initial entropy is low, then there will be fewer free protons, most of the protons being confined to the nuclei. With a smaller abundance of free protons there will be less electron capture during infall and a higher trapped lepton fraction at bounce. This results in a larger homologous core, so that the shock forms further out and has less of the remaining iron core to dissociate.

To summarize, the figure indicates that to *ensure* a successful explosion the initial model should come in the lower left hand corner, below all lines. None of the models evolved to date do. Marginal explosions result from the models lying lowest on the curve. Clearly the cores with better evolution should head towards the lower left hand corner if the prompt shock is to work. An iron core mass of 1.10 $M_\odot$ and low entropy would appear to be what is needed. This is a severe reduction from the models listed in Table 2, but it should be recalled that initial iron core masses have come down from 1.55-1.65 $M_\odot$ to those shown in the past ten years. Furthermore, all of the masses in Table 2.2 are substantially larger than the Chandrasekhar masses, Eq. (2.2), calculated with the average Y_e for each model. Clearly, the possibilities, consistent with present limits on nuclear burning rates, in

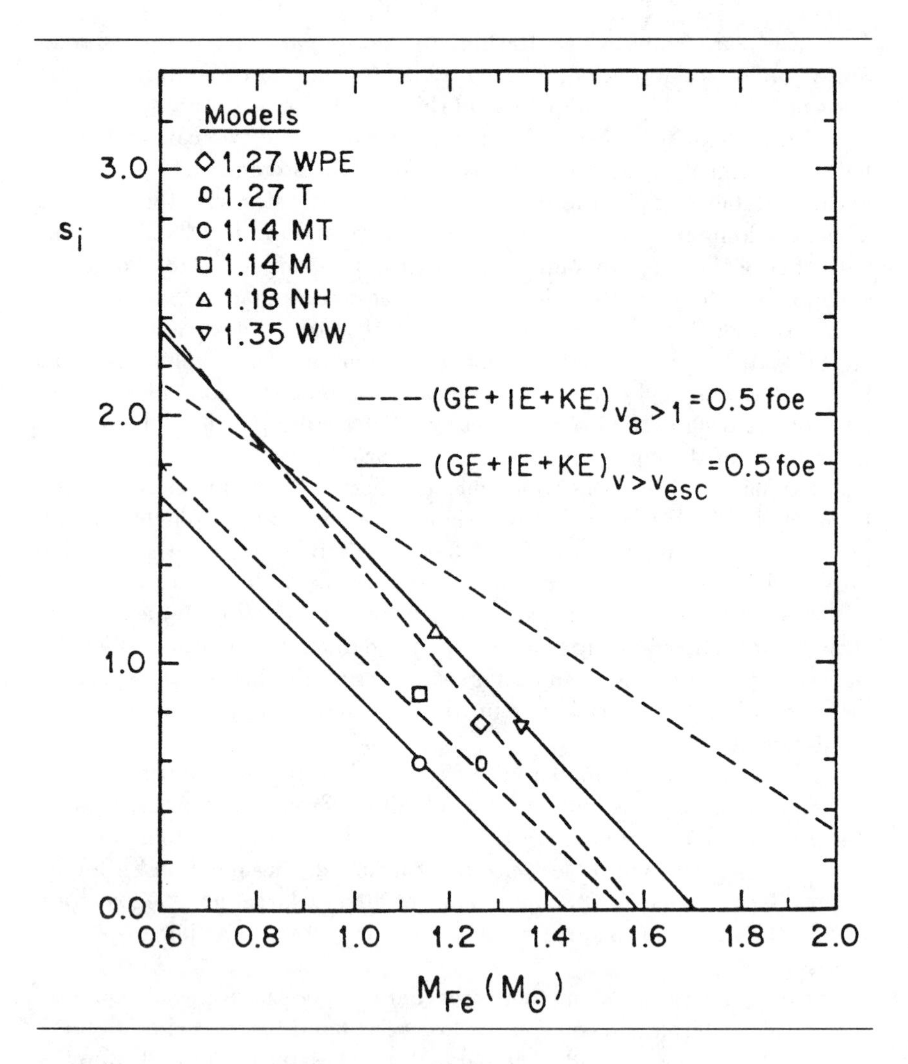

FIGURE 2.3
The central entropy - iron core mass plane ($s_i - M_{Fe}$ plane) for precollapse models. The straight lines are loci of points for which $E_{expl} = 0.5$ foe. Two different definitions of explosion energy are used. The first, denoted by $(GE + IE + KE)_{v>v_{esc}}$, is the sum of gravitational, internal (measured from iron) and kinetic energies of all mass shells having velocities greater than the escape velocity. The other is the sum of gravitational, internal and kinetic energies of all mass shells having positive velocities exceeding 10^8 cm/s.

TABLE 2.2
Precollapse Models

1.27 WPE	The 1.27 $M_\odot$ iron core of a presupernova model evolved from a 15 $M_\odot$ main sequence model by Woosley, Pinto, and Ensman [25].
1.35 WW	The 1.35 $M_\odot$ iron core of a presupernova model evolved from a 12 $M_\odot$ main sequence model by Woosley and Weaver [19].
1.18 NH	The 1.18 $M_\odot$ iron core of a presupernova model evolved from a 13 $M_\odot$ main sequence model by Nomoto and Hashimoto [22].
1.27 T	A 1.27 $M_\odot$ core obtained from model 1.27 WPE by reducing the temperature in each mass zone to 70% of its original value, then adjusting Y_e to maintain the original force imbalance of the zone.
1.14 M	1.14 $M_\odot$ core obtained from model 1.27 WPE by reducing the mass in each mass zone to 90% of its original value, then adjusting Y_e to maintain the original force imbalance of the zone.
1.14 MT	1.14 $M_\odot$ core obtained from model 1.27 WPE by reducing the mass and temperature in each mass zone to 90% and 70%, respectively, of their original values, then adjusting Y_e to maintain the original force imbalance of the zone.

evolution of initial iron cores must be explored. This seems to me to be the most important immediate problem. Unfortunately, this is an extremely complex business. Furthermore, as the remeasurement of the $^{12}C(\alpha,\gamma)^{16}O$ reaction cross section recently showed, there can be large uncertainties in this evolution.

2.3 Implications of Chiral Symmetry Restoration at High Densities for EOS

We have found that the nuclear matter EOS in the density range $\rho \lesssim 4\rho_0$ plays a critical role in the collapse and explosion of large stars. In Ref. 6 we motivated a low compression modulus K_0. What about the EOS in the range $\rho \sim 2 - 4\rho_0$? The collapse of large stars seems to be the only way to empirically get at the cold EOS of nuclear matter in this density range, although neutron stars give information on the EOS of neutron matter and the two EOS's are related. If the EOS is soft, however, the densities in neutron stars will, for the most part, be greater than $4\rho_0$; the information from them may apply chiefly to the higher density region $\rho \sim 4 - 8\rho_0$.

Most calculations to date give quite stiff EOS's for nuclear matter in the region $> 2\rho_0$, but we shall now show that a number of effects which

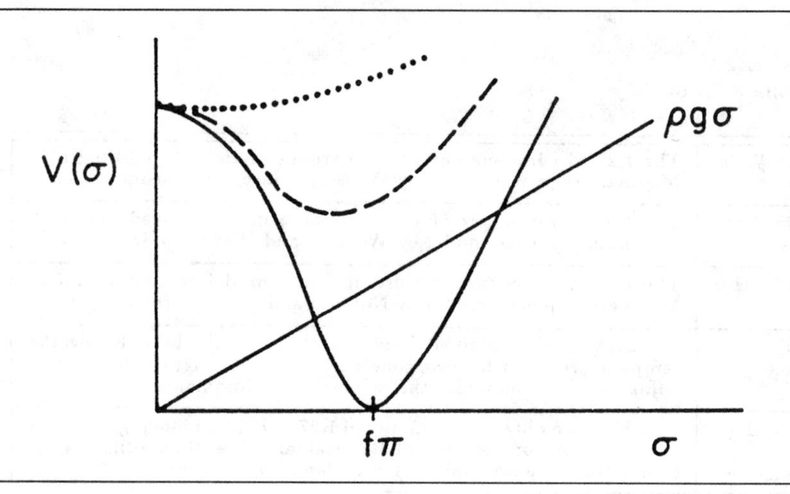

FIGURE 2.4
Behavior of the field energy $V(\sigma)$ vs σ : solid lines, with no matter present, and dashed line, in the presence of matter. The straight line gives the contribution from the matter. The dotted line shows the minimum having moved to $\sigma = 0$.

tend to soften the EOS have been left out of these calculations.

A number of phase transitions, π^0 condensation, K^- condensation, η condensation etc., are expected [11] to occur in the region of $2 - 5\rho_0$. Each of these will soften the EOS somewhat. Possibly the most interesting phase transition - at least the one that is discussed the most in nuclear physics - is the chiral restoration transition: e.g., the transition from nucleons and mesons to quarks and gluons. Although this chiral restoration transition may occur at only a very high density, $\sim 5 - 10\rho_0$, we see effects from it already at nuclear matter density. For example, the fact that the nucleon effective mass m_n^* is substantially less than m_n (Mahaux et al [26] find $m_n^*/m_n = 0.85 \pm .03$ for nuclear matter) can be interpreted as a step on the way towards chiral restoration where $m_n^* \to 0$.

In QCD one has only one scale Λ_{QCD}, equivalently, $f_\pi = 93$ MeV, the pion decay constant, which we find convenient to use as the scale in the Goldstone mode. The nucleon mass comes as

$$m_n = g f_\pi \tag{2.5}$$

in the no loop approximation of the σ-model. This comes about from the coupling

$$\delta H = g\overline{\psi}\sigma\psi, \tag{2.6}$$

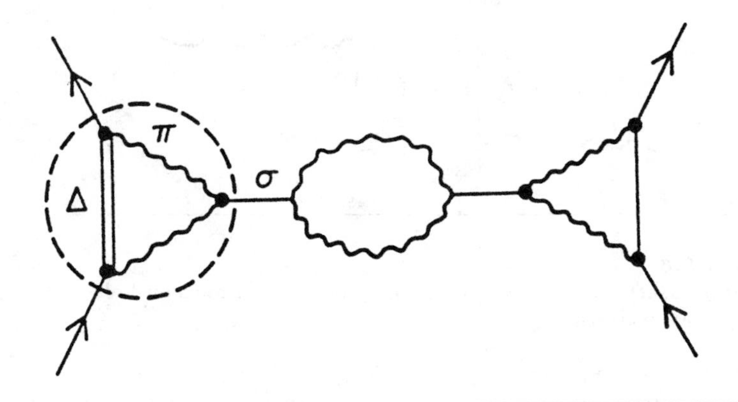

FIGURE 2.5
One particular contribution to the scalar channel in boson-exchange interactions. The encircled process on the left can be viewed as a vertex correction to the $NN\sigma$ coupling of Eq. (2.6); the pion bubbles as self-energy insertions in the σ.

where ψ is the nucleon field and σ is the scalar field. In the σ-model, the field energy $V(\sigma)$ has a minimum in the Goldstone vacuum for $\sigma = f_\pi$ and this produces the nucleon mass of Eq. (2.5).

As matter is added, the minimum in V(σ) moves to lower σ (see Fig. 2.4). If $m_n^* = 0.85 m_n$ at $\rho = \rho_0$, this means that the minimum comes at $f_\pi^* = 0.85 f_\pi$. In other words, it is trivial that the nucleon *in medium* mass scales with f_π^*. One suspects this to be true of the masses m_ρ and m_ω of the vector mesons, since these obey the KFSR [27] relation

$$m_\rho^2 = 2 f_\pi^2 g_g^2, \tag{2.7}$$

where g_g is the vector-meson gauge coupling. In principle, this coupling constant could change in medium, but, as we shall see, in general the coupling constants do not seem to be substantially modified by medium-dependent effects, so we expect the in-medium vector meson masses m_ω^* and m_ρ^* to scale as f_π^*.

The scalar attraction between nucleons which binds nuclei arises in a complicated fashion, from correlated pion pairs, coupled to angular momentum J=0. A particular contribution to this attraction is shown in Fig. 2.5. This effective scalar meson has, of course, a distributed mass spectrum, but can be reasonably represented in boson exchange models as having a "sharp" mass $m_\sigma \cong 560$ MeV. Roughly speaking, it is composed of two pions, each with kinetic energy about equal to their rest mass; thus,

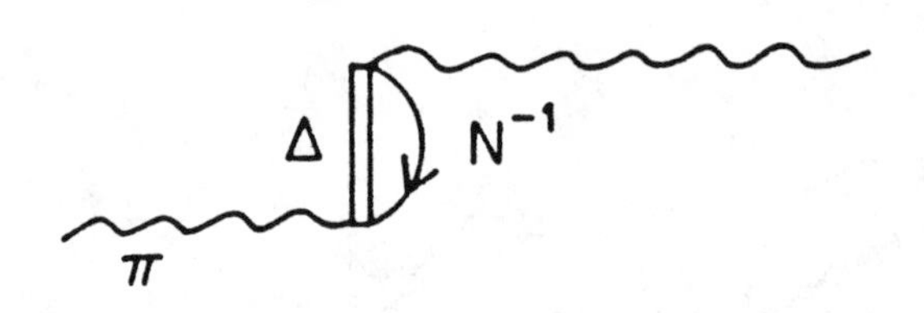

FIGURE 2.6
Self-energy insertion in the virtual pion, consisting of nucleon-hole and virtual isobar excitation.

$m_\sigma \cong 4m_\pi$. Although the discussion here may appear to be rough, it has a good grounding in the dispersion theoretical description [28] of the scalar attraction.

There will be modifications to the process, Fig. 2.5, in medium, the most important one concerning self-energy insertions in the virtual pions, Fig. 2.6. Estimation [10] of these medium-dependent effects produces an effective σ mass m_σ^* which varies as

$$(\frac{m_\sigma^*}{m_\sigma})^2 = 1 + \frac{(g'_{N\Delta} - \frac{k^2}{k^2+m_\pi^2})\frac{8}{9}\frac{\rho}{\rho_{nm}}}{1+\frac{8}{9}g'_{N\Delta}\frac{\rho}{\rho_{nm}}}. \tag{2.8}$$

Here the coupling constant $f_{\pi N\Delta} = 2$ has been used; $g'_{N\Delta}$ is, in pionic units, the Ericson-Ericson Lorentz-Lorenz correction [29], possibly supplemented by contributions from ρ-meson exchange.

With chiral restoration as $f_\pi^* \to 0$, the σ and $\vec{\pi}$ becomes degenerate so m_σ should decrease towards m_π. One would expect m_π, which arises from explicit chiral symmetry breaking, through the up and down current quark masses, to change much in the medium.

With the estimate [10] of $< k^2 >= 3m_\pi^2$, one sees that m_σ^*/m_σ will drop with density as long as $g'_{N\Delta} < 0.75$. In fact we [30] find $g'_{N\Delta}$ in the region $\frac{1}{3} < g'_{N\Delta} < 0.45 - 0.5$. This is in agreement with the detailed analyses of Arima et al [31], who find $g'_{N\Delta} \sim 0.4$. For this value, $m_\sigma^*/m_\sigma = 0.88$ at $\rho = \rho_0$, close to the nucleon effective mass $m_n^*/m_n = 0.85 \pm 0.03$. The values $g'_{N\Delta} = \frac{1}{3}$ and $g'_{N\Delta} = 0.45$ bracket the latter value.

After the above argumentation [10] concerning the decrease of m_σ^* with density ρ was carried out, it was realized that calculations in the Nambu-Jona-Lasinio formalism [32] must produce such a drop of m_σ^* with ρ. In Fig. 2.7 we show results of calculations by Bernard, Meissner and Zahed [33]. The NJL calculation is essentially a Hartree-Fock theory of the negative-energy quark sea, cut off at some Λ. The quarks exchange

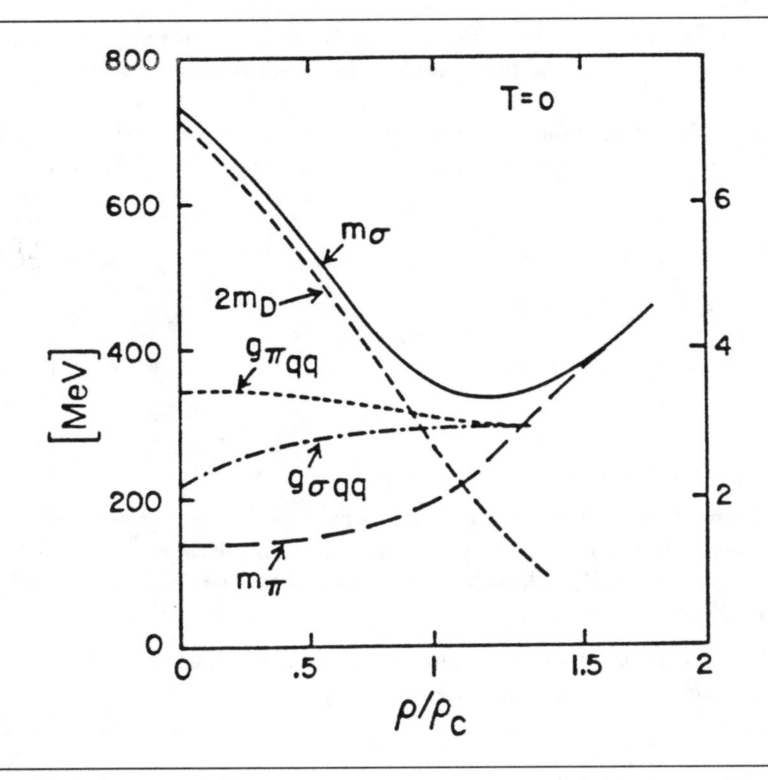

FIGURE 2.7
Results from a calculation in the Nambu-Jona-Lasinio model for the masses of σ and π mesons as function of density. Similar results have been obtained by Hatsuda and Kunihiro [34].

the same effective scalar meson as we discussed above. Above a certain strength of scalar field coupling, the Goldstone mode in which the quarks have dynamically generated masses lies below the massless Wigner mode. (We neglect current quark masses, since we shall be working here with up and down quarks where those masses are small.) One can take the dynamically generated quark mass m_Q to be

$$m_Q \cong m_n/3, \tag{2.9}$$

since three quarks make up a nucleon. (The actual situation may be more complicated than this, if a substantial part of the nucleon mass comes [12] from terms which explicitly break chiral symmetry.)

The NJL interaction is strong and coherent in the pion channel, producing a Goldstone boson. The pion mass comes from the explicit chiral symmetry breaking; i.e., from the up and down quark current masses. As can be seen from Fig. 2.7, this mass is reasonably constant as function of ρ.

The NJL interaction is not, however, coherent in the scalar channel, and the mass of the σ-particle comes out to be $m_\sigma \cong 2m_Q$. Thus we see that

$$m_\sigma^* \cong \frac{2}{3} m_n^* \tag{2.10}$$

in medium, or, alternatively,

$$m_\sigma^* \cong m_\sigma \frac{m_n^*}{m_n} = \lambda(\rho) m_\sigma , \tag{2.11}$$

where

$$\lambda(\rho) \equiv m_n^*/m_n . \tag{2.12}$$

Considering that [26] $m_n^* \cong 0.85 m_n$ at nuclear matter density, we can easily check that Eq. (2.11) gives about the same decrease as Eq. (2.8) with increasing density. Thus, we have obtained the decrease in σ-mass by two methods:

1. Phenomenology, using medium-dependent corrections, calculated in a language of nucleons and mesons.

2. A fundamental theory of dynamical generation of quark masses.

The results for the behavior of m_σ^* with ρ are similar. Although possibly less fundamental, it is useful to have procedure **1** above, because it can be phrased in terms of quantities we know. We note that effects such as the swelling of nucleons in medium, etc., result here from pionic interactions, and thus have to do with the behavior of the pion cloud of the nucleon, in agreement with considerations of Ericson et al [35]. Furthermore, we should note that the range of the medium-dependent effects is $\sim \hbar/m_\pi c$, just because they are connected with pions, and it will not, in general, be a good approximation to treat them in zero range [36]. It makes sense that the main effects will come from the lowest-mass particles.

Our result is that interaction in nuclear matter scale as

$$V_\sigma = \frac{g_{\sigma NN}}{4\pi} \frac{e^{-m_\sigma^* r}}{r} = \lambda(\rho) \frac{g_{\sigma NN}^2}{4\pi} \frac{e^{-\lambda m_\sigma r}}{\lambda r} \tag{2.13}$$

where λ is

$$\lambda(\rho) = \frac{m_n^*}{m_n} . \tag{2.14}$$

The ω-exchange potential behaves similarly. The kinetic energy goes as

$$T(r) = \frac{\hbar^2}{2m_n^*}\frac{\partial^2}{\partial r^2} \tag{2.15}$$

for each particle. If we now make the scale change

$$x = \lambda r \tag{2.16}$$

we see that

$$H(x) = \lambda H_0(x) \tag{2.17}$$

where H_0 is the Hamiltonian where nucleon and meson masses have their vacuum values m_σ, m_ω, m_n. The Schrödinger equation is

$$H(x)\psi(x) = \lambda H_0(x)\psi(x) = \lambda E_0\psi(x). \tag{2.18}$$

Again, E_0 is the energy for the Hamiltonian with vacuum values of the masses. The prescription is to calculate the energy for the vacuum values m_σ, m_ω and m_n, and then multiply by $\lambda(\rho)$ which now carries the full information about the effective masses. Note that form factors, etc., which can be thought of as containing meson masses, must be scaled in the same way.

One easily finds, using m_n rather than m_n^* in, say, relativistic mean field calculations [2] that E_0 is ridiculously low. Furthermore, the source of the scalar field is to be taken as $\rho_s(m_n)$, the scalar density calculated with m_n, rather than m_n^*. Thus, E_0 is small, essentially negligible compared with the large values of E found at high densities in the RMF calculations. With the σ mass m_σ which decreases with density, the scalar field no longer decouples with increasing density.

It is not clear that the ω-meson mass m_ω^* scales quite as fast as m_σ^* with density (Some of the ω energy comes from agencies other than the constituent quark masses.) We would view that decreasing m_ω^* at the same rate as m_σ^* maximizes the repulsion, or minimizes the attraction. Even with this minimal attraction, H_0 gives too much attraction at nuclear matter density and no tendency towards saturation, and since $\lambda(\rho)$ varies slowly and smoothly, this latter factor won't help much. The above is easy to understand, since the effective mass m_n^* has been replaced by m_n through our scaling. However, just the decrease of m_n^* with density ρ is one of the chief agents, if not the chief one, for producing saturation in essentially all nuclear matter calculations. Furthermore, nucleon loops cannot help the situation, since there are none in the x-variable calculations, the nucleon mass being the bare mass. The σ-meson loops are attractive, and also cannot help saturation. Saturation *is* helped by standard mechanisms

[2]Our scaling argument is quite general, not limited to mean field calculations; however, introduction of pion exchange involves special arguments, but the pion can be accomodated in a straightforward way

of the nuclear many-body problem not included in the relativistic mean field approach: Pauli blocking of the second-order pion exchange, dispersion corrections for off-shell effects in the interactions, etc., but it seems likely that at least second-order, if not higher, σ-meson loops are needed to produce saturation. The mechanisms used to date to produce saturation in the relativistic mean-field problem "evaporate", once the σ-meson mass is allowed to decrease with density. It must be admitted that we do not have a convincing scenario to date of how saturation is produced by higher-order loop terms, but we shall point out that possible mechanisms for this are well known.

The general problem of *too much* softness from the σ-mass decreasing with density is by no means a new one. It was encountered by Lee and Margulies [37] in the form that m_σ^2 went negative with increasing density, resulting in tachyonic behavior for the σ-meson. These authors cured this by introducing a matter-dependent two-loop term $\Delta U_\sigma(\rho)$ which brought about saturation. Jackson et al [38] implemented this in a mean field treatment of the many body problem. Recently Brown, Müther and Prakash [39] showed that with inclusion of all of the effects described above, a sensible description of the nuclear matter problem could be achieved in the relativistic Dirac-Brueckner approach. At high densities, the EOS became rather stiffer than the soft EOS's discussed earlier, however.

Recent developments [40] based on work of Richard Ball [41], which in turn is essentially carrying out the program of Weinberg to enforce local chiral invariance on effective Lagrangians, shows that the ρ-meson is not as fully clothed with soft pions as the σ-meson. Presumably the ω, which is strongly coupled to the $\rho - \pi$ system, behaves more like the ρ than like the σ. Thus, one would not expect the vector meson masses to drop as rapidly as the σ mass with increasing density, and attraction would have to be added back in the BMP work [39].

These questions are under active investigation.

2.4 Conclusions

Although Supernova 1987A (Shelton) was only one event, occuring 383 years after the preceding "nearby" Supernova 1604 (Kepler), it gives us important information about the equation of state of nuclear matter at high densities; namely, that the equation of state is quite soft, with $\Gamma \sim 2.5$ in the density range $\rho \cong 2 - 4\rho_0$. Also the compression modulus K_0 should be low.

Implications from the fact that nucleons must merge to quarks and gluons at higher densities, the chiral symmetry restoration transition, can be formulated in the language of medium-dependent effects. Inclusion of these substantially soften the equation of state, indicating that theory can be made consistent with the needs of Supernova 1987A (Shelton).

Acknowledgements

This work was supported in part by the U.S. Dept. of Energy, Grant No. DE-FG02-88ER40388

Bibliography

[1] E. Baron, J. Cooperstein and S. Kahana, Phys. Rev. Lett. **55** (1985) 126.

[2] J.R. Wilson, in *Numerical Astrophysics*, ed. J. Centrella, J. Le Blanc and R. Bowers (Jones and Bartlett, Boston, 1985)p. 374 ; H.A. Bethe and J.R. Wilson, Ap. J. **295** (1985) 14.

[3] B. Friedman and V.R. Pandharipande, Nucl. Phys. **A399** (1983) 51.

[4] J.P. Blaizot, D. Gogny and B. Grammaticos, Nucl. Phys. **A265** (1976) 315.

[5] M. Prakash and K.S. Bedell, Phys. Rev. **C32** (1985) 1118.

[6] G.E. Brown, *Proc. Third International Conference on Nucleus Nucleus Collisions* (Saint-Malo, France, June 6-11, 1988); Nucl. Phys. **A488** (1988) 689c.

[7] J.J. Molitoris and H. Stöcker, Phys. Rev. **C32** (1985) 346; see also, J.W. Harris *et al*, Phys. Rev. Lett. **58** (1987) 463.

[8] T.L. Ainsworth, E. Baron, G.E. Brown, J. Cooperstein and M. Prakash, Nucl. Phys. **A464** (1987) 740.

[9] J. Aichelin, A. Rosenhauer, G. Peilert, H. Stöcker and W. Greiner, Phys. Rev. Lett. **58** (1987) 1926.

[10] T.L. Ainsworth, G.E. Brown, M. Prakash and W. Weise, Phys. Lett. **200** (1987) 413.

[11] G.E. Brown, Phys. Rep. **163** (1988) 167.

[12] D.B. Kaplan and A.E. Nelson, Phys. Lett. **175B** (1986) 57.

[13] M. Prakash, T.L. Ainsworth and J.M. Lattimer, Phys. Rev. Lett. **61** (1988) 2517.

[14] S.E. Woosley, Ap. J. **330** (1988) 218.

[15] H.A. Bethe, G.E. Brown, J. Applegate and J.M. Lattimer, Nucl. Phys. **A334** (1979) 487.

[16] P. Goldreich and S. Weber, Ap. J. **238** (1980) 991.

[17] G.E. Brown, H.A. Bethe and G. Baym, Nucl. Phys. **A375** (1982) 481.

[18] J. Cooperstein, H.A. Bethe and G.E. Brown, Nucl. Phys. **A429** (1984) 527.

[19] S.E. Woosley and T.A. Weaver, Bull. Am. Astr. Soc. **10** (1984) 971.

[20] S. Bruenn, Ap. J. **341** (1989) 385.

[21] R.M. Kremer et al, Phys. Rev. Lett. **60** (1988) 1475.

[22] K. Nomoto and M. Hashimoto, Phys. Rep. **163** (1988) 13.

[23] S. Woosley and T.A. Weaver, Phys. Rep. **163** (1988) 79.

[24] K. van Riper, Los Alamos Natl. Lab., Preprint LA-UR-87-432.

[25] S.E. Woosley, P.A. Pinto and L. Ensman, Ap. J. **324** (1988) 466.

[26] C. Mahaux and R. Sartor, Nucl. Phys. **A475** (1987) 247; C.H. Johnson, D.J. Horen and C. Mahaux, Phys. Rev. **C36** (1987) 2252; C. Mahaux, priv. comm.

[27] K. Kawarabayashi and M. Suzuki, Phys. Rev. Lett. **16** (1966) 255; Riazuddin and Fayvazuddin, Phys. Rev. **147** (1966) 1071.

[28] G.E. Brown, Chiral Symmetry and the Nucleon-Nucleon Interaction, in *Mesons in Nuclei*, ed. M. Rho and D.H. Wilkinson (North Holland, Amsterdam, 1975).

[29] M. Ericson and T. Ericson, Ann. Phys. **36** (1966) 232.

[30] G.E. Brown, Mannque Rho and W. Weise, Z. Physik (in press).

[31] A. Arima, T. Cheon, K. Shimizu, H. Hyuga and T. Suzuki, Phys. Lett. **122B** (1983) 126.

[32] Y. Nambu and G. Jona-Lasinio, Phys. Rev. **122** (1961) 345; *ibid* **124** (1961) 246.

[33] V. Bernard, U.-G. Meissner and I. Zahed, Phys. Rev. Lett. **59** (1987) 966.

[34] T. Hatsuda and T. Kunihiro, Phys. Rev. Lett. **55** (1985) 158; Prog. Theor. Phys. Supplement **91** (1987) 284.

[35] M. Ericson and M. Rosa Clot, Z. Physik **A324** (1986) 373.

[36] G. van der Steenhoven *et al*, Phys. Rev. Lett. **58** (1987) 1727.

[37] T.D. Lee and M. Margulies, Phys. Rev. **D11** (1975) 1591.

[38] A.D. Jackson, M. Rho and E. Krotscheck, Nucl. Phys. **A407** (1983) 495.

[39] G.E. Brown, H. Müther and M. Prakash, Stony Brook Preprint (1988).

[40] G.E. Brown and Mannque Rho, in progress.

[41] Richard Ball, *Proc. Conf. on Skyrmions and Anomalies* (Mogillany, Poland, Feb. 21-23, 1987) and to be pub.

CHAPTER APPENDIX
Epilogue, March 1989

Since my talk in November 1988, much has happened. I summarize it briefly here. On a visit to Santa Cruz in January of this year (1989), Hans Bethe and I found that the electron capture matrix used by Weaver and Woosley terminated with A=60, whereas BBAL [15] showed that larger A dominated as the nuclei became neutron rich. Preliminary results show the transitions $^{70}Ga + e^- \rightarrow {}^{70}Zn$ and $^{68}Cu + e^- \rightarrow {}^{68}Ni$ to be larger than the rates Weaver and Woosley included for these neutron rich nuclei. Consequently, we expect substantially more electron capture, once these nuclei are included. This should lower the size of the core, which goes roughly as Y_e^2. Furthermore, very "fat" beta decays have not been included. For example, $^{68}Cu \rightarrow {}^{68}Zn + e^-$ has a half life of only 31 s, very small compared with the Kelvin-Helmholtz contraction time of $\sim 10^6$ s during which the core of the star cools before going into collapse. Inclusion of such processes will bring down the entropy substantially, decreasing the thermal additions to the collapsing core. We believe that inclusion of these effects may result in a substantially smaller core, more favorable for the explosion mechanism I discussed here.

Most exciting is the report in the New York Times of 10 February 1989 that "Astronomers View Creation of a Pulsar Amid Stellar Debris" and, furthermore, that it is spinning nearly 2,000 times per second inside the debris of the recent stellar explosion.

The rapid rotation, if confirmed, is very interesting. Friedman et al, in Nature, 8 December 1988 (p.560) suggested that PSR 1957 + 20 and PSR 1937 + 214, rotating at only $\sim$ 1/3 the above speed, were rotating near their critical velocity, namely, near the velocity where matter would fly off their surface. They suggested that this was no coincidence, that possibly pulsars rotating faster than these two don't exist because the equation of state is quite stiff.

In News and Views of that same issue of Nature, I suggested that finding a pulsar which rotated faster than these two would invalidate their argument.

Indeed, the equation of state would have to be very soft for a neutron star of gravitational mass 1.4 $M_\odot$ to rotate at 2000 times per second. If the EOS is soft, the star can be quite compact, so that gravity is stronger and the velocities at the surface somewhat less. In this way, gravity can keep the star from flying apart. Estimates show that the central density of the star has to be greater than ten times nuclear matter density. Quantitative calculations take some time to carry out, but it is clear that it will be difficult to obtain a stable star rotating at 2000 times per second without a very soft EOS.

Of course, the kinetic energy in rotation is tremendous, $\sim$ 100 foe. This may well wreak havoc with our one-dimensional radially symmetrical calculations discussed earlier.

3

Detecting the Quark-Gluon Plasma

G. Bertsch
Dept. of Physics and Cyclotron Lab.
Michigan State University
East Lansing, MI 48824

3.1 Introduction

What are the properties of matter at high temperature and density? This question is the focus of a large program of research in high energy heavy ion physics. Besides general interest, there are two particular reasons for this study. First, the theory of Quantum Chromodynamics (QCD), which is the accepted theory of strong interaction physics, is capable in principle of predicting these properties. Lattice calculations suggest that a phase transition may take place at a temperature of about 200 MeV.

Another reason for interest in matter at high temperature is that the universe in the Big Bang is thought to have originated at high density and cooled through these temperatures. If there were a first order phase transition, it could have important consequences for the later evolution. For example, baryonic matter might become more concentrated as a result of the phase transition; this would affect the later nucleosynthesis into elements.

To study hot matter, we first need a way to produce it. The only possibility in the laboratory is to collide heavy nuclei at high energy. Obviously high energy is needed, but why use nuclei instead of nucleons or other hadrons? It is important that the hot matter reaches a local equilibrium under the conditions of study. Only then does the concept of temperature have meaning. If we were condensed matter physicists, and we doubted the equilibrium, we could just leave the sample in the oven longer. That isn't possible here; the high temperature lasts only a short time. The conditions for equilibration become more favorable by making the region of the high energy density as large as possible. And that is done using heavy nuclei as projectiles rather than hadrons.

There are two accelerators involved in this program. In Brookhaven, the AGS has accelerated heavy nuclei up to 15 GeV/n in experiments starting from 1987. This program is now in full operation, and interesting results are presently emerging. In Geneva at CERN the large accelerator SPS accelerates ions up to 200 GeV/n, in an experimental program that started in 1986. Both CERN and Brookhaven use ions of lighter mass elements, mainly oxygen and sulfur. The CERN program had a major running period in 1987, and the main results that I discuss come from this run. These results were presented in a conference in 1987, and have since been published [1].

As I will try to explain, the results are interesting and consistent with the expected phase transition, but proof is still lacking. Future experiments starting in 1990 should do better; I have particular expectations for measurements of pion correlations. Further experiments are under discussion at CERN using beams of accelerated lead nuclei. This would allow substantially higher energy density to be reached. Finally, there are plans in the USA to build a heavy ion collider at Brookhaven, to collide 200 GeV/n nuclei against each other. This might take the system out of the mixed state of hadronic matter and plasma, into a state of superheated plasma.

My own interest in this subject started from the point of view of nonrelativistic heavy ion physics. There we deal with a system that is extremely complicated to describe in detail. Nevertheless one is able to calculate with fair confidence certain kinds of observables, particularly those involving conserved quantities such as linear momentum. Classical simulations of the collision turn out to be very useful providing the quantum effects of the Pauli principle are included. So it is natural to try analogous simulations of the ultrarelativistic collisions. There are two important differences between these regimes, however. In lower energy heavy ion collisions, we believe we understand the important degrees of freedom and how they interact. One can then make a complete dynamical calculation, starting from the cold nuclei approaching the collision region. In the ultrarelativistic domain, we do not know in enough detail how the important degrees of freedom, the partons in the nucleons, interact to make reliable calculations. On the other hand, the dynamics in the low energy domain is not so favorable for determining the equation of state. The equilibration process takes a considerable fraction of the total collision time, so that the observables are sensitive to the details of the equilibration as well as to the equation of state. It has so far not been possible to extract equation of state information from these experiments. The equilibration should be more favorable in the ultrarelativistic domain. Simple estimates based on expected densities of partons indicate that the system would quickly come to local equilibrium. It is thus reasonable to pursue models in which the details of the formation of the hot matter are ignored, and instead assume a local equilibrium at a certain time. This is my attitude and all inferences are based on this assumption. But we must keep in mind that it is essentially untested as of yet.

3.2 Initial Conditions

Before we can model the dynamics of the expanding hot matter, we need to know how the system starts out. The appearance of the hot zone for a 200 GeV/n Oxygen nucleus colliding with a Gold target nucleus is shown pictorially in Fig. 3.1. We view the collision in the midrapidity frame, so both nuclei are Lorentz contracted. The left figure shows the mass distributions at the point when the surfaces just touch. The right figure shows the nuclei and the hot zone between them at the time 1 fm/c later. The excitation is in a roughly cylindrical zone; it expands longitudinally at the projectile velocity, i.e. close to the velocity of light. The transverse motion is less clear and depends on the details of the dynamics that we wish to study.

For our initial conditions, we assume some distribution of energy in this cylindrical interaction zone. Experiments already tell us quite a bit about that energy distribution. Essentially, we only have to look at the projected mass overlap between the projectile and the target to determine the transverse distribution of energy. This is the conclusion of the experiments that measure the so-called transverse energy produced in the collision. The transverse energy E_T is defined as the sum over final state particles

$$E_T = \sum_i E_i \, sin\theta_i \ , \tag{3.1}$$

where E_i and θ_i are the energies and angles of the emerging particles. The probability distribution of this quantity follows quite closely the behavior expected if the energy is produced according to the geometric overlap of the projectile and target nucleons. This is illustrated in Fig. 3.2. On the horizontal axis is the transverse energy, and the differential cross section in this variable is plotted. The two graphs show data from Brookhaven [2] and from CERN [3]. A peak in the distribution is seen at small transverse energy. This is due to peripheral collisions. At higher transverse energy the

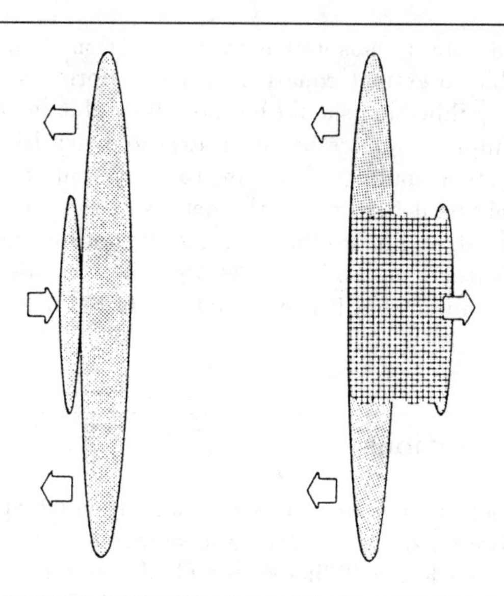

FIGURE 3.1
Visualization of a central collision between oxygen and gold nuclei. The oxygen beam has an energy of 200 GeV/n, i.e. a rapidity of $y = \cosh^{-1}(200/0.931) = 6.06$, but the system is viewed in the midrapidity frame, $y = 3.03$. Thus both target and projectile are Lorentz-contracted by a factor $\gamma = \cosh 3.03 = 10.4$. The right drawing shows the positions of the nuclei 1.5 fm/c after they first make contact.

cross section has a plateau, corresponding to more central impact parameters. Finally, the distribution falls exponentially above a certain point.

The shape of the transverse energy distribution is very easy to understand geometrically. One calculates the number of collisions between nucleons, assuming that the nucleons are distributed in (Lorentz-contracted) spheres having the volume of the nucleus, and they move only in the longitudinal direction. In Fig. 3.2a, each nucleon-nucleon collision is assumed to produce a certain distribution of transverse energy. These contributions are then convoluted with the distribution of NN collisions. The theoretical distribution for the Brookhaven experiment is shown in the figure, with the contributions from different numbers of NN collisions separately graphed. The overall curve agrees very well with experiment.

I still have to explain how the distribution of transverse energy for individual NN collisions is determined. In the analysis of Fig. 3.2a it was parameterized to fit the data. A more fundamental analysis, modelling the

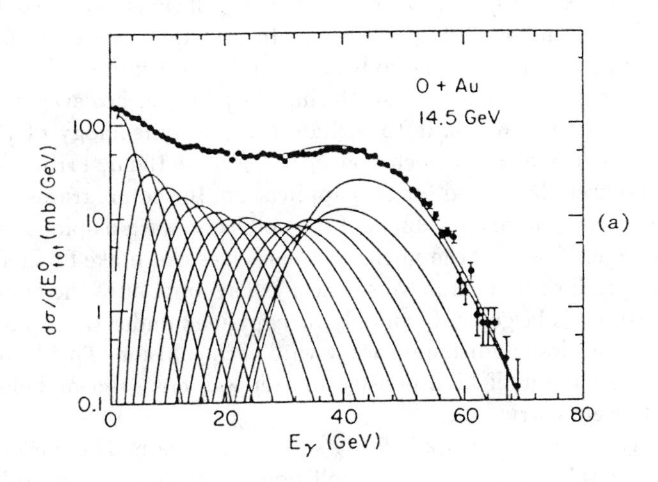

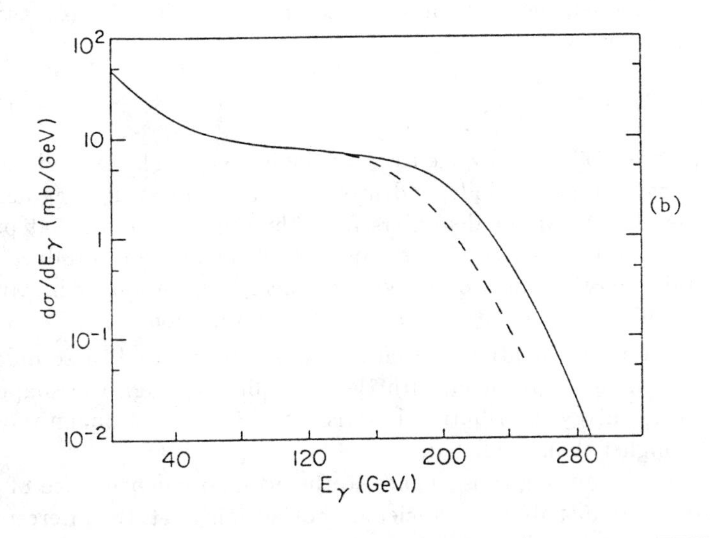

FIGURE 3.2
Transverse energy distribution. a) is from 14.5 GeV/n collisions of O + Au, measured at Brookhaven. The experimental data is fit by a sum of functions representing convoluted NN transverse energy distributions. b) is from 200 GeV/n collisions of O + Au, measured at CERN (dashed line). The solid line is the prediction of the VENUS code, which assumes independent parton collisions in the target and projectile nucleons.

underlying NN dynamics, has been applied to interpret the higher energy data from CERN. To calculate from the NN collisions, we need to distinguish primary and secondary collisions. In principle, the distribution of transverse energy from primary collisions could be taken directly from the empirical systematics of the NN scattering. In practice, however, one uses computer codes that are built to include the phenomenology of the NN scattering. There are many such codes, but two of them, Fritiof [4] and Venus [5], are well developed for this application. In the programs, it is assumed that the secondary products of the collisions, such as pions, play no subsequent role. However, the nucleons themselves can make further interactions and produce more transverse energy. The amount of the transverse energy measured is large, and can only be explained under the assumption that the nucleons lose additional energy with each collision. This important finding in the pursuit of high density matter was not so clear before the experiments were carried out.

The way these codes work is fairly simple to explain. The nucleons are treated as excitable particles. Each collision gives the nucleon additional excitation energy, according to some probability distribution. In the case of the Fritiof model, the probability to excite each nucleon from mass M to M' is given by

$$dP \sim \theta(M' - M)\frac{dM'}{M'} , \tag{3.2}$$

with an upper cutoff given by the total available energy. The excited nucleons then decay completely independently, according to a string fragmentation scheme. The Venus model differs from the Lund model in that pairs and triples of partons are the basic objects that interact and produce the strings. Multiple collisions produce greater energy deposition by involving more of the nucleon's constituents in the string formation.

It may be worth mentioning that at a quantitative level these models are not in complete agreement with the data [6]. Although the shape of the transverse energy distribution is satisfactory, its overall magnitude is about 20% higher than predicted.

There is an amusing consequence to the geometric dependence of the transverse energy distribution. Nuclei are not all spherical; the intercepted mass of a deformed nucleus will depend on its orientation. The maximum mass is presented when the projectile goes along the major axis of a deformed target. Conversely, when the target is oriented with the major axis transverse to the beam, the overlap will be smaller. Thus the shoulder on the distribution of the transverse energy should be less sharp in the case of deformed nuclei. This can be seen in Fig. 3.3, taken from data by the Helios collaboration [7]. The two targets compared are tungsten (W) and platinum (Pt). These differ in mass by only about 10 units out of nearly

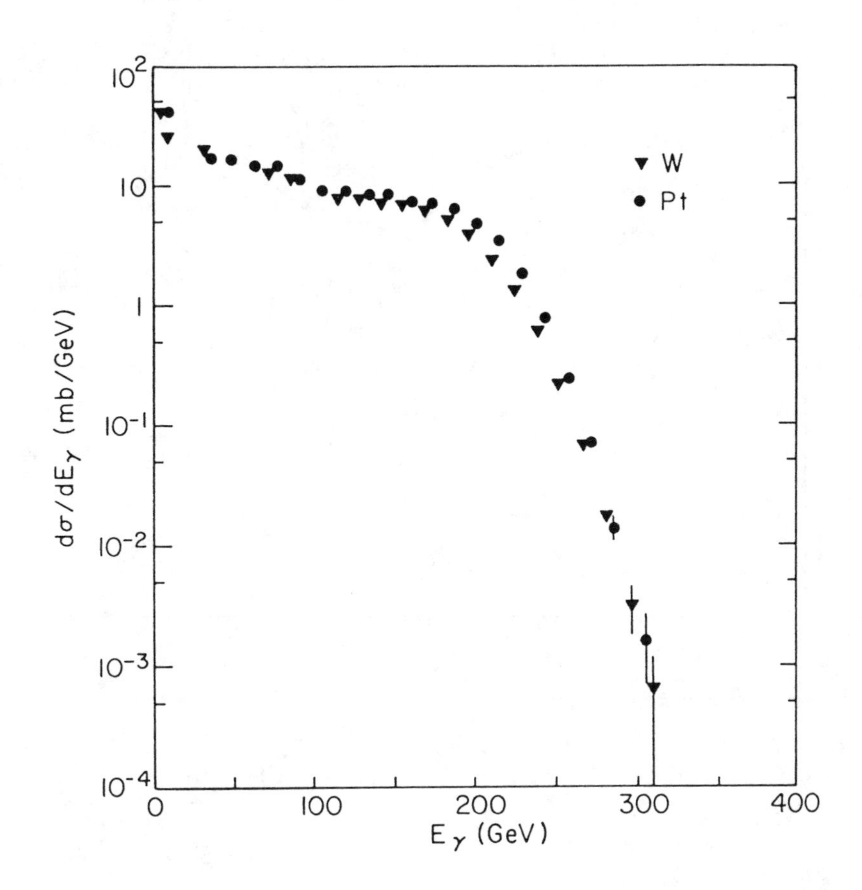

FIGURE 3.3
Transverse energy distribution comparing tungsten and platinum targets. The softer shoulder of the tungsten data is explained by nuclear deformation effects.

200, but W is deformed while Pt is spherical. It may be seen that the shoulder is broader in the case of W, as expected.

Having established how the energy is distributed spatially in the direction transverse to the beam, we next ask about the longitudinal distribution of energy. There are two extreme models which have been explored. One of these, proposed by Bjorken [8], is quite simple from an analytic point

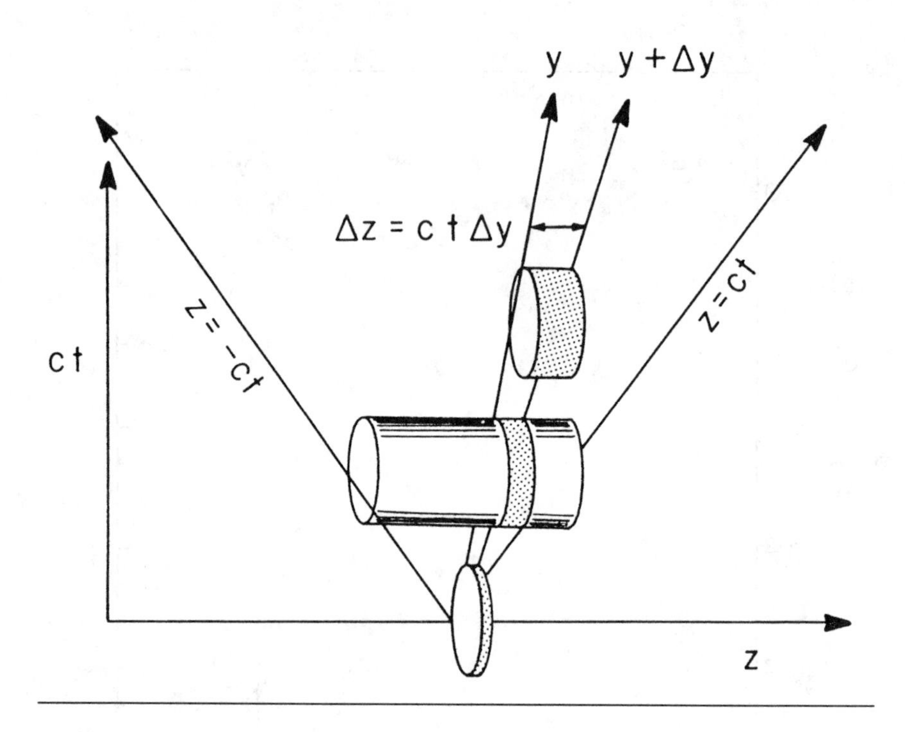

FIGURE 3.4
Time development of the hot matter in Bjorken's boost invariant expansion.

of view. The longitudinal distribution of energy is assumed to be boost-invariant. That is, the hot zone in Fig. 3.1 should look the same in other frames besides the one we chose. For nearby frames, we see the target and projectile discs receding from the collision point at a velocity close to c. The hot matter is locally at rest at the collision point, i.e. it has no net longitudinal motion at the collision point. Thus, the average rapidity of the matter does not change in a comoving frame. This is illustrated in Fig. 3.4. Consider a small slice of the cylinder, bounded by planes moving with rapidities y and $y + \Delta y$. From the Lorentz kinematics, $z = \tau \sinh y$ and $t = \tau \cosh y$, we find that the volume in the slice varies with time as $\Delta V = \Delta z A = c\tau \Delta y A$, where A is the cross sectional area of the cylinder. In the Bjorken model, the flux of entropy across any transverse surface is zero and there is no transport of entropy from one region of rapidity to another. Thus, in the absence of entropy-producing processes, a measurement of entropy in the final state can be extrapolated back to the collision time. Specifically, if dS/dy is the entropy per unit rapidity in the final state, and σ is the entropy density, then the relation is

$$dS/dy = \sigma Act \ . \tag{3.3}$$

In Bjorken's model the final state observables should be independent of rapidity. This is very nearly the case for the average transverse momentum of the final state particles. However, the number of particles per unit rapidity is not constant over a broad range. The rapidity distribution is better described empirically by a Gaussian,

$$dn/dy \approx exp(-\frac{(y-y_0)^2}{2y^{*2}}) \ . \tag{3.4}$$

The variance of the Gaussian is about $y^* = 1.5$ for the CERN experiments. This kind of behavior is predicted by Landau's hydrodynamic model [9]. In the Landau model one assumes that the collision brings the matter to rest in some intermediate rapidity frame, and the spatial distribution of matter is a disc, due to the Lorentz contraction. The highest gradient of the density is along the beam axis, so the matter is accelerated most along that direction. The resulting rapidity distribution resembles a Gaussian.

Of these two extreme models, the Bjorken model is probably more realistic. According to the more microscopic pictures such as contained in the Fritiof model, the energy deposition is via string fragmentation. If the strings were of infinite extent, they would break in a boost-invariant way. The lack of boost invariance in the final state reflects the finite size effects of the ends of the strings and the initial distribution of strings in rapidity.

In the calculations that I will describe below, my collaborators and I used the Bjorken assumption. I don't believe that the results would be much different if a more realistic longitudinal distribution of energy were assumed. The main effect would be that the system would evolve from a higher entropy density, since matter would be accelerated out of the midrapidity region.

3.3 Equation of State

I have mentioned entropy without explaining how it is measured. In fact we cannot measure it in the thermodynamic sense, because there is no adiabatic energy transfer. If the system is in local thermal equilibrium, the entropy density can be calculated from the temperature. For a gas of massless particles, the relation from statistical mechanics is [10]

$$\sigma = \frac{4\pi^2}{90}(n_B + \frac{7}{8}n_F)T^3 \ , \tag{3.5}$$

where n_B and n_F are the number of species of bosons and fermions, respectively. We pretend that the final state is a gas of massless pions, with $n_\pi = 3$, so the formula for the entropy density in the hadronic phase is

$$\sigma_\pi = \frac{2\pi^2}{15} T^3 . \tag{3.6}$$

The formula for the number density of particles has the same T-dependence,

$$\rho_\pi = n_\pi \int \frac{d^3p}{8\pi^3} \frac{1}{exp(p/T) - 1} = \frac{3\zeta(3)}{\pi^2} T^3 \tag{3.7}$$

The entropy per particle is then given by the ratio of Eq. (3.6) to Eq. (3.7), $S_\pi = \sigma_\pi / \rho_\pi \approx 3.6$. So the way we estimate the entropy of the final state is to count the number of pions and multiply by 3.6.

We now need some dynamics to evolve the initial state that was constructed. Our goal is to measure the equation of state, but the best we can do is to try out various models and compare with experiment. The motivation for modelling in general was explained by Weisskopf in the following anecdote:

> "A model is like an Austrian railroad timetable. Austrian trains are always late. A foreign visitor asked the conductor on an Austrian train, why they even bother printing the timetables. The conductor's reply: How else would we know how late the trains are?"

In our model of the equation of state the low density phase is assumed to be an ideal gas of massless pions. Then the entropy function is given above in Eq. (3.6), and the pressure is

$$P_\pi = \sigma_\pi T/4 . \tag{3.8}$$

The quark-gluon phase is also treated as a gas of massless particles. The entropy density is determined from Eq. (3.5). To count the number of species of particle, note that each kind of quark is described by a Dirac equation, having 4 degrees of freedom. We consider two flavors of quark, "u" and "d", and three colors. Thus $n_F = 4 \cdot 2 \cdot 3 = 24$. The gluons have two helicities and 8 colors, making $n_B = 16$. The coefficient in parentheses in the entropy formula is thus $16 + 24 \cdot 7/8 = 37$, which is more than an order of magnitude higher than in the lower density phase. Thus at the transition temperature the two phases will have very different entropy densities. This is also the case for the energy densities in the two phases.

The only difference from the massless particle thermodynamics in the model of the quark-gluon phase is in the pressure and energy formulas. The pressure consists of two terms, the first of which is the pressure due to the particle motion. This has the same form as Eq. (3.5). The second term is

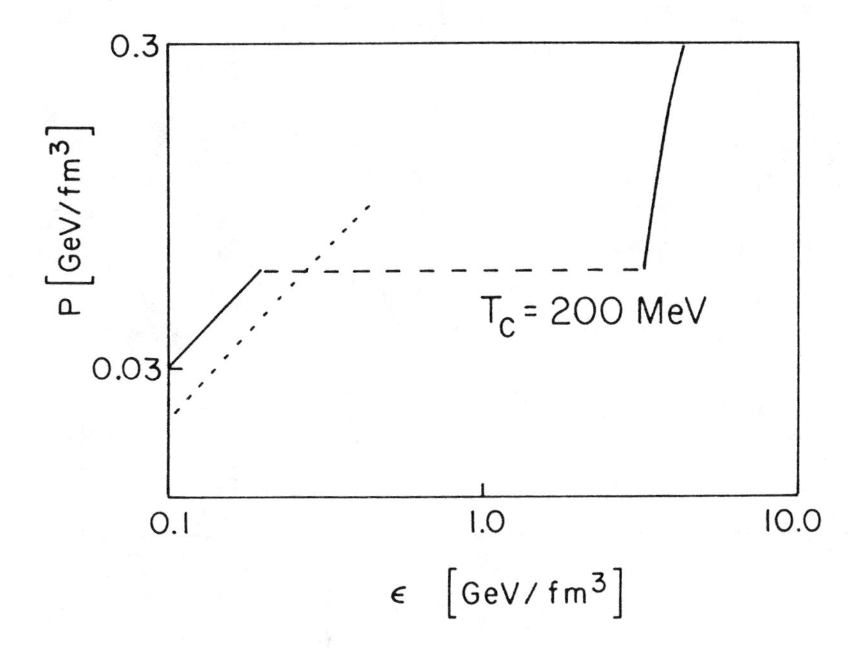

FIGURE 3.5
The equation of state in the simple pion gas/quark-gluon bag model, as a function of energy density. The short-dashed line shows the hadronic equation of state when all light-quark mesons are included.

a constant, the so-called "bag" pressure. The total is

$$P_{qg} = \frac{\sigma_{qg} T}{4} - B \, . \tag{3.9}$$

At the transition temperature the two pressures are equal, which gives a condition relating it to B. We consider the transition temperature (or B) as a parameter of the model, to be determined if possible by the data. A graph of the equation of state is shown in Fig. 3.5, plotting pressure as a function of energy density [11]. On the left is the pure pion gas phase, on the right the pure plasma phase, and the flat region in the middle is the mixed phase. Note that the mixed phase extends over an order of magnitude in density, for the reason given above. In the figure is also shown the equation of state of a hadronic gas consisting of the light-quark mesons: π, ρ, ω, and η. The

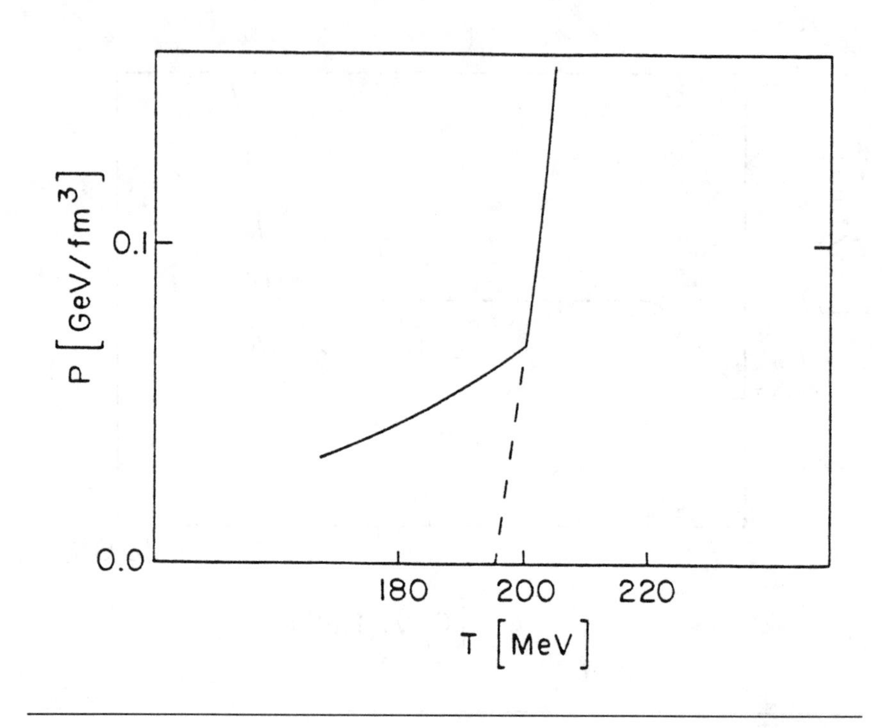

FIGURE 3.6
Equation of state in the simple model, as a function of temperature. The dashed line shows a possible metastable region of the quark-gluon phase.

more particles that are included, the lower will be the pressure for a given energy density. In fact, it was suggested a long time ago that there might be so many particles that any amount of energy could be put into the system at a certain limiting temperature. The pressure would then also approach a limiting value. The only way to distinguish these models qualitatively is to go the quark-gluon phase, where QCD predicts a renewed increase in pressure.

In Fig. 3.6 the equation of state is shown with temperature as the independent variable. Note that the pressure changes much more rapidly with temperature in the plasma phase than in the pion phase. We can imagine cooling the system from the plasma phase, creating a metastable plasma state of lower pressure. If the phase transition is first order, some metastability is expected.

The lattice gauge calculations of QCD are not yet settled enough to

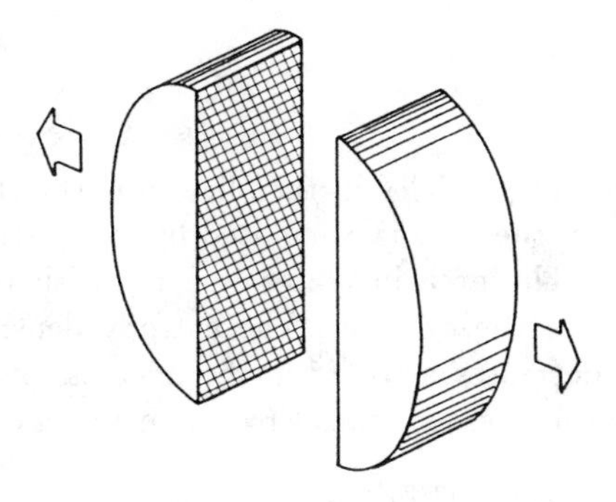

FIGURE 3.7
Geometry for the argument relating pressure to transverse momentum in Eqs. (3.10) and (3.11).

replace simple models of high density such as I described above. The magnitude of the entropy jump from one phase to the other is consistent with the simple model, at least in some calculations [12]. In calculations with only gluon fields, the transition appears to be first order, although the full entropy density of the plasma phase may not be present immediately above the transition temperature. Technical difficulties associated with quarks on the lattice prevent any definite statements about the phase transition when both quark and gluon degrees of freedom are included.

3.4 Observables

Assuming now that we have an equation of state and can calculate the dynamics of the expanding hot matter, what should we measure? I believe two observables are particularly relevant to the study of the equation of state. One quantity is the average transverse momentum of the particles in the final state. As emphasized by Van Hove [13], the transverse momentum directly reflects the pressure history of the expansion. We can see this in a semiquantitative way by examining a slice through the cylinder of hot expanding matter. We cut the slice in half, as in Fig. 3.7, and ask how much force is exerted on the cut. This force transfers momentum between the two halves, and thus produces transverse momentum in the final state particles. Suppose the two halves have momentum p_{slice} and $-p_{slice}$ and that the average motion respects the axial symmetry of the cylinder.. Then the average transverse momentum of the particles contained in the slice is given by

$$< p_t >= \frac{2p_{slice}}{\pi N} , \tag{3.10}$$

where N is the number of particles in the slice, counting both sides. The transverse momentum of the slice is calculated by integrating the pressure P over area and time. The area increases linearly with time due to the longitudinal expansion of the cylinder. The entropy density decreases as t^{-1}, and the pressure decreases as $t^{-4/3}$ in the massless pion model of the hadronic phase. Then the average transverse momentum is given by

$$< p_t >= \frac{2}{\pi N} \int P dA dt \sim t^{2/3}\Big|_{\tau_0}^{\tau_f} . \tag{3.11}$$

The dependence on the equation of state is obvious in this relation. A stiff equation of state has more pressure at a given entropy density and produces a higher average transverse momentum. Conversely, if a phase transition makes the equation of state very soft, there would not be more transverse momentum produced, despite a higher energy density.

However, note that the integral in Eq. (3.11) is not very well behaved. It diverges at the upper limit. The integral must be cut off at the freezeout time τ_f, when interactions cease. The dependence on the initial time is less important, but still could be significant. For these reasons, much more detailed calculations are required to interpret the experiments.

The time duration and spatial extent of the strongly interacting stage of the collision are also important characteristics which might indicate a possible phase transition. The space-time distribution of the interacting zone can be measured by pion correlations, as will be explained later. The time measurement is particularly interesting. In the bag model of the phase transition, the matter must be rarified by more than an order of magnitude to convert it completely to the other phase. If there is an interface separating the two phases, this severely limits the rate at which the phase transition can take place. The interface travels into the plasma, converting dense plasma into pion gas. However, the flux of entropy in the pion gas is bounded by c times its entropy density. Thus the interface cannot move faster than c times the ratio of entropy densities. This slow phase conversion produces a long-lived plasma droplet. In Table 3.1, I quote some of the estimates of this phase conversion that may be found in the literature. Assuming that the phase conversion rate is c/10, a droplet of initial radius 2 fm would last 20 fm/c. This is a very long time compared to duration times based on collision rates in an expanding pion gas.

TABLE 3.1
Estimates of phase interface velocity

$v \leq \frac{1}{6}c$	Ref. [14]
$v = \frac{2}{3\sqrt{3}}\frac{\epsilon_\pi}{\epsilon_{qg}}c$	Ref. [15]
$v = \frac{1}{4}\frac{\epsilon_\pi}{\epsilon_{qg}}c$	Ref. [16]

3.5 Detailed Models

I now turn to the specific analytic and numerical models for dealing with the expansion dynamics. The simplest model is Bjorken's, which assumes pure longitudinal expansion. As mentioned before, the boost invariance and the assumption of entropy conservation imply that the entropy per unit rapidity is constant and the entropy density decreases as t^{-1}. Since all other quantities can be expressed in terms of the entropy density as the independent variable, the entire evolution of the system is fixed. The model exhibits a long-lived mixed state because of the limited expansion rate from purely longitudinal motion. Clearly the neglect of transverse expansion is a drastic approximation, requiring better modeling. We saw in Eq. (3.11) that the freezeout time is an important parameter of the evolution; in the Bjorken model the system never freezes out [17]. To remedy this deficiency it is necessary to calculate the transverse expansion explicitly. The evolution can then no longer be determined analytically; the price one pays for the more accurate modelling is that the results are strictly numerical. Hydrodynamic calculations have been reported in Refs. [18,19]. In these calculations, one still must assume something about the freezeout, typically that it occurs at a certain density or a certain temperature. Also, the phase transition dynamics in hydrodynamics is rather complicated, with the possibility of rarifaction shocks and entropy production. The hydrodynamic conservation laws are insufficient to specify the details of the transition under these conditions.

For these reasons a more detailed model was made by McLerran together with myself and others [16]. To avoid the uncertainties of freezeout, we decided to simulate the evolution of the pionic final state. By tracking coordinates of the individual pions, we could see how they collide and determine the spatial and temporal size of the collision region as well as the transverse momentum distribution. This model needs pi-pi scattering cross sections as input. Even though nobody has measured the scattering process, we were able to extract adequate information about the scattering from the literature.

We simulated the dense phase of matter by a collection of droplets of appropriate density. The droplets are assumed to have the same initial size, and this is taken as a parameter of the model. However, the range of this parameter is somewhat limited. A droplet radius smaller than about 1 fm would make the concept of a separate phase doubtful, since the QCD size scale is of the order of 1 fm. Very large droplets, say 2 fm radius or larger, would produce obvious fluctuations in the rapidity distribution of the final state particles.

The conversion of matter from one phase to the other is treated by a model very similar to the compound nucleus model of neutron evaporation from a nucleus [20]. The hadronic state is produced by emission of pions from the droplet surface. The rate of emission is determined by detailed balance, assuming that the droplet absorbs all incident pions. The resulting formula for the emission rate W is

$$W = \frac{c}{4}\pi R^2 \rho_\pi(T) , \tag{3.12}$$

where $\rho_\pi(T)$ is the density of pions at temperature T, given by Eq. (3.7). Our model neglects the interaction between the droplets; because the droplets are quite heavy their thermal motion is small.

With these ingredients we can simulate the evolution of the system from the start of the phase transition to the final state. The time at which the phase conversion begins of course varies for different longitudinal rapidities due to time dilation. We fill the cylinder with randomly placed droplets, which begin emitting pions at the proper time corresponding to the beginning of the phase transition. The evolution of the system proceeds by time steps; in each step pions may be emitted or absorbed from droplets, they may scatter from each other, and they of course change position due to their velocity. After a long enough time, all interactions cease and the properties of the final state pions are saved in a file. The information about the pions that we need to save is their momenta and the coordinates of their last interaction point. That could be be where they last collided with another pion, or where they were emitted from a droplet.

In general, our results from the cascade modelling confirm the previous hydrodynamic calculations. In hydrodynamics, the dense phase stays within the original zone of creation. The droplets in our model also tend to remain close to their point of origin, because they are so massive. Very little entropy production was found in the hydrodynamic phase transition, and the droplets also do not produce much entropy when they evaporate pions.

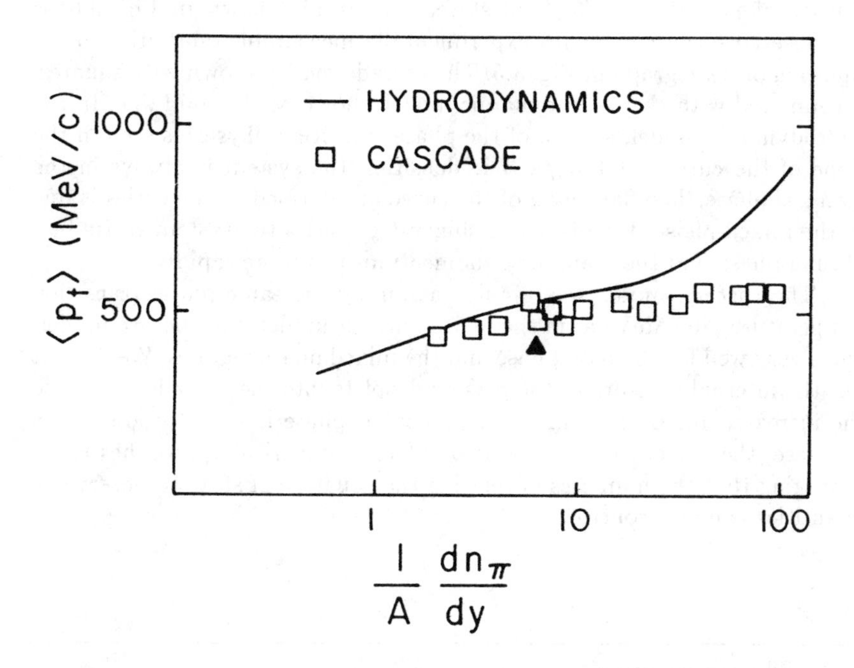

FIGURE 3.8
Final state transverse momentum as a function of the rapidity density of the produced pions. The solid line is the hydrodynamic prediction of Ref. [19], and the squares are the results of the cascade calculation of Ref. [16]. The experimentally measured point from O + Au collisions at CERN is shown by the triangle.

It can be shown that an isolated droplet radiating pions would produce 30% more pions than one converting to the pion phase under adiabatic conditions. This difference is not very much, when we compare with the order of magnitude difference in entropy densities expected for the two phases.

One more qualitative conclusion is worth mentioning. We found relatively little interaction between the pions in the hadronic state. On the average, each pion makes about one collision on the way out. These collisions do not affect the momentum distribution of the pions very much, so the final state pions almost provide a direct view of the plasma droplet surface.

Turning to the quantitative results, the model predicts the average transverse momentum, which can be displayed as a function of the rapidity density of particles in the final state. The graph, shown in Fig. 3.8, is the closest we can come with experimentally measurable quantities to the equation of state graph in Fig. 3.5. The cascade model, shown with squares, is compared with the hydrodynamic model, shown as the solid line. In the hydrodynamic model, a trace of the phase transition physics is seen in the slope of the curve. At low particle densities, the system is always in the pion gas phase. The flattening of the curve at intermediate densities is due to the mixed phase. Finally, at the highest densities the system starts in a plasma phase and the transverse momentum rises more rapidly.

The cascade model was applied assuming the same phase transition temperature (200 MeV) as in the hydrodynamic model, and the two models agree very well for the pion phase and the mixed phase regimes. We did not put the internal pressure of the plasma droplets into the cascade model, so the increased momentum in the superheated phase is not reproduced. In any case, the agreement of these two different numerical approaches is encouraging that the numerics of relating the equation of state to observable quantities is under control.

3.6 Experiment

Let us now take a closer look at the experimental data. I divide the experiments into two categories. In one class are the experiments that measure qualitative effects, such as the suppression or enhancement of flavored meson production. This kind of experiment is valuable to show the degree of interaction and equilibration in the hot zone. However, measurement of the equation of state requires experiments that determine dimensional quantities such as momentum. The relevant data here can be summarized in a few sentences. I take as a typical collision ^{16}O on a heavy target, with a projectile momentum of 200 MeV/n. The density of pions in the final state goes to about 120 per unit of rapidity. The data used to obtain this number is shown in Fig. 3.9. Only π^- were measured, so the number from this graph is multiplied by 3 to get the total pion yield. As mentioned earlier, to get so many pions in the final state requires that both primary and secondary collisions deposit energy into the central region. We see on the graph that the central region extends from about 1.5 to 3.5 units of rapidity, where the pion density has a broad peak.

The average transverse momentum of these pions is 400 MeV/c, which is only 10% higher than in NN collisions. Experimentalists were probably disappointed when they found such a small difference between heavy ion

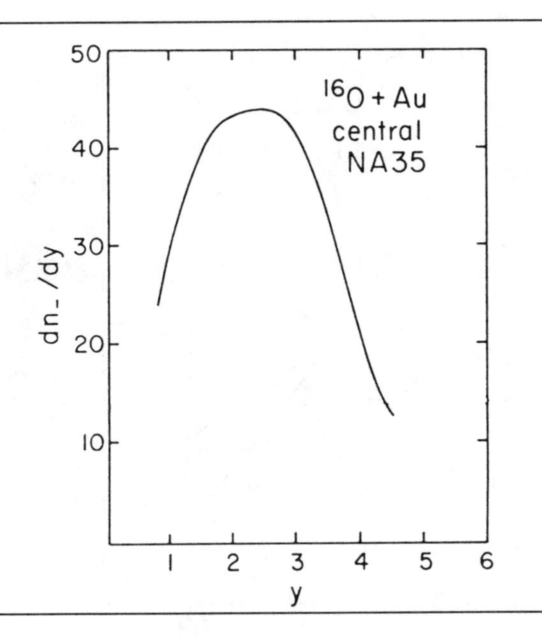

FIGURE 3.9
Experimental pion rapidity distribution for central O + Au collisions at 200 GeV/n, from Ref.[1], p. 92.

and NN collisions. But this is exactly what to expect if the system goes into a mixed phase. I have put the data point from these measurements on Fig. 3.8, the experimentalist's equation-of-state graph. The transverse momentum is in fact slightly lower than predicted for a phase transition at T=200 MeV.

At this point I should mention a feature of the pion momentum spectrum that is not explained so far and is very puzzling. The transverse momentum distribution of the pions, shown in Fig. 3.10, has an excess of low momentum pions that are not present in NN collisions. Comparing NN and heavy ion distributions, one finds that about 1/4 of the final state pions are in this low momentum group. A suggestion has been made that the effect may be due to a cooling of the pions by radial expansion [21], but this idea is not supported by the cascade or hydrodynamic calculations. Another possibility is that the pions are somehow associated with slow-moving baryons; again no detailed estimates supporting this idea have been published. Obviously, it is necessary to understand this data before any firm conclusions can be made that are based on the assumption of local equilibrium.

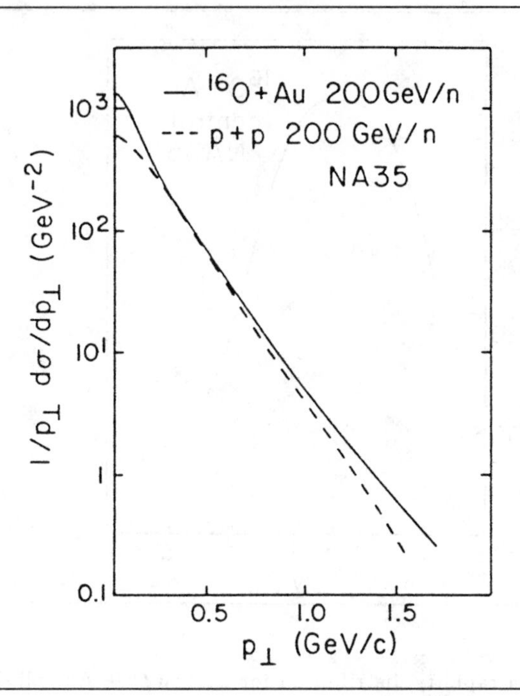

FIGURE 3.10
Momentum spectrum of the pions from central O + Au collisions at 200 GeV/n, from Ref. [1], p. 96.

3.7 Pion Interferometry

I now want to discuss a completely different kind of measurement, called pion interferometry or HBT interferometry [22]. Identically charged pions have correlations due to their Bose statistics, and these correlations reflect the distribution in space and time of the source of the pions. The theory of this effect may be shown in a few steps starting from the wave function for a pair of pions,

$$\begin{aligned} \psi_{\pi\pi} &= \frac{1}{\sqrt{2}}\left(e^{ip_1\cdot r_1+ip_2\cdot r_2}+e^{ip_1\cdot r_2+ip_2\cdot r_1}\right) \\ &= \sqrt{2}e^{\frac{1}{2}iK\cdot(r_1+r_2)}cos(\frac{1}{2}q\cdot(r_1-r_2))\ , \end{aligned} \tag{3.13}$$

where $q = p_1 - p_2$ is the difference in momentum between the pions and $K = p_1 + p_2$ is their sum. The amplitude to create the pions is given by the overlap with some source function $s(r)$,

$$A \approx \int dr_1 dr_2 s_1(r_1) s_2(r_2) \psi_{\pi\pi} \, . \tag{3.14}$$

The modulus squared of this overlap is the probability to produce the pions with the given momenta. It simplifies under the assumption of an incoherent source, by which is meant

$$< s_i(r_1) s_i(r_2) > \approx G(r_1)\delta(r_1 - r_2) \, . \tag{3.15}$$

Then the probability integral becomes

$$P(q) \approx 1 + \int dr_1 dr_2 G(r_1) G(r_2) cos(q \cdot (r_1 - r_2)) \, . \tag{3.16}$$

Thus the measurement of $P(q)$ gives direct information about the Fourier transform of the source function $G(r)$. In making this argument, we assumed that the pions are created at the same time. Eq. (3.16) is applicable to the more general case of different creation times if $q \cdot r$ is interpreted as the 4-vector dot product. It is useful to use the equal time wave function, however, because interactions such as the Coulomb interaction can be included by replacing the free wave function by the appropriate scattering wave function in a potential.

In this analysis, the assumption of an incoherent source restricts one to large spatial sources. The quality of the data tends also to be limited, so that one can only extract one or two parameters describing the source size. In any case, the results of this analysis are very different for hadronic sources and for the heavy ion collisions. For a hadronic source the correlation tends to be independent of the direction of the correlation; the source size is small, about 1 fm [23]. The correlations between pions in the CERN heavy ion collisions show a much larger source size, of the order of 6 fm [24].

The expectations for the source parameters are very different, depending on whether or not there are long-lived heavy particles produced as intermediates. The source for the two extreme scenarios, hot pion gas or plasma droplet initial state, are depicted in Fig. 3.11. We imagine observing a slice of the cylinder in a frame in which the pions come out with zero longitudinal motion. The two transverse directions to the axis of the cylinder we denote as outward and sideward. The outward direction is along the direction of motion of the pions, and the sideward is perpendicular to the pion direction and the cylinder axis. If the pions are formed immediately in a hot gas, their density is very hot and they make many collisions before the freezeout. These collisions take place as the gas expands, and the source is much larger radially than the initial cylinder. However, as pointed out by Pratt [25], these pions will be moving radially outward so that the pions on

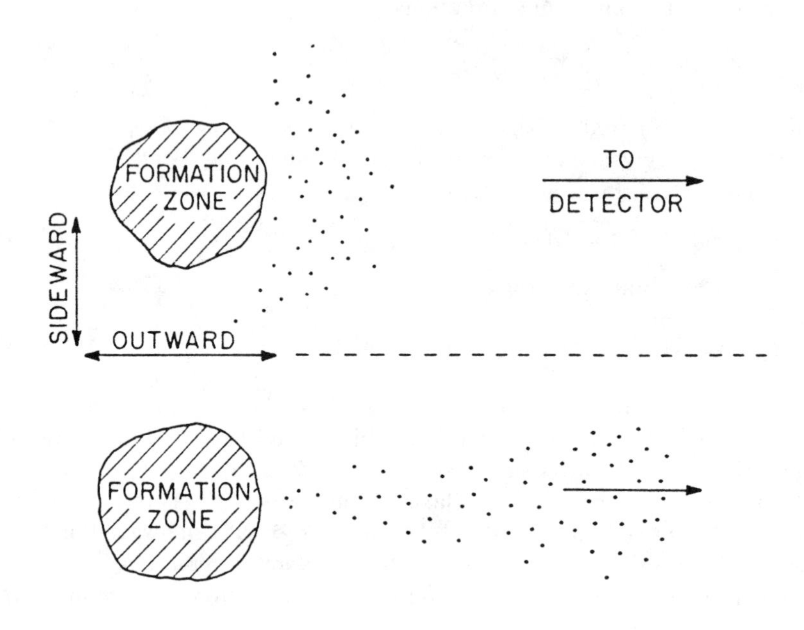

FIGURE 3.11
Pion source distribution in transverse direction, for two scenarios of the evolution. The upper diagram depicts the source points for a hot pion gas, showing only pions that are moving in the direction of the detector. The lower diagram depicts the corresponding pions from a decaying plasma phase. The horizontal elongation is due to the time delay converting high density matter to the low density phase.

the opposite side of the cylinder will not go into the detector. Effectively, the source seen by the detector just covers a region in front of the formation zone. This is depicted in the figure. In the plasma droplet scenario, the pions evaporate rather slowly so the spatial source size in the outward direction becomes long, as depicted in the figure. So the characteristic of heavy droplet formation is a long outward source and a moderately small sidewards source.

To test this idea quantitatively, I generated a pion source function $G(r)$ from the cascade simulation. But first an important technical point: the pions' momenta as well as their positions is determined by the source function, so we must explicitly include a dependence on p as well as r in G. We took the source function to be [26]

$$G(r,p) = \sum_i \delta^4(r - r_i) exp(-\frac{(p - p_i)^2}{p_0^2}) , \tag{3.17}$$

where the sum runs over the final state pions produced by the cascade simulation. The p_0 is a smoothing parameter. The expression we evaluate to get the correlation is then the same as Pratt's formula [25]

$$P(q,K) = 1 + \frac{\int d^4r d^4r' G(r, K/2) G(r', K/2) cos(q \cdot (r - r'))}{\int d^4r d^4r' G(r, p_1) G(r', p_2)} . \tag{3.18}$$

It is conventional to parameterize the source function by a Gaussian, in which case the correlation is also Gaussian. If we write the source function as

$$G(r,p) = exp(-(\frac{r_l}{R_l})^2 - (\frac{r_{side}}{R_{side}})^2 - (\frac{r_{out}}{R_{out}})^2) exp(-(\frac{p}{p_0})^2) , \tag{3.19}$$

then the correlation function is simply

$$P = 1 + exp(-\frac{1}{2}((q_l R_l)^2 + (q_{side} R_{side})^2 + (q_{out} R_{out})^2) . \tag{3.20}$$

The sideward and outward dimensions resulting from the cascade simulation are shown in Fig. 3.12. In the lower corner on the left side are the correlation lengths obtained for the hot pion gas model, *i.e.* all the pions were assumed to be already formed at the initial time (taken as $\tau_0 = 1$ fm/c). One of the points shows the correlations if the pions freely traveled from this point without interaction. The source size then reflects only the transverse size of the projectile nucleus, which is 2.2 fm in the case of ^{16}O. The nearby point on the graph shows the results with pion scattering included. The physical source size is much larger, but hardly any effect is seen in the correlation, for the reason explained above.

A dramatically different correlation function is obtained in the plasma droplet scenario. The outward dimension becomes much larger. For droplets of radius 1 fm, the outward dimension is about 10 fm/c. This increases for droplets with larger initial radii; smaller initial radii do not have much effect. The large source radius found experimentally [24] could arise from the droplets. To confirm this, it is necessary to separate the outward and sideward dimensions, which unfortunately is not so easy with the present experimental data. However, there are indications that the large dimension is associated with the outward direction [27]. One possible problem connected with this interpretation should be mentioned [28]. More massive mesons than the pion are formed in the collision, and their decay lifetimes will contribute to a larger source size. In particular, the ω meson with a mean life of 20 fm/c produces decay pions that masquerade as a long-lived source. The effect is substantial when meson production rates of the Lund model are used in the calculation. However, the Lund model predicts a

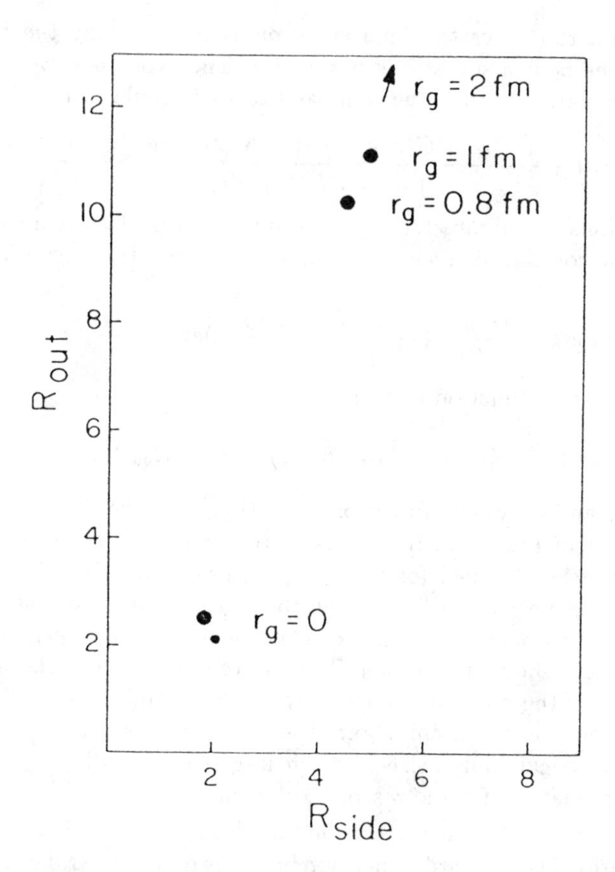

FIGURE 3.12
Comparison of outward and sideward source sizes from the cascade simulation. The formation zone is a boost-invariant cylinder with a Gaussian radius parameter 1.8 fm, producing hot matter that evolves to a final state containing 120 pions per unit rapidity. The small point near the origin shows the dimensions if the pions are formed immediately and do not interact. Collisions between points in the hot pion gas hardly change the source size, as shown by the nearby larger point. Models with plasma droplet formation give very large outward dimensions. For initial droplet radius 0.8 fm and 1.0 fm, the outward dimension is about 10 fm/c. Increasing the droplet radius to 2.0 fm doubles the outward dimension.

rather high proportion of ω mesons in the final state. If the collisions allow local thermal equilibrium, the proportion would be much lower. It is also found that in NN collisions the proportion of ρ mesons, while substantial, is less than assumed in the Lund model.

3.8 Outlook

Clearly, the data looks promising, suggesting that high density matter has been created and that it disperses with moderate pressure on a moderately long time scale. However, we need to know more about the equilibration. On the experimental side, one can imagine more sophisticated experiments that would measure proportions of higher mesons such as the ρ and the η. The strange mesons are interesting also, although we do not expect as rapid an equilibration in strangeness as in the degrees of freedom associated with ordinary quarks.

It is also important for theorists to estimate the time scales for equilibration using more fundamental descriptions of the hadronization process. One study reported an equilibration time of 2.5 fm/c, using a parton model of the nucleons [29]. On the other hand, hadronization times are much shorter in the Lund model [30]. On a fundamental level, the QCD parton structure of a nucleus may differ in important respects from a collection of independent nucleons. Collisions between the partons with the modified probability distribution may take place on rather short time scales, of the order of 0.3 fm/c [31,32]. Clearly, this is an area where more work needs to be done.

Acknowledgements

The author acknowledges conversations with L. McLerran, L. Van Hove, and W. Willis. This work was supported by the National Science Foundation under Grant PHY 87-14432.

Bibliography

[1] Z. Physik **C38** (1988), published as a separate book by Springer Verlag.

[2] *ibid*, p. 39.

[3] T. Akesson *et al.*, Phys. Lett., to be pub. This is the Helios collaboration.

[4] B. Andersson, G. Gustafson, and B. Nilsson-Almqvist, Nucl. Phys. **B281** (1987) 289.

[5] K. Werner, Phys. Lett. **B208** (1988) 520.

[6] T. Awes, priv. comm.

[7] T. Akesson *et al.*, Phys. Lett., to be pub.

[8] J.D. Bjorken, Phys. Rev. **D27** (1983) 140.

[9] L.D. Landau, *Collected Works*, p. 569-585; Izv. Akad. Nauk Ser. Fiz. **17** (1953) 51.

[10] For the statistical mechanics of massless bosons, see any text on black-body radiation, *e.g.*, L. Landau and E. Lifshitz, *Statistical Physics* (Addison Wesley, Reading, MA, 1958) p. 165.

[11] The energy densities of the two phases are given by: $e_{pi} = 3\sigma_{pi}/4$ and $e_{qg} = 3\sigma_{qg}/4 + B$.

[12] T. Celik *et al.*, Nucl. Phys. **B256** (1985) 670.

[13] L. Van Hove, Phys. Lett. **B118** (1982) 138.

[14] L. Van Hove, Z. Physik **C27** (1986) 135.

[15] P. Danielewicz and P.V. Ruuskanen, Phys. Rev. **D35** (1987) 344.

[16] G. Bertsch, M. Gong, L. McLerran, V. Ruuskanen and E. Sarkkinen, Phys. Rev. **D37** (1988) 1202.

[17] An exercise for the student: show that the one-dimensional hydrodynamics never freezes out by finding the total number of collisions a particle makes. This is estimated as $\int dt < \sigma v \rho >$ where σ is a cross section, $v \approx c$ is a relative velocity. The particle density, ρ, has a time dependence given by the considerations in the text.

[18] G. Baym, B. Friman, J. Blaizot, M. Soyeur, and W. Czyz, Nucl. Phys. **A407** (1983) 541.

[19] H. von Gersdorff, L. McLerran, M. Kataja, and P. Ruuskanen, Phys. Rev. **D34** (1986) 794; M. Kataja, P. Ruuskanen, L. McLerran, and H. von Gersdorff, Phys. Rev. **D34** (1986) 2755.

[20] V. Weisskopf, Phys. Rev. **52** (1937) 295.

[21] T.W. Atwater, P.S. Freier and J. Kapusta, Phys. Lett. **199** (1987) 30; K. Lee and U. Heinz, Regensburg preprint, 1988.

[22] The initials HBT stand for the authors of the paper where the technique of using boson correlations to infer a source size was first proposed, R. Hanbury-Brown and R. Twiss, Nature **177** (1956) 27. The technique was introduced to hadronic physics using pion correlations by G. Goldhaber, S. Goldhaber, W. Lee and A. Pais, Phys. Rev. **120** (1960) 300.

[23] T. Akesson,*et al.*, Z. Phys. **C36** (1987) 517.

[24] A. Bamberger *et NA35 al.*, Phys. Lett. **B203** (1988) 320.

[25] S. Pratt, Phys. Rev. Lett. **53** (1984) 1219.

[26] G. Bertsch, M. Gong, and M. Tohyama, Phys. Rev. **C37** (1988) 1896.

[27] J. Harris, *Proc. Quark Matter '88* (Lenox, MA), Nucl. Phys., to be pub.

[28] M. Gyulassy and S. Padula, *ibid.*

[29] D. Boal, Phys. Rev. **C33** (1986) 2206.

[30] T. Csorgo, J. Zimanyi, J. Bondorf and H. Heiselberg, NORDITA preprint (1988).

[31] A.H. Mueller and J. Qiu, Nucl. Phys. **B268** (1986) 427.

[32] J.P. Blaizot and A.H. Mueller, Nucl. Phys. **B289** (1987) 847.

4

The Standard Model and Beyond

Mary K. Gaillard
Department of Physics and Center for Particle Astrophysics
University of California at Berkeley
and
Theoretical Physics Group, Physics Division
Lawrence Berkeley Laboratory
Berkeley, CA 94720

4.1 Introduction

The field of elementary particle, or high energy, physics seeks to identify the most elementary constituents of nature and to study the forces that govern their interactions. Increasing the energy of a probe in a laboratory experiment increases its power as an effective microscope for discerning increasingly smaller structures of matter. Thus we have learned that matter is composed of molecules that are in turn composed of atoms, that the atom consists of a nucleus surrounded by a cloud of electrons, and that the atomic nucleus is a collection of protons and neutrons.

The more powerful probes provided by high energy particle accelerators have taught us that a nucleon (proton or neutron) is itself made of objects called quarks. Different quarks (q) are distinguished by attributes known as "flavor" and "color". The flavor quantum numbers of nucleons, such as electric charge, are determined by the flavor quantum numbers of the three "valence quarks" that are their constituents. The proton p (with electric charge $Q_p = +1$ in units of the positron charge) is composed of two "up" quarks u ($Q_u = +\frac{2}{3}$) and one "down" quark d ($Q_d = -\frac{1}{3}$); the neutron n (electric charge $Q_n = 0$) is composed of two down quarks and one up quark, as illustrated schematically in Fig. 4.1. In both cases the

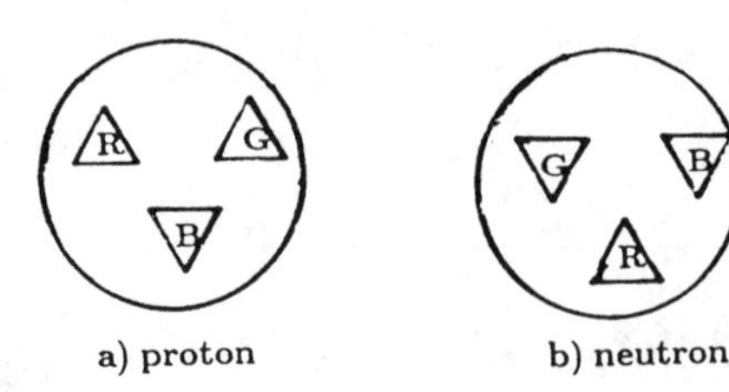

FIGURE 4.1
Schematic representation of a) a proton containing two up quarks and one down quark, and b) a neutron containing two down quarks and one up quark; each of the three quarks in a nucleon carries a different color charge.

three valence quarks are each of a different color, i.e. the the nucleons are "white", or color neutral, as are all known bound states of quarks.

The forces among quarks and electrons are understood within a general theoretical framework called the "standard model", that accounts for all interactions observed in high energy laboratory experiments to date. These are commonly categorized as the "strong", "weak" and "electromagnetic" interactions. In this lecture I will describe the standard model, and point out some of its limitations.

Probing for deeper structures in quarks and electrons defines the present frontier of particle physics. I will discuss some speculative ideas about extensions of the standard model and/or yet more fundamental forces that may underlie our present picture.

4.2 Elementary Glue: QED and QCD

It is well known that electrons are bound to the nucleus in an atom by the exchange of photons (γ) between the negatively charged electrons and the positively charged protons in the atomic nucleus. Photons are the quantum excitations of the electromagnetic field: a ray of light or a radio wave is a beam of photons. The exchange of photons creates an attractive force between particles of opposite electric charge. This is the electromagnetic force; its effects are described by the quantum field theory known as quantum electrodynamics or QED.

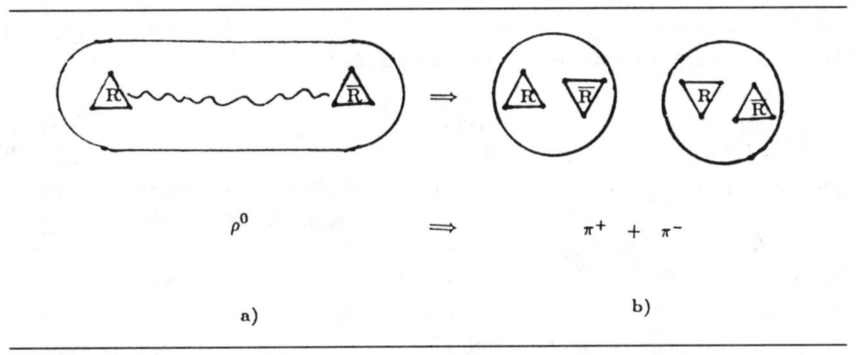

FIGURE 4.2
The neutral ρ–meson can be depicted as a quark and an antiquark attached to the ends of a string. At large distances the string tension becomes infinite; the string breaks, resulting in the creation of a color– and flavor–neutral $q\bar{q}$ pair. Thus the ρ breaks up into a pair of π–mesons, rather than a (confined) $q\bar{q}$ pair.

In a similar way, protons and neutrons are bound together within a nucleus by the exchange of particles called π-mesons, or pions. Pions are not elementary particles, but are themselves quark–antiquark bound states. Like quark bound states they are color neutral, that is, a red quark is bound to an "anti–red" antiquark, etc. In other words, unlike the "atomic glue" (the photon), which is elementary (as far as we know today), the nuclear glue is not. Pion exchange is one manifestation of the strong nuclear force.

According to the standard model, quarks are in turn bound to one another inside nucleons by the emission and absorption of elementary quanta called gluons (g), believed to be the fundamental mediators of the strong force. Their effects are described by a quantum field theory similar to QED that goes by the name of quantum chromodynamics, or QCD.

However, gluons differ from the photon in that they are themselves carriers of the strong "color charge" to which they couple; in contrast, the photon, which couples to electromagnetic charge, is itself electrically neutral. Since gluons are carriers as well as mediators of the color charge, the theory predicts the existence of "glueballs", that is, bound states of gluons with no valence quarks or antiquarks. Such states are difficult to establish experimentally, but evidence for glueballs is actively being sought among the debris produced in elementary particle collisions.

Another consequence of this self–coupling of the gluon is that the color force between two colored particles increases in strength with increasing distance between the particles. As a result, quarks and gluons do not appear as free particles in nature. They exist only inside composite particles like nucleons and pions, generically known as "hadrons". The name is derived from the Greek *hadros* (thick, heavy); all quark bound states, or hadrons, are subject to the strong force.

When one tries to use a very high energy probe to split a hadron into its constituents, one finds instead that quark–antiquark pairs (as well as gluons) are created in the strong color field and rearrange themselves together with the original constituents to form more hadrons. One way to model this phenomenon is to imagine a *meson* (the generic name for quark–antiquark bound states), for example, as a string whose ends carry the valence quark quantum numbers that determine the flavor properties of the meson. When the string is stretched over a sufficiently large distance (in practice, a distance that is large in comparison with the range of the nuclear force, which is about one fermi, or 10^{-13} cm) the string tension becomes infinite, and the string breaks because it cannot be stretched further: the two pieces of the broken string correspond to two mesons (see Fig. 4.2).

In spite of the fact that quarks and gluons do not themselves traverse particle detectors in an unbound state, the elementary interactions among them can nevertheless be studied in high energy laboratory experiments. This is because the hadronic debris resulting from a hard collision involving quarks and/or gluons arranges itself into collimated jets of particles. These jets leave visible tracks or electronic signals in detectors. The measured energy and direction of each high energy jet reflects the original energy and direction of an elementary quark or gluon from which the jet emanated. The hadronic jet is formed by a Bremsstrahlung process through which the primary quark or gluon radiates gluons and quark–antiquark pairs that ultimately (i.e. within a distance less than the "confinement radius" of about one fermi) convert to hadrons (mostly mesons). The collimated jet topology that results can be understood from the peaking towards small angles that is characteristic of a Bremsstrahlung distribution.

Jet production in e^+e^- collisions is illustrated in Fig. 4.3 which shows events recorded in the JADE detector at the DESY colliding ring facility in Hamburg, Germany, where electron and positron beams collide head on, each with an energy of 18 GeV (1 GeV = 10^9 electron Volts (eV), roughly the rest energy of a proton). The events shown are a) a collinear two–jet event resulting from the elementary process

$$e^+ + e^- \rightarrow \gamma \rightarrow q\bar{q} \; , \tag{4.1}$$

b) an event with two acollinear jets together with an energetic photon produced by Bremsstrahlung from one of the electrons or quarks in the process of Eq. (4.1), and c) a three–jet event interpreted a hard gluon Bremsstrahlung by one of the quarks.

Fig. 4.4 shows a "Lego plot" of the energy deposited in the Collider Detector at Fermilab (CDF) in Batavia, Illinois, by two quark and/or gluon

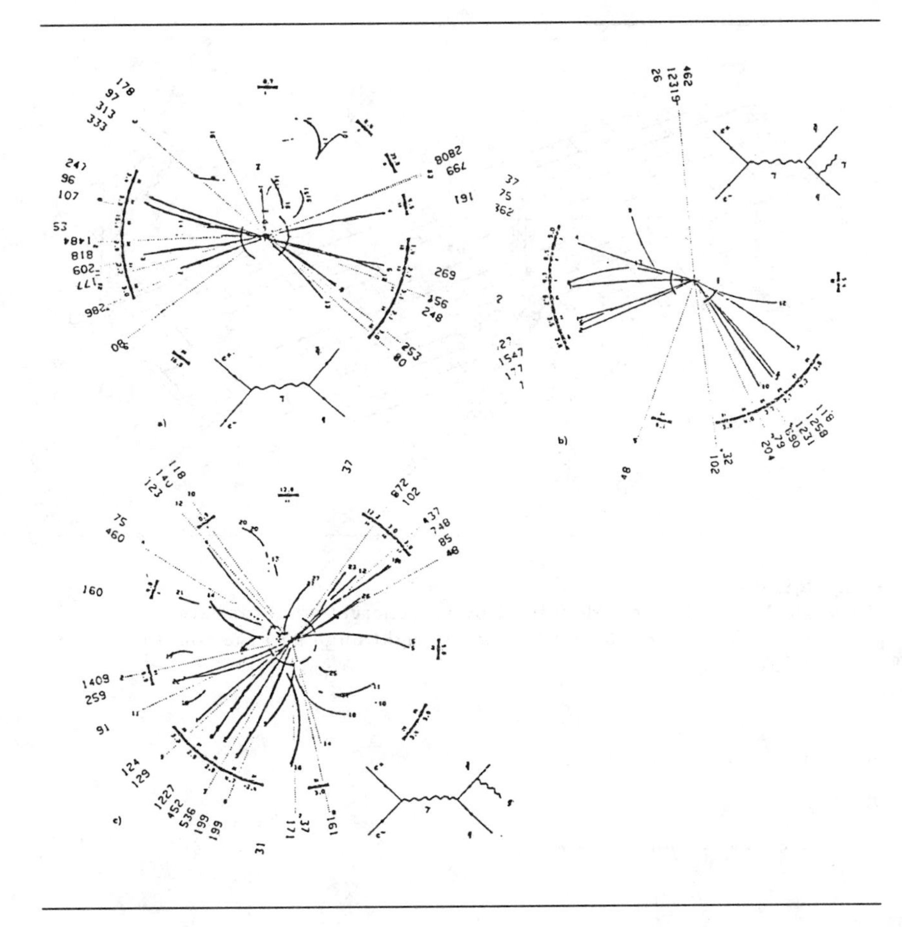

FIGURE 4.3
Two– and three–jet events observed in e^+e^- collisions at the DESY collider in Hamburg.

jets emitted back to back at about a 90° angle with respect to colliding proton and antiproton beams, each with an energy close to a TeV, that is, 10^{12} eV. A highly accelerated proton is in fact an ensemble of constituents including a colorless and flavorless "sea" of gluons and of quark and antiquark ($\bar{q}$) pairs, as well as its valence quarks. The proton's energy is shared among these constituents, each of which carries on average about a tenth to a sixth of the proton energy. The event shown in Fig. 4.4 is interpreted as resulting from a head–on collision of one of the proton's constituents with a constituent of the antiproton, *e.g.*:

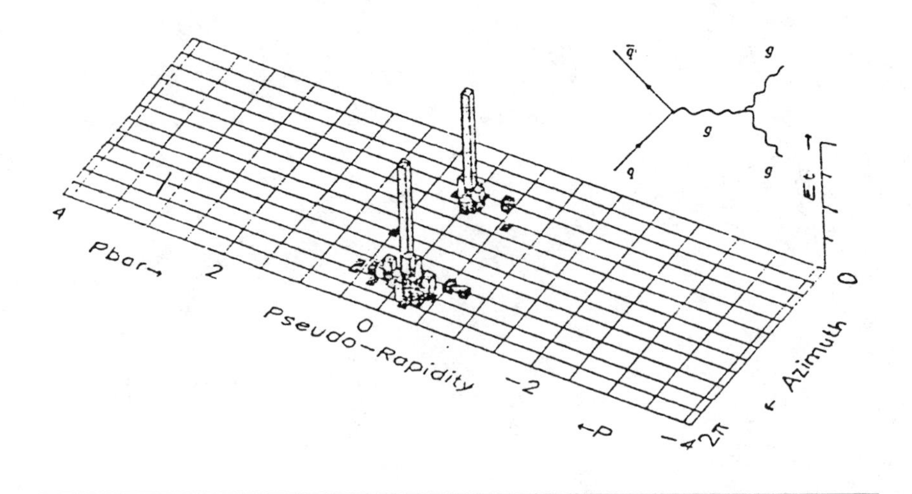

FIGURE 4.4
"Lego plot" of the energy deposited by two energetic hadron jets produced back-to-back, at approximately right angles to the colliding $p\bar{p}$ beams at Fermilab.

$$q + \bar{q} \rightarrow g \rightarrow q + \bar{q} \text{ or } g + g$$

or

$$g + g \rightarrow g \rightarrow q + \bar{q} \text{ or } g + g \,. \tag{4.2}$$

4.3 Ungluing Nuclear Matter: The Weak Force

The third elementary force of the standard model is the weak force, responsible for the radioactive β–decay of unstable nuclei. This decay occurs when, for example, one of the down quarks in a nucleon converts into an up quark emitting an electrically charged quantum known as a W–particle (Fig. 4.5). The W rapidly converts into an electron and a neutrino, which is an electrically neutral, apparently massless particle. Together with the electrically neutral Z–particles, the W's are the mediators of the weak force. Like gluons, they couple to one another—that is, they are carriers of the "weak charge", but unlike either the gluons or the photon, they are massive: the W and Z weigh, respectively, about 80 and 90 times the proton mass. They were discovered in 1983 in experiments using proton–antiproton ($p\overline{p}$) colliding beams at the European Center for Particle Physics, CERN, in

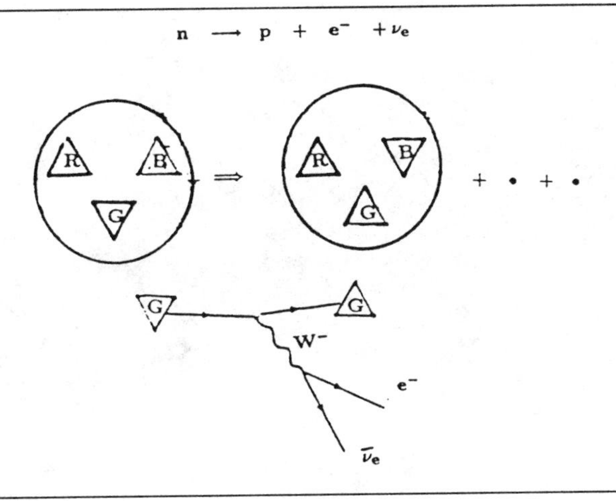

FIGURE 4.5
Schematic representation of neutron β–decay: one of the down quarks in the neutron converts into an up quark of the same color, emitting an $e^-\bar{\nu}_e$ pair by the exchange of a quantum of the W field.

Geneva, Switzerland. Their properties are being studied in detail both at CERN and at the higher energy $p\bar{p}$ collider at Fermilab. Fig. 4.6 shows an example of an event recorded by the CDF detector at Fermilab which is interpreted as the production and subsequent decay of a W particle, *e.g.*,

$$\bar{u} + d \to W^-, \quad W^- \to e^- + \bar{\nu}_e \, . \tag{4.3}$$

The figure shows the electron energy which is deposited in the detector via electromagnetic interactions. However, the weakly interacting neutrino leaves no trace of its passage, resulting in a large apparent imbalance of energy and momentum in the recorded event. This striking signal is very characteristic of W production and decay.

The properties of the Z will soon be studied in even greater detail at facilities where electrons collide with their antiparticles, positrons: a linear collider, SLC, that has recently begun operation at Stanford University, and a higher energy, circular collider, LEP, whose construction has just been completed at CERN.

Ordinary matter is composed of electrons (e) and of up and down quarks (u and d). Quarks of each flavor carry one of three possible color charges. Together with the electron neutrino (ν_e), the neutral, massless particle emitted in radioactive β–decay, these particles are the members of what is referred to as the first family of matter particles. Two other

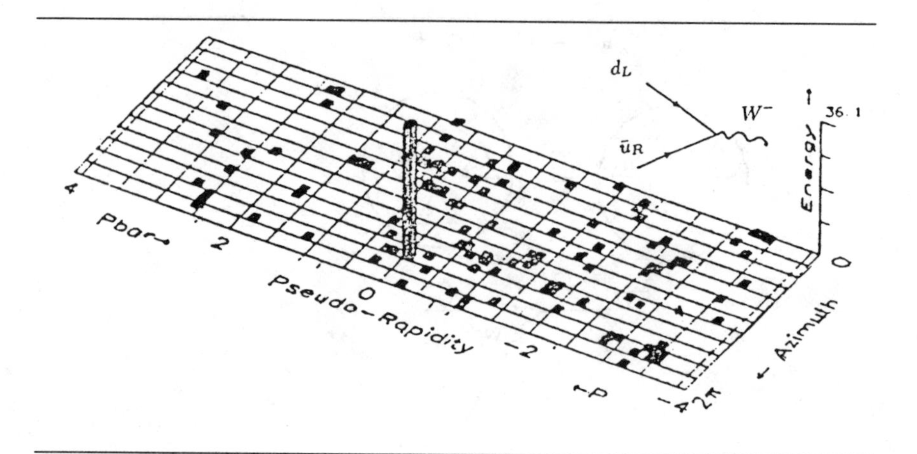

FIGURE 4.6
"Lego plot" of the energy deposited by an energetic, wide angle electron, interpreted as a decay product of a W–boson produced in $p\bar{p}$ collisions at Fermilab.

families of matter particles are known to exist. Their members (Table 4.1) have properties identical to those of the first family, except that they are more massive (aside, possibly, from neutrinos, which have all been found to be massless within the experimental errors).

The quark flavors analogous to (u, d) for the second and third families, respectively, are called "charm" and "strangeness" (c, s) and "top" and "bottom" (t, b). The top quark is in fact so heavy that it has not yet been established experimentally, but few particle physicists doubt its existence. These heavy quarks live only briefly, decaying rapidly to lighter quarks by a mechanism similar to the $d \to u$ transition of Fig. 4.5 that induces β–decay. Thus the heaviest quark t can decay via a chain:

$$\begin{array}{llllllll} t \to & b & + & \ell^+ & + & \nu^\ell & & \\ & \hookrightarrow & c & + & \ell^- & + & \overline{\nu_\ell} & \\ & & \hookrightarrow & s & + & \ell^+ & + & \nu_\ell \\ & & & \hookrightarrow & \nu & + & \ell^- & + \ \overline{\nu_\ell} \end{array} \tag{4.4}$$

where ℓ stands for e, μ, or τ, except for the last two steps in the chain, in which there is insufficient energy release for production of the more massive τ. Each link in the chain occurs via the exchange of a W particle, as in Fig. 4.5.

TABLE 4.1
Elementary fermions are grouped into three families, each with leptons (ℓ) of two flavors and quarks (q) of two flavors and three colors (taken here as Red, Blue and Green). The known families are listed below with the fermion masses (m) and electric charges Q indicated. (1 GeV= 10^3 MeV = 10^9 eV)

Quarks	m_q	Q_q	Leptons	m_ℓ	Q_ℓ
$u_R u_B u_G$	a few	+2/3	ν_e	< 18 eV	0
$d_R d_B d_G$	MeV	−1/3	e	.5 eV	−1
$c_R c_B c_G$	≈ 1.5 GeV	+2/3	ν_μ	< $\frac{1}{4}$ MeV	0
$s_R s_B s_G$	∼ 100 MeV	−1/3	μ	106 MeV	−1
$t_R t_B t_G$	> 78 GeV	+2/3	ν_τ	< 35 MeV	0
$b_R b_B b_G$	≈ 5 GeV	−1/3	τ	1.8 GeV	−1

Each family also includes two types of leptons (ℓ, ν_ℓ), that is, particles that carry no color charge and thus do not interact with gluons, like (e, ν_e) of the first family. The electrically charged counterparts of the electron for the second and third families are called the muon (μ) and the τ-lepton, respectively. Like the heavy quarks, they are short-lived and decay to lighter leptons:

$$\tau^- \rightarrow \nu_\tau + e^- + \bar{\nu}_e \quad \text{or :} \quad \begin{array}{l} \tau^- \rightarrow \nu_\tau + \mu^- + \bar{\nu}_\mu \\ \qquad \hookrightarrow \nu_\mu + e^- + \bar{\nu}_e \end{array} \tag{4.5}$$

Their companion neutrinos (ν_μ and ν_τ) may be stable.

All of the above matter particles are "fermions", which means that they are tiny spinning tops, carrying a half a unit (in units of Planck's constant $\hbar = h/2\pi$) of "spin" or intrinsic angular momentum. The elementary glue, that is, the quanta of the electromagnetic, color and weak fields (generically called gauge fields) carry one unit of spin. These spin−1 particles, called "vector bosons", are, respectively, the photon (γ), eight colored gluons (g), and the weak bosons W and Z. The interactions among the various particles are described by quantum field theories collectively known as "gauge theories." The W particles carry electromagnetic charge ($Q_{W^\pm} = \pm 1$) as well as weak charge, that is, they also couple to the photon. Thus in the standard model the weak and electromagnetic forces are described together by what is referred to as the "electroweak" gauge theory.

4.4 Symmetry and Broken Symmetry: Gauge Theories

The strong color interactions are characterized by a high degree of symmetry. Particles with the same spin and flavor, but different color charges, have identical masses; in fact there is no way to distinguish experimentally among particles that differ only in their color charge. The weak and electromagnetic interactions are understood in the context of an "electroweak" theory according to which the laws of nature are such that members of a fermion family with the same color charge, but different electric charge (e.g., a red up quark and a red down quark, or the electron and its neutrino) should be similarly indistinguishable. Clearly this is not what we observe: the symmetry of the elementary laws of nature is not reflected in the world around us.

To understand symmetry and symmetry breaking, imagine that the earth is a perfect sphere with no magnetic field. An ant crawling over the earth's surface would be completely lost; it could not distinguish one place from another. Now turn on the earth's magnetic field and give the ant a compass. The ant can now distinguish the north pole from the south pole, but it doesn't know its position along the equator: the presence of the magnetic field reduces the original spherical symmetry of the earth to a cylindrical symmetry (Fig. 4.7a).

Next consider a sphere in an abstract space, such that the north pole represents green color charge and the south pole red (Fig. 4.7b); the symmetry of the strong color interactions means that no position on this sphere is distinguishable from any other.

Finally, consider a sphere in another abstract space (Fig. 4.7c) where the north and south poles correspond to electric charges differing by one unit. The observed spectrum of particles and their interactions can be understood if we assume the existence of a field—known as the Higgs field—which distinguishes the north and south poles of this sphere, leaving a residual "cylindrical" symmetry.

Just as a photon passing through matter moves at a velocity v less than the speed of light, c, due to its interactions with atomic electrons, the W and Z bosons interact with the Higgs field that permeates all space, and they cannot propagate with the speed of light; they acquire an "index of refraction", $n = v/c$ or equivalently, a rest mass: the velocity is related to energy E and momentum p by $v = pc/E$, where for a particle of mass m, $E^2 = (cp)^2 + (c^2m)^2$, so the "index of refraction" is related to the mass by

$$n = \frac{p}{\sqrt{p^2 + m^2c^2}} \to \begin{cases} 0 & p^2 \ll m^2c^2, \\ 1 & p^2 \gg m^2c^2. \end{cases} \tag{4.6}$$

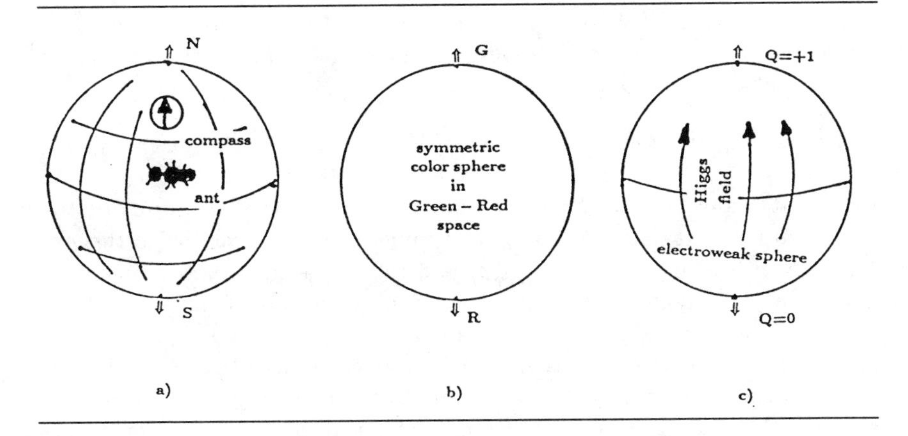

FIGURE 4.7
a) The spherical symmetry of the earth is broken by its magnetic field, leaving a residual cylindrical symmetry. b) The unbroken symmetry of QCD makes it impossible to distinguish one color from another. c) The conjectured "Higgs field" breaks the electroweak symmetry, making different electric charges distinguishable, but leaves a residual symmetry (conservation of total electric charge).

We see that at energies which are large compared with the boson rest masses, these masses become negligible, and the W and Z propagate essentially with the speed of light. The photon does not interact with the Higgs field (corresponding to the residual cylindrical symmetry of the imagined sphere), and hence is exactly massless.

The mass m_W acquired by the heavy gauge boson W is proportional to the strength g_W of its coupling to the Higgs field and to the strength v of that field. The field strength

$$v = \frac{2m_W}{g_W} \simeq 250 \text{ GeV} \tag{4.7}$$

is known from the measurement of the free neutron lifetime. To understand this, first recall that the force

$$\vec{F} = \frac{\partial V}{\partial \vec{r}} \tag{4.8}$$

that attracts two oppositely charged particles at rest, separated by a distance r and with electric charges $\pm e$, arises from the Coulomb potential

$$V = \frac{e^2}{4\pi r} \tag{4.9}$$

that is characteristic of a static electromagnetic field. A force mediated by a massive particle with mass m and coupling strength g is instead derived from a Yukawa potential

$$V = \frac{g^2}{4\pi r} e^{-mrc/\hbar}, \tag{4.10}$$

with finite range $r_{int} = \hbar c/m$. This can be understood in terms of the Heisenberg uncertainty principle: The range of the interaction is the uncertainty in distance: $r_{int} = \hbar/\Delta p$, and mc is the minimum momentum uncertainty Δp required for the emission of a particle of mass m_W from a much lighter particle. For a force mediated by W–exchange, $m_W c = 80$ GeV/c, and so $r_{int} \simeq 10^{-16}$ cm, which is much smaller than the effective radius of the nucleon: $r_{\text{nucl.}} \simeq 10^{-13}$ cm. As a consequence, the probability amplitude for neutron β–decay, which is determined as the Fourier transform of the potential ,Eq. (4.10), is suppressed by a factor proportional to the area $r_{int}^2 \propto m_W^{-2}$ over which the weak interaction is effective, and is therefore proportional to $v^{-2} = (g_W/2m_W)^2$. The coupling g_W has been independently determined by a series of measurements of weak transitions, including neutrino–induced interactions, and it was thus possible to predict accurately the mass of the W—as well as that of the Z—before they were directly produced in high energy $p\bar{p}$ collisions.

In a gauge field theory, each spin–one particle, or gauge boson, is identified with a symmetry of nature. In the unbroken symmetry phase (i.e. in the absence of a Higgs field) of the electroweak gauge theory, there are three gauge bosons $W_{1,2,3}$, with couplings of strength $g \equiv g_W$ to one another and to fermions. These three bosons correspond to the axes of rotation in an abstract three–dimensional space that represents flavor quantum numbers. In fact, the W's couple only to *left–spinning* fermions f_L, that is, fermions whose spin is anti–parallel to their momentum (as for a left–handed screw). Left spinning fermions are said to have one–half unit of *negative helicity*. Thus, for example, a rotation of 180° about the first or second axis (Fig. 4.8) in this abstract space effects the transformations

$$u_L \leftrightarrow d_L, \quad \nu_{eL} \leftrightarrow e_L^- \tag{4.11}$$

(i.e. a rotation north pole ↔ south pole in the sphere of Fig. 4.7c). In the symmetric phase, the laws of nature are unchanged by arbitrary rotations about these axes.

In addition there is a gauge boson B with couplings of strength g' to both left– and right–spinning fermions, and with no self–coupling (like the photon of QCD). The symmetry associated with B is a phase transformation on fermions:

$$f_i \rightarrow e^{i\alpha_i} f_i, \tag{4.12}$$

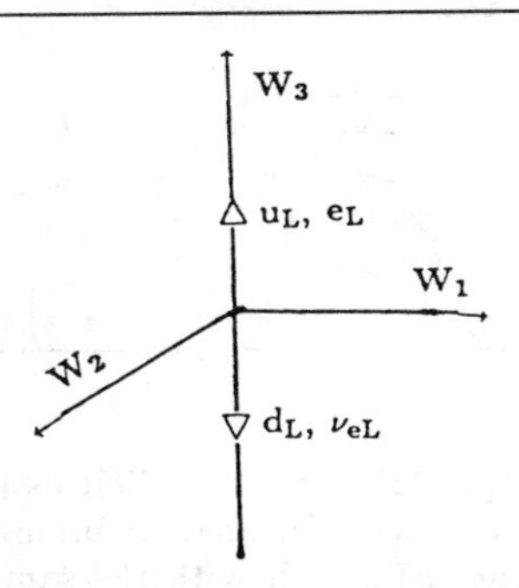

FIGURE 4.8
The three electroweak gauge bosons W_i correspond to three axes in an imaginary space. Fermions differing in electric charge by one unit correspond to equal and opposite positions along the third axis.

where f_i can be any right–spinning (f_R—possessing one half unit of *positive helicity*) or left–spinning fermion; the phase α_i depends on both the helicity and the flavor quantum numbers of the fermion f_i.

In the broken symmetry phase (i.e., in the presence of a background Higgs field), the physical states, that is, those states of well–defined mass, are linear combinations of the above, namely,

$$W^+ = \frac{1}{\sqrt{2}}(W_1 + iW_2) \quad , \quad W^- = \frac{1}{\sqrt{2}}(W_1 - iW_2) \, ,$$
$$Z = \cos\theta_w W_3 - \sin\theta_w B \quad , \quad \gamma = \cos\theta_w B + \sin\theta_w W_3 \, . \tag{4.13}$$

The known QED couplings of the photon γ — which remains massless because the background Higgs field is electrically neutral — makes it possible to relate the mixing angle θ_w to the electromagnetic (e) and weak (g, g') coupling strengths:

$$\sin\theta_w = \frac{e}{g}, \quad \cos\theta_w = \frac{e}{g'} \, . \tag{4.14}$$

This mixing angle also determines the relative rates of various transitions mediated by Z–exchange, such as

$$\nu + f_i \to \nu + f_i \, . \tag{4.15}$$

Thus measurements of a variety of neutrino–induced processes allowed the determination of the parameter θ_w and consequently, via Eq. (4.14), the couplings g and g', which in turn gave predictions for the W and Z masses through Eq. (4.7) and an analogous relation for the Z–particle. The Z–mass turns out to be related to the W–mass by:

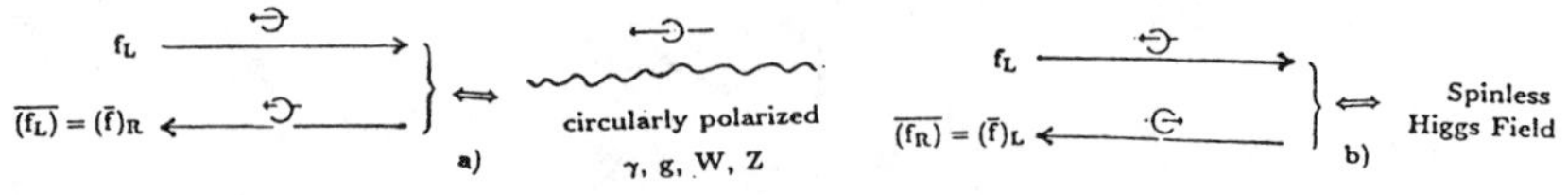

f_L
$\overline{(f_L)} = (\bar{f})_R$
circularly polarized
γ, g, W, Z
a)
f_L
$\overline{(f_R)} = (\bar{f})_L$
Spinless
Higgs Field
b)

FIGURE 4.9
a) A fermion–antifermion pair of opposite helicities (and opposite momentum directions) has the correct angular momentum to emit a spin–1 gauge field quantum . b) A pair with the same helicities can couple to the spinless Higgs field.

$$m_Z = \frac{m_W}{cos\theta_w} \,. \tag{4.16}$$

A characteristic feature of the couplings of gauge bosons to fermions is that emission or absorption of a gauge boson by a fermion leaves the helicity of the fermion unchanged. Equivalently, consider, a difermion system composed of a left–spinning fermion f_L and the antiparticle $\overline{(f'_L)}$ (which is right–spinning: $\overline{(f'_L)} \equiv (\bar{f}')_R$) of another left–spinning fermion, with no net orbital angular momentum (see Fig. 4.9a). This difermion system has net helicity -1, which is the same as that of a circularly polarized photon or other gauge boson. It therefore has the correct spin quantum numbers to couple to spin–1 bosons, as in the processes (4.1) and (4.2).

On the other hand a difermion system composed of a left–spinning fermion f_L and the antiparticle $\overline{(f'_R)} \equiv (\bar{f}')_L$ of a right–spinning fermion, and with no net orbital angular momentum (Fig. 4.9b), has net helicity 0, i.e., that of a spin–zero particle. In other words, emission or absorption of a spin–zero particle flips fermion helicity. Electrically charged quarks and leptons couple to the background Higgs field, and therefore propagate with a velocity less than the speed of light, that is, they acquire masses. As they propagate their helicity oscillates, which is precisely the property of a massive fermion. As far as we know, only left–spinning neutrinos occur in nature, and they may by exactly massless.

In the symmetric phase (no Higgs field), the fermions are necessarily massless because fermions of opposite helicity have different transformation properties under the symmetries of the gauge theory (e.g., Eqs. (4.11) and (4.12)) and therefore left $\leftrightarrow$ right oscillations are forbidden. However, in contrast to the case of the massive gauge bosons, the coupling strengths of fermions to the Higgs field cannot be independently determined, and therefore the mass of the top quark, for example, cannot be predicted. This is one of the weaknesses of the standard model.

4.5 Open Questions: Physics at the TeV Scale

While the standard model successfully accounts for a very large body of experimental data, the origin of electroweak symmetry breaking and the nature of the Higgs field are not understood. Another puzzle is the replication of fermion families. Why are there three families? Are there more? There is no understanding of the pattern of fermion masses, nor of the flavor mixing parameters that determine how fast heavy quarks and leptons decay to lighter ones via the chains of Eqs. (4.4) and (4.5). All of these parameters are determined by the couplings of fermions to the Higgs field; put another way, we do not understand why these couplings take the values that are observed. In particular, the couplings of fermions to the Higgs field seem to be the only couplings in nature that are not invariant under the CP operation which turns a fermion spinning parallel to its direction of motion into an antifermion spinning antiparallel. (C is the charge conjugation operation that turns a particle into an antiparticle of the same spin and momentum; P is the parity, or mirror reflection, operation that inverts momentum direction but not spin, and thus inverts helicity.) If CP were an exact symmetry of nature we would have no way of understanding the predominance of matter over antimatter in our universe, which made possible the formation of galaxies, planets, DNA, ... and life.

Many theorists believe that these questions can be answered only by probes with energies higher than those accessible at facilities now in operation. A signpost for the requisite hard collision energy is provided by the scale of electroweak symmetry breaking which is fixed by the value of the Higgs field strength, $v = 2g_W/m_W$, corresponding to an energy of a fourth of a TeV. Since, among the known particles, the massive W and Z couple most strongly to the Higgs field, their collisions are expected to provide the most efficient probe of its nature and origin. W's and Z's are present in colliding hadrons or leptons because, like photons, they can be radiated from energetic quarks and leptons—but with reduced energy. Calculations indicate that W and Z collisions with up to one or two TeV of total energy will be probed at the Superconducting Super–Collider (SSC) that is planned for construction in Waxahachie, Texas. In the SSC protons of 20 TeV energy will collide head on. This will permit the observation of quark interactions with total collision energies of up to about 4 TeV, and of collisions between radiated bosons with a total energy of 1 or 2 TeV.

As regards understanding electroweak symmetry breaking, the simplest possibility is the existence of an elementary spin–zero (Higgs) field that has a potential energy–density which has a local maximum for vanishing field strength, and a global minimum at a constant value v of field

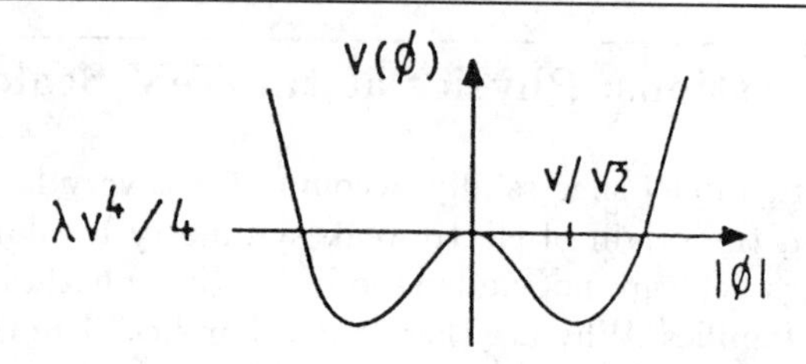

FIGURE 4.10
A scalar field potential energy density that has its minimum away from the symmetric point $|\phi| = 0$.

strength, as illustrated by the "Mexican hat potential" of Fig. 4.10. If this Higgs field is a vector on the imagined "electroweak sphere" of Fig. 4.7c, and if it is nonvanishing, it must be oriented. The direction of its orientation defines the "north" and "south" poles on that sphere, that is, it distinguishes between particles of different charges and thus breaks the electroweak symmetry.

Consider, for example a complex scalar field which can be parametrized as

$$\varphi(x) = \frac{\rho(x)}{\sqrt{2}} e^{i\theta(x)/v} , \tag{4.17}$$

where ρ and θ are real functions of the space–time point x. The potential energy–density

$$V(\varphi) \equiv V(\rho) = \frac{\lambda}{4}(\rho^2 - v^2)^2 \tag{4.18}$$

is minimized for $\rho = v$ or $|\varphi| = v/\sqrt{2}$. However, since V is invariant under phase transformations

$$\theta(x) \to \theta(x) + \alpha , \tag{4.19}$$

where α is a constant, the vacuum is degenerate. The background Higgs field takes on a constant value

$$\langle\varphi\rangle_{vac} = \frac{v}{\sqrt{2}} e^{i\theta_0/v}$$

so as to minimize the energy density, but quantum excitations that change the value of θ cost no energy; these excitations correspond to states of massless particles with zero energy and momentum. Quite generally, for every degree of degeneracy of the vacuum with respect to a continuous transformation on the fields, there is a massless, spinless particle, called a

Goldstone boson.

In a gauge theory in which the scalar fields couple to gauge boson fields, there is a larger symmetry, called a "gauge symmetry". For example if the spinless field φ defined in Eq. (4.17) is coupled to the electromagnetic field A_μ ($\mu = 0, \ldots, 3$ is a Lorentz index), the theory is invariant under phase transformations that depend on the space–time variable x, provided that the field A_μ is shifted by a total divergence:

$$\theta(x) \to \theta(x) + \alpha(x), \quad A_\mu \to A_\mu - \frac{1}{ev}\frac{\partial_\mu \alpha(x)}{\partial x^\mu}, \tag{4.20}$$

where e is the electric charge of φ. This gauge invariance allows one to make the choice $\alpha(x) = -\theta(x)$ with no observable consequences; with this choice the θ degree of freedom is apparently removed from the theory. It reappears as the longitudinally polarized component of the now massive "photon". Again quite generally, for every degree of vacuum degeneracy with respect to a *gauged* continuous transformation on the fields, the corresponding gauge boson acquires a mass. The electroweak symmetry is broken by the introduction of four real spinless fields, with a potential energy–density that has three degenerate directions in the space of field variables. The three associated massless particles are "eaten" by the massive $W^\pm$ and Z bosons to become their longitudinally polarized components.

However, there is always at least one massive scalar particle left over; in the above example, it corresponds to quantum excitations that move the field ρ away from its ground state value. The mass is determined by the curvature of the potential at its minimum; this gives

$$m_H^2 = 2\lambda v^2 = \frac{\lambda}{2}\,(\mathrm{TeV})^2\,, \tag{4.21}$$

for the mass of the "Higgs particle" $H \equiv \rho - v$, where I used the result (4.7) for the value of v in the electroweak theory, as inferred from β–decay data.

We see that the Higgs mass is determined by the parameter λ that also governs scalar self–interactions, that is, the terms quartic in the field ρ in the potential of Eq. (4.18). A Higgs particle with a rest energy as high as a TeV could be discovered at the SSC. If it were more massive than that, W and Z collisions would grow rapidly in rate when their total collision energy exceeds a TeV. This is because their longitudinal components, which are really eaten scalar particles, also have strong couplings if the parameter λ is of order unity or larger. The study of these new W and Z interactions with enhanced rates would provide an alternative probe of the physics associated with electroweak symmetry breaking. These interactions could be observable at the SSC if it operates at its maximum design energy and luminosity.

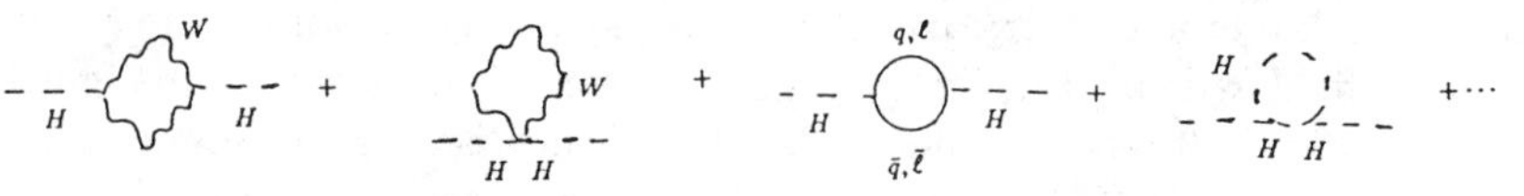

FIGURE 4.11
Quantum "loop" effects can contribute large corrections to the classical value of the scalar mass.

However, the existence of an elementary Higgs particle is problematic because the rest energy of a spinless particle gets large contributions from quantum fluctuations; these are illustrated by the "loop diagrams" in Fig. 4.11 that represent the emission and reabsorption of bosons and fermions by the Higgs field. When these effects are taken into account, it becomes difficult to understand how the Higgs mass can be as small as a TeV or so, as is required by the data and by the consistency of the theory. One can in fact show that allowing the quantum corrected value of the Higgs mass, and therefore the quantum corrected value of the coupling strength λ, to become arbitrarily large is not a viable option.

An alternative possibility is that the Higgs field is a composite field induced by fermion–antifermion pairs. It is believed that the effects of strong coupling at large distance in QCD lead to "quark condensation" (somewhat analogous to the formation of Cooper pairs), that is, nonvanishing vacuum expectation values of bilinear quark fields:

$$\langle \overline{q_L} q_R \rangle \equiv (v_\pi)^3 \approx (100 \text{ MeV})^3 \neq 0 \,. \tag{4.22}$$

The condensate (4.22) has the effect of breaking the so–called chiral symmetry of QCD. If quarks were massless, QCD would be invariant under separate (constant) phase transformations on right– and left–spinning quarks:

$$q_L \rightarrow e^{i\alpha} q_L, \quad q_R \rightarrow e^{i\beta} q_R \tag{4.23}$$

because the couplings of quarks to gluons does not mix quarks of different helicity. Since the up and down quarks have very tiny masses (a few MeV), QCD is to a very good approximation invariant under the transformations of Eq. (4.23) for $q = u, d$. The nonvanishing of Eq. (4.22) at the QCD vacuum, *i.e.* the state of lowest energy, breaks invariance under the chiral transformations of Eq. (4.23) and implies a vacuum degeneracy. This phenomenon explains many features of the observed mass spectrum of hadrons, in particular the massiveness of the nucleons N: $m_N \gg 3m_q$, and the very small pion masses: $m_\pi \approx \frac{1}{7} m_N$. According to this picture, if u and d were exactly massless, so would be the pions: they would be the Goldstone bosons associated with the vacuum degeneracies of the chiral symmetric theory.

The condensate (4.22) also breaks the electroweak gauge symmetry, which involves separate transformations among left– and right–spinning quarks. If there were no other source of electroweak symmetry breaking, the pions would be eaten by the $W^\pm$ and Z particles which would acquire masses of order

$$m_W = \cos\theta_w m_Z = \frac{g_W v_\pi}{2} \approx 30 \text{ MeV} . \tag{4.24}$$

To get a value of v as large the observed one (in other words to get a sufficiently large W mass) from the fermion condensation mechanism requires the introduction of new fermions, called "technifermions" that interact very strongly via a new force, "technicolor", transmitted by "technigluons". These new techniparticles, like quarks and gluons, would exist in bound states that could be produced only at collision energies as high as those to be made available at the SSC.

Alternatively, the potentially large quantum corrections to the Higgs rest energy could be damped if the theory possessed a larger symmetry, called supersymmetry, which relates fermions (particles with half integral spin) to bosons (particles with integral spin). This is because fermion loops and boson loops in the quantum corrections of Fig. 4.11 contribute with opposite signs. If supersymmetry were an exact symmetry of nature, there would be equal numbers of fermions and bosons occurring in pairs with equal mass, and their couplings to the Higgs particle would be related in such a way that the quantum corrections from boson loops would exactly cancel those from fermion loops. We know that supersymmetry is not an exact symmetry of nature: fermion and bosons are not observed in equal mass pairs. However if this lack of apparent supersymmetry is due to some symmetry breaking mechanism (similar to the electroweak gauge symmetry breaking) that gives large masses to the partners of the observed particles, the quantum corrections would still be damped by a factor proportional to the boson–fermion mass splitting ΔM. In other words, quantum effects would contribute a correction to the Higgs mass of order:

$$\Delta m_H^2 \sim \frac{g^2}{16\pi^2} \Delta M^2 . \tag{4.25}$$

A Higgs mass of a TeV or less could be understood if $\Delta M \leq 1$ TeV.

According to this conjecture, every known particle has a "superpartner" or companion "sparticle" with identical properties (mass, electric charge, ...), except that it differs in spin by one half unit. Thus the spin–$\frac{1}{2}$ particles of the standard model imply the existence of partner spin–0 particles, and conversely:

Spin = 1/2	$\Longleftrightarrow$	Spin = 0
quarks(q)	$\Longrightarrow$	squarks($\tilde{q}$)
leptons(ℓ)	$\Longrightarrow$	sleptons($\tilde{\ell}$)
neutrinos(ν)	$\Longrightarrow$	sneutrinos($\tilde{\nu}$)
Higgsino($\widetilde{H}$)	$\Longleftarrow$	Higgs(H)

Similarly, the spin–1 gauge bosons of the this model should be accompanied by spins–$\frac{1}{2}$ "gauginos": gluinos, the photino, the Wino, ...

Spin = 1	$\Longleftrightarrow$	Spin = 1/2
g	$\Longrightarrow$	$\tilde{g}$
γ	$\Longrightarrow$	$\tilde{\gamma}$
$W^{\pm}$	$\Longrightarrow$	$\widetilde{W}^{\pm}$
Z	$\Longrightarrow$	$\widetilde{Z}$

These sparticles are being sought, so far unsuccessfully, at accelerator facilities throughout the world. It is not possible to make precise predictions about their masses, but if supersymmetry plays a role in determining the scale of electroweak gauge symmetry breaking, at least some of them should have masses less that a TeV, and therefore accessible to planned, if not currently operating, machines. Squarks and gluinos would be abundantly produced at $p\bar{p}$ or pp collider facilities of sufficient energy, and any electrically charged sparticle would be produced together with its "anti-sparticle" in sufficiently energetic e^+e^- collisions.

4.6 Speculations: Physics Near the Planck Scale

The technicolor hypothesis replaces elementary spin–zero fields with fermion–antifermion composite spinless fields. Yet another possibility is that quarks and leptons are themselves composite and that some substructure would by revealed by analysing data from high energy scattering experiments. Quarks and/or leptons that have common constituents are expected to have additional, hitherto unobserved, very short range interactions arising from the elementary couplings of their constituents. At energies E that are small in comparison with the inverse range r_c of these new interactions: $(r_c E/\hbar c)^2 \ll 1$, these interactions are suppressed in rate relative to the standard model gauge interactions by a factor $(r_c E/\hbar c)^4$. However their effects can be observable at energies that are considerably smaller than $\hbar c r_c^{-1}$. The absence of such effects in present experimental data suggests that

$$r_c < \hbar c \,(\mathrm{TeV})^{-1} \approx 10^{-17}\mathrm{cm}\,, \tag{4.26}$$

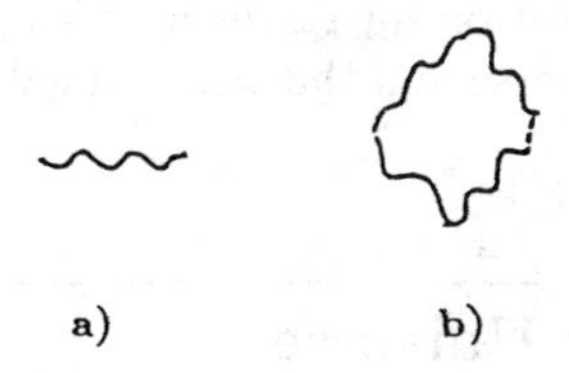

FIGURE 4.12
According to superstring theory the observed particles are the lowest vibrational modes of a) open strings (gauge bosons) and b) closed strings (graviton).

a result which implies that the scale at which quark and/or lepton compositeness becomes manifest is probably too large to be a relevant factor in determining the scale of electroweak symmetry breaking.

A more fashionable idea at present is that elementary "particles" are not particles at all, but rather the lowest vibrational modes of tiny strings—open strings or closed strings, see Fig. 4.12—with an extension of the order of the Planck length, about 10^{-33} cm. The excited string modes would have masses $M^2 \sim \hbar T/c^3$, where T is the string tension that is assumed to be governed by the value G_N of Newton's gravitational constant:

$$\hbar c T \sim G_N^{-1} \sim (10^{19}\,\mathrm{GeV})^2\,. \tag{4.27}$$

When supersymmetry is included, this "superstring" theory provides the only known possibility for a consistent quantum theory of gravity. It suggests that space–time is actually ten dimensional, but with six dimensions curled up with a radius comparable to the Planck length. In the context of superstring theory, many new exotic particles (in addition to the superpartners of ordinary particles) are predicted. Some of these could be produced at the present or planned accelerator facilities. Others would interact with ordinary matter only with couplings of gravitational strength, that is, too weakly to be produced in laboratory experiments. They could, however, have implications for cosmology.

It is also conjectured that higher symmetries entail couplings among quarks and leptons that can induce the proton to decay to leptons and pions or other mesons. Together with CP violation, proton instability could account for the existence of matter in today's universe. Evidence for proton decay has been sought unsuccessfully in deep mine experiments. If they exist, the interactions responsible for proton decay are probably far too weak to be detected in accelerator experiments. However new particles discovered at or below the TeV scale could give clues to the underlying theory, which in turn could shed light on questions such as proton instability, the density of matter in the universe and the origin of galaxy formation, among many others.

4.7 The Very Early Universe

According to the Big Bang theory, the universe began with an explosion yielding a hot dense gas of elementary particles that subsequently expanded and cooled. Its evolution to the presently observed universe was determined by the total number of particle species and by the properties of their interactions at very high energies. The growing field of "particle astrophysics" involves the search for relic particles from the Big Bang, in particular, particle species that might not have been observed in high energy accelerator experiments. Observational cosmology has in fact provided evidence for the presence of nonluminous matter in galactic halos. In addition, there are theoretical arguments which suggest that nuclear and atomic matter make up only a small fraction of the total matter in the universe. Perhaps the presence of superstring–predicted states with couplings to ordinary particles only through interactions of gravitational strength could account for this nonluminous matter. Other candidates for "dark matter" include massive neutrinos, stable sparticles, black holes, ... The determination of the nature of dark matter is in turn a crucial element that is needed for our understanding of the process through which galaxies were formed from density fluctuations in an otherwise homogenous early universe.

The evolution of the very early universe may have involved phase transitions analogous to the condensation of a gas or the freezing of a liquid. The Higgs field is similar to a ferromagnetic material, in which the lowest energy state is one with all electron spins aligned. Spin alignment implies a direction; the choice of this direction breaks the rotational symmetry of the laws of nature, just as the choice of a Higgs field orientation breaks the electroweak symmetry. In the ferromagnet, rotational symmetry can be restored by heating: the hot, energetic electrons become randomly oriented and there is no longer a special, preferred direction. Similarly, the electroweak symmetry should have been restored in the hot early universe. If the universe supercooled in the false, symmetric vacuum, this would have created a constant energy density that would have caused the universe to expand exponentially until the transition to the lower energy, asymmetric phase occurred. Such a period of exponential expansion could explain cosmological puzzles like the homogeneity, isotropy and flatness of the present universe. A consequence of this scenario is that the universe is closed, which in turn implies that the mass density of the observed universe is much greater that the density that can be accounted for by ordinary matter.

This is the theoretical argument in support of the existence of dark matter that I alluded to above.

Returning to the standard model of elementary particles, it is in fact known that the specific phase transition associated with electroweak symmetry breaking cannot have led to the inflationary epoch just described. However if there is a more fundamental theory underlying the standard model, and if that theory possesses still higher degrees of symmetry, there may have been many other, earlier phase transitions in the cosmological evolution. Understanding the origin of one such phase transition, which has an energy scale that appears to be accessible for study at achievable accelerator energies, would have profound implications for cosmology as well as for particle physics.

5

Tunneling in Many–Fermion Systems: From Instantons to Fission

John W. Negele
Center for Theoretical Physics
Laboratory for Nuclear Science and Department of Physics
Massachusetts Institute of Technology
Cambridge, MA 02139

A microscopic theory of barrier penetration is described in which a Feynman path integral is used to provide a physical picture of the collective path. The idea of an instanton in the case of a Boson field theory is generalized to many Fermions and used to determine the dynamical path self-consistently. The essential roles of Fermion nodal surfaces and symmetry breaking are emphasized. The theory is applied to a solvable pedagogical model, which demonstrates its quantitative accuracy and the importance of solving for the optimal collective path. Similar features are observed in a more realistic calculation of a nuclear system in three dimensions.

5.1 Introduction

One of the great contributions of Feynman to theoretical physics is the formulation of quantum mechanics in terms of path integrals. This way of thinking about physics provides both a physical picture in terms of time histories and a powerful framework for systematic calculations. Although the path integral may be viewed as an optional reformulation of what is already known for some problems, for other applications such as quantization of gauge theories or quantum cosmology, it is a fundamental and essential tool.

In this talk, I will describe another problem for which an appropriate path integral provides an essential tool — understanding the tunneling of many-Fermion systems. This problem exhibits fruitful interplay between several fields of physics. Many of the basic ideas, such as the instanton or bounce, have their foundations in statistical mechanics and field theory. However, in the past, these ideas had been restricted to Boson fields and the simplest geometries, and it required the rich phenomenology and detailed microscopic studies of nuclear fission to stimulate the generalizations to many Fermions and the spatial geometry of self-bound finite systems. Although the discussion will be framed in the context of spontaneous fission of atomic nuclei, the ideas are applicable to a variety of systems, ranging from induced fission and exotic radioactivity to alkali metal clusters.

The fundamental problem in formulating a microscopic theory of collective motion is to establish a general framework in which the Hamiltonian and the process specify the collective path. Physically, we expect some form of mean-field theory in which the optimal collective path through the classically forbidden region is determined self-consistently. From our experience with the Nilsson model and constrained Hartree-Fock calculations, we know that the rearrangement of the nodal structure of single-particle wave functions, and the associated symmetry breaking, will play a crucial role in the dynamics. This physical mean-field picture emerges simply and naturally from path integrals, which provide a powerful and convenient framework for a consistent quantum formulation of the problem. I will briefly review the general theory, details of which may be found in the literature [1–5] and then address an illuminating pedagogical model and the fission of a nucleus in three dimensions.

5.2 Mean Field Theory

The physical foundation of mean-field theory is the idea that the behavior of each particle is governed by the mean field generated by interactions with all the other particles. This mean field, or equivalently, the one-body density matrix, is the obvious candidate to communicate collective information and we seek a formulation of the quantum many-body problem in which it emerges naturally. A path integral may be expressed in terms of evolution in any convenient complete or overcomplete set of states, and when evaluated in the stationary-phase approximation, yields a path in this space. In this way, a many-body path integral in the space of Boson coherent states yields a self-consistent time-dependent Hartree theory, and a many-Fermion path integral in the space of Slater determinants yields a time-dependent Hartree-Fock theory. To show the basic idea, I will first review the case of a path integral with a single degree of freedom, and then generalize to the many-body problem.

5.2.1 Path Integral with a Single Degree of Freedom

A path integral deals with the non-commutativity of operators in quantum mechanics by expressing the evolution operator as a product of infinitesimal evolution operators for which non-commutativity may be ignored. Thus, one writes

$$\begin{aligned}\langle q_f|e^{-iHT}|q_i\rangle &= \langle q_f|e^{-i\epsilon H}\int dq_n|q_n\rangle\langle q_n|e^{-i\epsilon H}\int dq_{n-1}|q_{n-1}\rangle \\ &\quad \times\langle q_{n-1}|e^{-i\epsilon H}\cdots\int dq_1|q_1\rangle\langle q_1|e^{-i\epsilon H}|q_i\rangle \qquad (5.1)\end{aligned}$$

and

$$e^{-i\epsilon(T+V)} = e^{-i\epsilon T}\, e^{-i\epsilon V} + \mathcal{O}(\epsilon^2)$$

with the result

$$\begin{aligned}\langle q_f|e^{-iHT}|q_i\rangle &= \int D(q_1\cdots q_n)e^{i\epsilon\sum_k\left[\frac{m}{2}\left(\frac{q_{k+1}-q_k}{\epsilon}\right)^2-V(q_k)\right]} \\ &\to \int D\left[q(t)\right]\, e^{i\int S(q(t))}\,, \qquad (5.2)\end{aligned}$$

where

$$S\left(q(t)\right) = \int_0^T dt\left[\frac{m}{2}\dot{q}(t)^2 - V\left(q(t)\right)\right]$$

is the classical action.

The eigenvalues of the one-dimensional potential sketched in (a) of Fig. 5.1 are obtained by calculating the poles of the resolvent

$$\begin{aligned}\mathrm{Tr}\frac{1}{E-H+i\eta} &= -i\int_0^\infty dT\, e^{iET}\int dq\langle q|e^{-iHT}|q\rangle \\ &= -i\int_0^\infty dT\, e^{iET}\int dq\int D\left[q(t)\right]\, e^{iS(q(t))}\,, \qquad (5.3)\end{aligned}$$

where it is understood that each trajectory $q(t)$ is periodic with end point q. The stationary-phase approximation is now applied in turn to each of the three integrals in Eq. (5.3). Variation of the trajectory $q(t)$ yields the Euler–Lagrange equations for the stationary solution

$$m\frac{d^2}{dt^2}q_0 = -\nabla V(q_0) \qquad (5.4)$$

and variation of the endpoint yields conservation of p and thus $\dot{q}$ at the

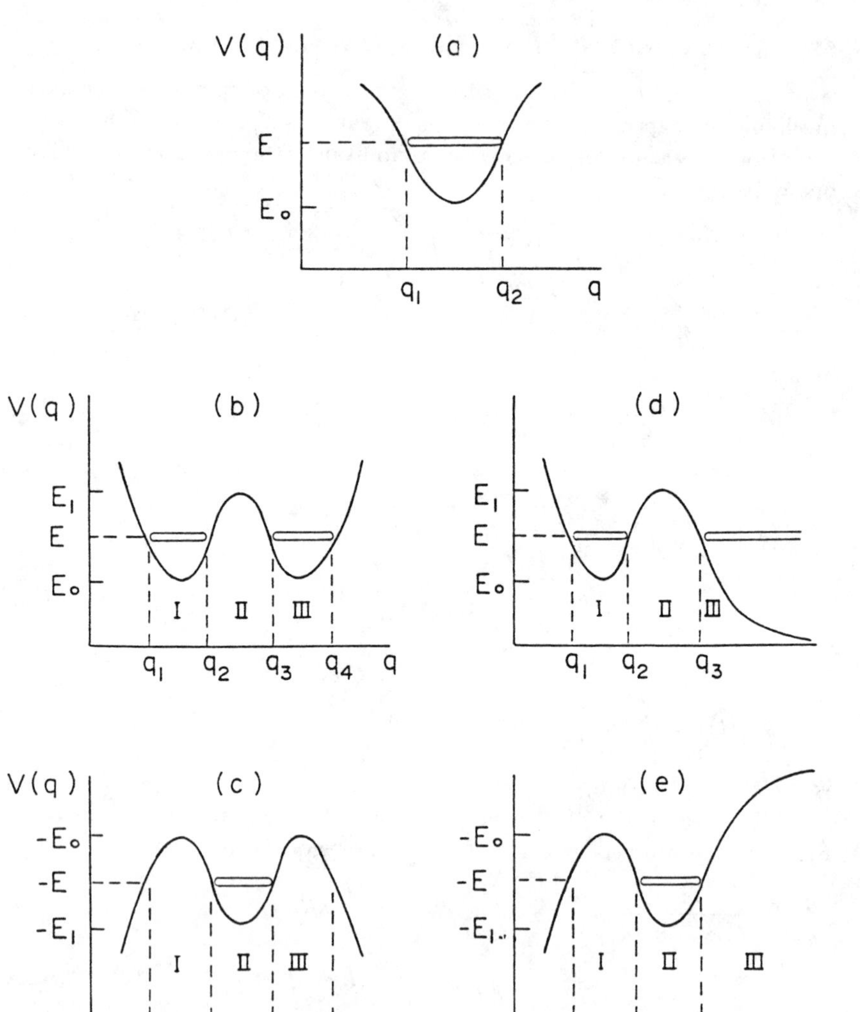

FIGURE 5.1
Sketches of the potentials $V(q)$, periodic trajectories, and turning points as described in the text.

endpoint. Finally, stationarity with respect to T yields

$$E = -\frac{\partial S(q_0, T)}{\partial T} = E(T) \,, \tag{5.5}$$

where $E(T)$ is the energy of a classical periodic orbit of period T,

$$T(E) = \int_{q_1}^{q_2} \sqrt{\frac{m}{2\,(E - V(q))}}\, dq \;. \tag{5.6}$$

Summing over all multiples T_n of the fundamental period, we obtain

$$\mathrm{Tr}\frac{1}{E - H + i\eta} = A \sum_N f_n\, e^{iW(T_n)} \,, \tag{5.7}$$

where

$$W(T_n) = ET_n + \int_0^{T_n} (p\dot{q} - H)\, dt = n \oint p\dot{q}dt$$

and A and f_n are factors arising from integrating quadratic fluctuations around the stationary points. Ignoring the factor f_n, the geometric series of Eq. (5.7) yields poles for energies satisfying the quantization condition

$$\oint p\dot{q}\, dt = 2n\pi \tag{5.8}$$

for any integer n. Inclusion in f_n of the $\frac{\pi}{2}$ phase at each turning point replaces $2n\pi$ by the Bohr-Sommerfeld condition $(2n+1)\pi$.

An interesting new feature arises in the analogous treatment of the double well in (b) of Fig. 5.1. In addition to periodic solutions of the form just discussed in regions I and III, there are stationary points in the complex T plane corresponding to periodic solutions in region II. A picturesque way to think of these solutions is to continue the classical equation of motion, Eq. (5.4), to imaginary time by replacing (it) by τ, with the result

$$m\frac{d^2}{d\tau^2} q_0 = -\nabla\left(-V(q_0)\right) \;. \tag{5.9}$$

That is, the trajectory which dominates the action in the stationary-phase sense corresponds to the classical solution in the inverted well. Such solutions were first introduced by Langer [6] in the context of bubble formation and correspond to the so-called instantons [7,8] or bounces [9] in field theory. These trajectories give rise to the period

$$T_{\mathrm{II}}(E) = 2i \int_{q_2}^{q_3} \sqrt{\frac{m}{2\,(V(q) - E)}}\, dq \tag{5.10}$$

and a contribution to the trace of $e^{-W_{\mathrm{II}}}$, where

$$W_{\rm II} = 2\int_{q_2}^{q_3} \sqrt{2m\,(V(q)-E)}\,dq\ . \tag{5.11}$$

A general periodic trajectory in the double well, Fig. 5.1b, is composed of all possible combinations of periodic orbits in each of the three regions, giving rise to the multiple geometric series:

$$\begin{aligned}\mathrm{Tr}\frac{1}{E-H+i\eta} &= A\sum_{k\ell m} f_{k\ell m}\, e^{ikW_{\rm I}(E)-\ell W_{\rm II}(E)+imW_{\rm III}(E)} \\ &= \frac{-2e^{i(W_{\rm I}+W_{\rm III})}-e^{iW_{\rm I}}-e^{-W_{\rm II}}-e^{iW_{\rm III}}}{(1+e^{iW_{\rm I}})\,(1+e^{iW_{\rm III}})+e^{-W_{\rm II}}}\ .\end{aligned} \tag{5.12}$$

Expansion of the denominator and the relations $\frac{\partial W}{\partial E} = T(E) = \frac{2\pi}{\omega}$ yield the familiar WKB energy splitting

$$E_n = E_n^0 \pm \frac{\omega}{2\pi}\, e^{-\frac{1}{2}W_{\rm II}}\ . \tag{5.13}$$

The lifetime of a metastable state is obtained by distorting the right-hand well to extend to the edge of an arbitrarily large normalization box, yielding the potential sketched in Fig. 5.1d. The lifetime is obtained by evaluating the smoothed level density, defined as the imaginary part of Eq. (5.3) with the infinitesimal η replaced by a finite width γ, such that γ is smaller than any physical width but larger than the level spacing in the normalization box. In this case, we obtain periodic stationary solutions in region I as before, and in lowest approximation these yield the result Eq. (5.8) for the energies of the quasi-stable states. In addition, periodic imaginary-time trajectories are obtained in region II, corresponding to solution of the classical equations of motion in the inverted potential sketched in Fig. 5.1e. The role of periodic solutions in region III is quite different than for the double well, and one may show that $e^{iW_{III}}$ yields negligibly small contributions.

Thus, the smoothed density of states has poles at complex energies $E_n^0 + \Delta E_n$ satisfying $1 + e^{iW_I(E_n^0+\Delta E_n)} = -e^{-W_{II}(E_n^0+\Delta E_n)}$ and expansion to first order in ΔE_n yields

$$\Delta E_n = -\frac{i\Gamma_n}{2}\ , \tag{5.14}$$

where

$$\Gamma_n = 2\frac{\omega(E_n)}{2\pi}\, e^{-W_2(E_n)}\ .$$

The level density is therefore proportional to $\left[(E-E_n)^2 + (\frac{1}{2}\Gamma_n)^2\right]^{-1}$ near E_n, so that Γ_n is the inverse lifetime of the metastable state. To within the factor of 2 which is presumably corrected by a careful evaluation of all corrections to the stationary-phase approximation, Eq. (5.14) is recognized as the familiar WKB formula for tunneling decay of a metastable state.

This simple example from one-dimensional quantum mechanics turns out to embody all the essential features of our subsequent treatment of the eigenstates of large-amplitude collective motion and tunneling decay of quantum many-body systems.

5.2.2 Many Particles

There are many alternative functional integrals for many-particle systems, each with its own advantages and limitations. The present objective is to obtain mean-field physics in the stationary-phase approximation. Hence, the many-particle Feynman path integral is inappropriate, since it yields the classical equations of motion. Rather, we will use overcomplete sets of coherent states and Slater determinants, respectively, for Bosons and Fermions.

The essential idea is most simply displayed for Bosons. Using coherent states $|\phi\rangle = e^{\int d\vec{x}\phi(\vec{x})\hat{\psi}^\dagger(\vec{x})}|0\rangle$ and inserting the completeness relation

$$\int \mathcal{D}[\phi^*(\vec{x}), \phi(\vec{x})]\, e^{-\int dx\, \phi^*(\vec{x})\phi(\vec{x})}|\phi\rangle\langle\phi| = 1 \tag{5.15}$$

between each factor $e^{-\epsilon H}$ in the evolution operator, we obtain [1]

$$\langle\phi_f|e^{-iHT}|\phi_i\rangle = \int \mathcal{D}\left[\phi^*(\vec{x},t), \phi(\vec{x},t)\right] e^{iS(\phi^*,\phi)}\;, \tag{5.16}$$

where

$$\begin{aligned} S(\phi^*,\phi) \;&=\; \int_0^T dt \int d\vec{x}\, \phi^*(\vec{x},t) \qquad\qquad (5.17)\\ &\times\; \left[-\frac{d}{dt} + \frac{\nabla^2}{2m} - \frac{1}{2}\int d\vec{x}'\, \phi^*(\vec{x}',t)\phi(\vec{x}',t)v(\vec{x}-\vec{x}')\right]\phi(\vec{x},t) \end{aligned}$$

plus a boundary term which is not relevant to the present discussion. The action, Eq. (5.17), is of the Hartree form and $\phi(x,t)$ has the physical interpretation of the condensate wave function. If we make the stationary-phase approximation and represent the evolution operator by a single optimal stationary trajectory, we obtain the time-dependent Hartree equation for the condensate wave function

$$i\frac{\partial}{\partial t}\phi(\vec{x},t) = \left[-\frac{\nabla^2}{2m} + \int d\vec{x}'\, \phi^*(\vec{x}',t)\phi(\vec{x}',t)v(\vec{x}-\vec{x}')\right]\phi(\vec{x},t)\;. \tag{5.18}$$

This equation describes the evolution of a drop of liquid ^{4}He in terms of a simple classical, function $\phi(\vec{x},t)$, which for low energy phenomena is nodeless. Although we have not stressed it, all the physics associated with short-range correlations, including the "hole" which the repulsive core of the Helium-Helium potential punches in the relative wave function, is subsumed in an effective interaction v. If one could attach repulsive charges of arbitrary strength to each Helium atom, one could generate a fission problem free of any of the Fermion nodal physics familiar in the nuclear case, and the bounce solution would smoothly evolve from a single Helium drop to two separated fragments.

The analogous result for Fermions is obtained economically using the completeness relation in the space of Slater determinants

$$\prod_{k,j} \int \mathcal{D}\left[\phi_k^*(\vec{x})\phi_j(\vec{x})\right] \delta\left(\int d\vec{x}\phi_k^*(\vec{x})\phi_j(\vec{x}) - \delta_{kj}\right) \times \mathcal{N}|\phi_1\phi_2\cdots\phi_N\rangle\langle\phi_1\phi_2\cdots\phi_N| = 1 , \tag{5.19}$$

where $|\phi_1\phi_2\cdots\phi_N\rangle$ denotes a Slater determinant which is composed of single-particle functions $\phi_1\cdots\phi_N$ and $\mathcal{N}$ is an irrelevant normalization factor [1,10]. The evolution operator becomes

$$\begin{aligned} \langle\Phi_f|e^{-iHT}|\Phi_i\rangle &= \int \mathcal{D}\left[\phi_1^*\cdots\phi_N^*\phi_1\cdots\phi_N\right]\prod_{kj} \qquad (5.20) \\ &\quad \times\delta\left(\int dx\, \phi_k^*(\vec{x},t)\phi_j(\vec{x},t) - \delta_{kj}\right) e^{iS(\phi^*\phi)} , \end{aligned}$$

where the action is

$$S(\phi^*\phi) = \int_0^T dt \left[\int dx \sum_k \phi_k^*(\vec{x},t) i\frac{d}{dt}\phi(\vec{x},t) - \mathcal{H}(\phi^*,\phi)\right] \tag{5.21}$$

and the Hartree-Fock energy functional is defined

$$\begin{aligned} \mathcal{H}(\phi^*,\phi) &= \sum_k \int d\vec{x}\phi_k^*(x)\frac{-\nabla^2}{2m}\phi_k(x) \\ &\quad + \frac{1}{2}\sum_{k,j}\int d\vec{x}\, d\vec{x}' \left[\phi_k^*(\vec{x},t)\phi_j^*(\vec{x}',t)v(\vec{x}-\vec{x}')\right. \\ &\quad \left. \times\left\{\phi_k(\vec{x},t)\phi_j(\vec{x}',t) - \phi_k(\vec{x}',t)\phi_j(\vec{x},t)\right\}\right] . \qquad (5.22) \end{aligned}$$

Note that the action Eq. (5.21) has the form of a classical field theory [11] with conjugate variables $i\phi_k^*$ and ϕ_k

$$\frac{\partial}{\partial t}\phi_k = \frac{\partial\mathcal{H}}{\partial(i\phi_k^*)} \quad ; \quad \frac{\partial}{\partial t}(i\phi_k^*) = -\frac{\partial\mathcal{H}}{\partial\phi_k} \tag{5.23}$$

so that we might naively expect the results for one degree of freedom to be generalized with $q \to \phi_k$, $p \to i\phi_k^*$.

Application of the stationary-phase approximation to $S(\phi^*\phi)$ with the orthonormality constraint in Eq. (5.20) enforced by Lagrange multipliers yields the following four-dimensional generalization of the usual three-dimensional Hartree-Fock problem

$$\left[i\frac{\partial}{\partial t}+\frac{\nabla^2}{2m}-\int d\vec{x}'\sum_k \phi_k^*(\vec{x}',t)\phi_k(\vec{x}',t)v(\vec{x},\vec{x}')+\text{exch}\right]\phi_j(\vec{x},t)$$
$$=\lambda_j\phi_j(\vec{x},t) \tag{5.24}$$

with the boundary conditions that the $\phi_k(\vec{x},t)$ vanish on the spatial boundary and are periodic with period T. Evaluating Eq. (5.20) using the periodic solutions to Eq. (5.24) yields the stationary phase result

$$\mathrm{Tr}\frac{1}{E-H+i\zeta}\propto\int_0^\infty dT\ e^{i[ET+S(T)]}\equiv\int_0^\infty dT\ e^{iW(T)}\ , \tag{5.25}$$

where $S(T)$ is the action $S[\phi^*,\phi]$, Eq. (5.21), evaluated with the periodic solutions of period T. The stationary-phase approximation to the T integral in Eq. (5.25) is performed precisely as in Eq. (5.17) for the one-dimensional problem, with the result that

$$\mathrm{Tr}\frac{1}{E-H+i\zeta}\approx\frac{e^{iW(T_0)}}{1-e^{iW(T_0)}}\ , \tag{5.26}$$

yielding the quantization condition

$$W(T_0)=\sum_k i\int dx\int_0^T dt\phi_k^*(x,t)\frac{\partial}{\partial t}\phi_k(x,t)=n2\pi\ . \tag{5.27}$$

As expected, this quantization condition has the structure of Eq. (5.8), with q,p replaced by $i\phi_k^*$ and ϕ_k.

The problem of barrier penetration can also be treated analogously to the one-dimensional example in Subsec. 5.2.1. The first step is to continue from real time to imaginary time. As before, we let $it\rightarrow\tau$ and it is convenient to use a symmetric time interval $(\frac{T}{2},\ -\frac{T}{2})$. One then finds [3] that the single-particle wave function $\phi_k(\vec{x},t)\rightarrow\tilde{\phi}_k(\vec{x},\tau)\equiv\phi_k(\vec{x},t=\frac{\tau}{i})$ which is purely real. The proper identification of the adjoint wave functions for analytic continuation is $\phi_k(\vec{x},t^*)^*\rightarrow\tilde{\phi}_k(\vec{x},-\tau)$. Note that the replacement $\phi^*(t)\phi(t)$ by $\tilde{\phi}(-\tau)\phi(\tau)$ systematically incorporates necessary physical properties, such as maintaining normalization of the density, and cancelling time-dependent factors of the form $\tilde{\phi}_k(\vec{x},\tau)=e^{\lambda\tau}\phi_k(x)$.

The action, Eq. (5.21), thus becomes

$$
\begin{aligned}
\tilde{S}[\tilde{\phi}(-\tau),\tilde{\phi}(\tau)] &= \int d\tau\Big[\Big[\int d\vec{x}\sum_k \tilde{\phi}_k(\vec{x},-\tau)[-\frac{\partial}{\partial\tau}+\frac{\nabla^2}{2m}]\tilde{\phi}_k(\vec{x},\tau) \\
&\quad -\frac{1}{2}\int\int d\vec{x}d\vec{x}'\tilde{\phi}_k(\vec{x},-\tau)\tilde{\phi}_j(\vec{x}',-\tau)v(\vec{x}-\vec{x}') \\
&\quad \times\{\tilde{\phi}_k(\vec{x},\tau)\tilde{\phi}_j(\vec{x}',\tau)-\tilde{\phi}_k(\vec{x}',\tau)\tilde{\phi}_j(\vec{x},\tau)\}\Big] \,. \qquad (5.28)
\end{aligned}
$$

To see the analog to the inverted potential of Eq. (5.9) it is useful to transform from the conjugate variables ϕ^* and ϕ to time-even and time-odd variables which correspond to coordinates and moments. We will seek to express the problem in the general form of a Lagrangian with a position-dependent mass [12]:

$$
L = \frac{1}{2}M(Q)\,\dot{Q}^2 - V(Q) \qquad (5.29)
$$

so that

$$
P = \frac{\partial L}{\partial \dot{Q}} = M(Q)\,\dot{Q} \qquad (5.30)
$$

$$
H = P\dot{Q} - L = \frac{P^2}{2M(Q)} + V(Q) \qquad (5.31)
$$

and thus

$$
L = P\dot{Q} - P\frac{1}{2M(Q)}P - V(Q)\,. \qquad (5.32)
$$

Now, consider the real time action Eq. (5.21), neglecting exchange terms for convenience, and change variables [3] such that

$$
\phi_k(\vec{x},t) = \sqrt{\rho_k(\vec{x},t)}e^{i\chi_k(\vec{x},t)}\,, \qquad (5.33)
$$

where $\rho(\vec{x},t)=\rho(\vec{x},-t)$ and $\chi(\vec{x},t)=-\chi(\vec{x},-t)$. Then, the action may be written as

$$
\begin{aligned}
S &= \int dt\left[i\sum_k \phi_k^*\dot{\phi}_k - \mathcal{H}(\phi_k^*,\phi_k)\right] \\
&= \int dt\int d\vec{x}\,\sum_k\left[\chi_k\frac{\partial}{\partial t}\rho_k - \frac{1}{2m}\rho_k(\nabla\chi_k)^2\right] - V(\rho_k)\,, \qquad (5.34)
\end{aligned}
$$

where

$$
V(\rho_k) = H\left(\sqrt{\rho_k},\sqrt{\rho_k}\right),
$$

that is, $V(\rho_k)$ is the Hartree-Fock energy functional Eq. (5.5) with a time-even determinant composed of real wave functions $\sqrt{\rho_k}$. The form of Eq. (5.34) is completely analogous to Eq. (5.28) where χ_k corresponds to a momentum, ρ_k corresponds to a coordinate, $\overleftarrow{\nabla}\frac{\rho}{2m}\overrightarrow{\nabla}$ is the inverse coordinate-dependent mass parameter and the potential is given by the HF energy functional.

In the imaginary-time case,the appropriate change of variables is

$$\tilde{\phi}_k(\vec{x},\tau) = \sqrt{\tilde{\rho}_k(\vec{x},\tau)}\, e^{-\tilde{\chi}_k(\vec{x},\tau)}\ , \tag{5.35}$$

where

$$\tilde{\rho}_k(\vec{x},\tau) \equiv \rho_k\left(\vec{x}, t = \tfrac{\tau}{i}\right) = \tilde{\rho}_k(\vec{x},-\tau)$$

and

$$\tilde{\chi}_k(\vec{x},\tau) = -i\chi_k\left(\vec{x}, t = \tfrac{\tau}{i}\right) = \tilde{\chi}_k(\vec{x},-\tau).$$

In this case, the action is

$$\tilde{S} = -\int d\tau \left(\int d\vec{x} \sum_k \left[\tilde{\chi}_k \frac{\partial}{\partial\tau}\tilde{\rho}_k - \frac{1}{2m}\tilde{\rho}_k \left(\nabla\tilde{\chi}_k\right)^2 \right] + V(\rho_k) \right)\ . \tag{5.36}$$

Comparing Eqs. (5.34) and (5.36) we see that relative to the terms involving χ, $V(\rho)$ undergoes a relative sign change, so that the stationary solutions simply correspond to solution in the inverted potential $V(\rho)$. Note that $V(\rho)$ is the multidimensional HF energy surface, the properties of which are known essentially only through constrained HF calculations.

In one respect, the problem of spontaneous decay in many dimension differs essentially from that in one dimension. Whereas in Fig. 5.1 parts d and e there was no problem in joining the solutions in the classically allowed and forbidden regions at the classical turning point, to obtain the geometric series, Eq. (5.12), in the multi-dimensional case the complete wave function composed of single-particle functions $\phi_\kappa(r,t)$ and $\tilde{\phi}_\kappa(r,t)$ must be joined at the classical turning point. Whereas this joining can in fact be demonstrated at the HF minimum, it does not occur in general and one must, therefore, use the alternative dilute instanton gas approximation [1,4,13] to evaluate the premultiplying factor corresponding to $\frac{\omega}{2\pi}$ in Eq. (5.14). With this one deficiency, repetition of the steps leading to Eq. (5.14) yields the following result for the lifetime of a metastable state

$$\Gamma = \sum_c \lim_{T\to\infty} \Gamma^c(T) \tag{5.37}$$

as a sum of particle widths

$$\Gamma^c(T) = \mathcal{K}\, e^{-\int_{-T/2}^{T/2} d\tau \int d\vec{x} \sum_k \tilde{\phi}_k^c(\vec{x},-\tau)\frac{\partial}{\partial\tau}\tilde{\phi}_k^c(\vec{x},\tau)}\ , \tag{5.38}$$

where $\tilde{\phi}_k^c$ represents a periodic solutions to the equations

$$\left[\frac{\partial}{\partial\tau} - \frac{\nabla^2}{2m} + \int d\vec{x}' \sum_k \tilde{\phi}_k(x',\tau)\tilde{\phi}_k(x',\tau)v(\vec{x}-\vec{x}')\right] \tilde{\phi}_j(\vec{x},\tau)$$
$$= \lambda_j \tilde{\phi}_j(\vec{x},\tau) \qquad (5.39)$$

for a specific fission channel c.

5.2.3 Symmetry Breaking

The truly novel feature of the many-Fermion tunneling problem which sets it apart from familiar problems in field theory is the physics associated with the nodal structure of the wave functions. The bounce solutions entering into vacuum decay and instantons in gauge theory are bosonic and smoothly evolve between classically allowed domains. In contrast, the nodal structure of Fermionic wave functions must rearrange substantially, and accounts for much of the richness of the physics.

The simplest case is in one spatial dimension. The bounce solution to a simple model with four orbitals is shown in Ref. [2]. Whereas the density evolves smoothly through the barrier, the nodes of each of the individual single-particle wave functions are substantially rearranged. At the initial time, the wave functions are simply the zero-, one-, two-, and three-node eigenfunctions of the static HF potential. The elongation to form a neck at the origin at $\tau = 0$ changes the structure, so that the first two wave functions resemble even and odd combinations of nodeless wave functions localized on separate sides of the origin and the last two correspond to even and odd combinations of wave functions located on each side of the origin containing a single node. Hence, a more illuminating representation would be states corresponding to sums and differences of the first two and last two wave functions, in which the total wave function would approximately factorize into two nearly separated subsystems.

In higher dimensions, as the shape of the mean field changes, the relative energies of wave functions with different numbers of nodes in the wide and narrow directions change, and the theory must address the corresponding level crossings.

To be specific, consider a deformed harmonic oscillator potential with the s and p orbital filled to produce a ^{16}O core. As the potential is elongated in the z-direction, an unoccupied state ϕ_{002} with two nodes in the z-direction decreases in energy while the occupied state ϕ_{100} with one node in the x-direction increases in energy, and at some point these levels cross. If the self-consistent field is constrained to be axially and reflection symmetric, these two levels cross without mixing. One possibility is to introduce pairing, in which case the mean field retains its symmetries, the orbitals contribute to the density in the form $a^2|\phi_{100}|^2 + b^2|\phi_{002}|^2$ and the pairing interaction governs the dynamics of tunneling. The other possibility is to allow the mean field to break all symmetries. In this case, near the level crossing the single-particle wave function is some mixture of the two orbitals, the orbitals contribute to the density in the form $|a\phi_{100} + b\phi_{002}|^2$ and the self-consistent broken symmetry mean field governs the dynamics. Before exploring this symmetry breaking in a three-dimensional system, it is illuminating to examine a simple solvable model.

5.3 A Pedagogical Model

A simple solvable model exhibits the essential features of mean field symmetry breaking [14]. For motivation, we consider a Hartree-like Hamiltonian for a large number of particles which could be in either of two single-particle states of a two-dimensional well: ϕ_x with a node in the x-direction and ϕ_y with a node in the y-direction. Further, we let z_i characterize the deformation of the orbital containing the ith particle. As the single-particle well is elongated in the x-direction, the z_i-increase, the single-particle energy of ϕ_x decreases, and the energy of ϕ_y increases. Since there are only two states in the model, and we are characterizing the deformation of each orbital by the variable z_i, each particle may be represented by the coordinate z_i and a spin. The full Hamiltonian for N distinguishable particles is

$$H = \sum_{i=1}^{N} \left(-\frac{1}{2}\frac{d^2}{dz^2} + \frac{1}{2}z^2 \right)_i + \kappa \left(\sum_{i=1}^{N} z_i \right) \left(\sum_{i=1}^{N} \sigma_z(i) \right) + \lambda \left(\sum_{i=1}^{N} \sigma_x(i) \right)^2 . \tag{5.40}$$

The second term represents the dependence of the single-particle energies on the total deformation of the system. The last term is the two-body residual interaction, representing the fact that for either a δ-function interaction or a separable interaction of the form $\sum_i x_i y_i \sum_j x_j y_j$ the following two-body matrix elements are roughly equal: $\langle \phi_x \phi_x | v | \phi_z \phi_z \rangle = \langle \phi_x \phi_z | v | \phi_z \phi_x \rangle$. Note that by transforming to collective c.m. variables $z = \sum_{i=1}^{N} z_i$ and $(N-1)$ relative coordinates, the relative coordinates decouple and one has a single dynamical variable coupled to the total spin. In this sense, the model may be viewed as a generalization of the familiar Lipkin model extended to allow for barrier penetration [15].

When $\lambda = 0$, the total spin projection $M = \langle \sum_i \sigma_z(i) \rangle$ is a good quantum number, the Hamiltonian is a shifted harmonic oscillator, the wave function is a shifted oscillator state

$$\phi_i(z, M, \nu) \propto H_\nu \left(\frac{1}{\sqrt{N}}(z - \kappa M N) \right) e^{-\frac{1}{2N}(z-\kappa MN)^2} |M\rangle ,$$

and there are two degenerate ground states: all spin up at $z = \kappa NM$ and all spin down at $z = -\kappa NM$. For non-zero λ, the diagonal matrix elements of the shifted oscillators are supplemented by off-diagonal matrix elements connecting $|M-2, \nu\rangle$ with $|M+2, \nu'\rangle$, the two degenerate ground states are split by tunneling, and the exact solution may be obtained by numerical matrix diagonalization.

The parameters appropriate to nuclear fission have been determined to reproduce a single-particle frequency corresponding to a giant quadrupole frequency of 10 – 15 MeV, a barrier height of 5 MeV and mixing comparable to that obtained in fission with realistic residual interaction matrix elements of the order of 0.2 MeV. Because all N level crossings occur simultaneously, an additional factor of N occurs in the mixing due to coherence, and the strength of the residual interaction has been reduced accordingly. The sign of λ is important in the Hartree solution, and the physical choice is negative λ, corresponding to an attractive residual interaction. The final parameters we use are $N = 40$, $\kappa = 0.00603$ and $\lambda = -0.0005$.

It is instructive to compare two approximations to the exact ground state splitting; the conventional cranking model which is based on constrained Hartree-Fock solutions and the imaginary time-dependent mean-field approximation derived in the previous section. By the proof of Eq. (6.15) of Ref. [2], one may show that the stationary-phase approximation corresponds to an expansion in $\frac{1}{N}$ for this model and thus should be extremely accurate.

The static mean field solution is obtained in the usual way using a product of single-particle wave functions of the form $\phi_i = \phi(z_i)\binom{\cos\frac{\theta}{2}}{\sin\frac{\theta}{2}}$. Variation yields the single-particle Hamiltonian

$$H_{sp} = -\frac{1}{2}\frac{d^2}{dz^2} + \frac{1}{2}z^2 + \kappa\langle\sigma_z\rangle z + \kappa\langle z\rangle\sigma_z + 2\lambda\langle\sigma_x\rangle\sigma_x + fz , \qquad (5.41)$$

where f is a constraining field used to constrain $\langle z\rangle$ and $\phi(z)$ is a shifted oscillator. The expectation values of the spin operators in the wave function are $\langle\sigma_z\rangle = N\cos\theta$ and $\langle\sigma_x\rangle = N\sin\theta$, so that $\langle\sigma_x\rangle$ may be expressed in terms of $\langle\sigma_z\rangle$. From the coefficients of σ_x and σ_z in the single-particle Hamiltonian, we conclude that either $\langle\sigma_x\rangle = 0$, in which case all spins are either up or down and there is no symmetry breaking, or else $\langle\sigma_z\rangle = \frac{\kappa\langle z\rangle}{2\lambda}$, corresponding to mixing of the two spin states and symmetry breaking. Using these relations, the energy and $\langle\sigma_x\rangle$ may be expressed as functions of $\langle z\rangle$, with the results shown in Figs. 5.2 and 5.3. The salient feature is that when the collective variable $\langle z\rangle$ is constrained, the system retains its symmetry for much of the path. Only close to the level crossing is the

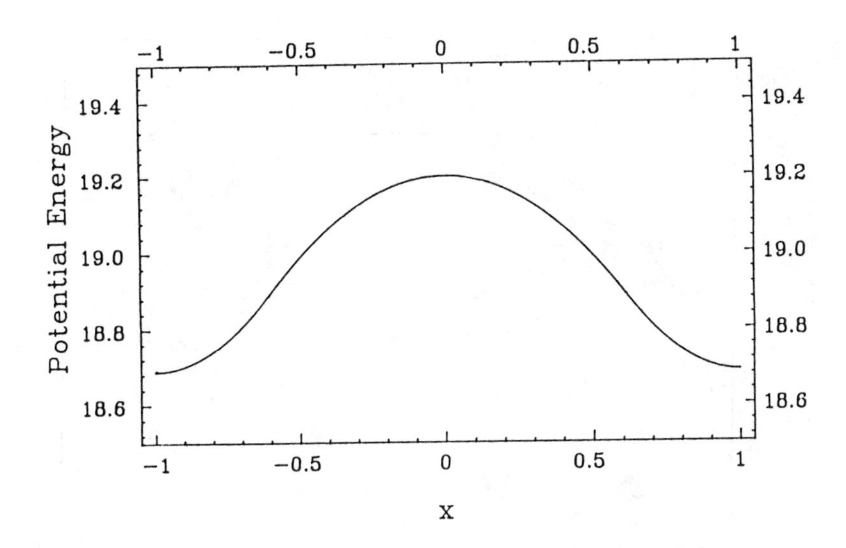

FIGURE 5.2
Collective potential in constrained mean-field theory.

symmetry broken with a significant expectation value $\langle\sigma_x\rangle \neq 0$. As derived in Ref. [14], the collective energy and cranking inertia in the symmetric regime are

$$\begin{aligned} E(\langle z\rangle) &= \frac{N}{2} + \frac{\langle z\rangle^2}{2N} - \kappa|\langle z\rangle|N \\ I &= \kappa^2 N^3 \end{aligned} \tag{5.42}$$

and in the broken symmetry regime are

$$\begin{aligned} E(\langle z\rangle) &= \frac{N}{2} + \frac{1}{2}\left(\frac{1}{N} - \frac{\kappa^2}{2|\lambda|}\right)\langle z\rangle^2 + \lambda N^2 \\ I(\langle z\rangle) &= \kappa^2 N^3 - \frac{1}{8\lambda N\left(\frac{4\lambda^2}{\kappa^4 N^2} - \frac{\langle z\rangle^2}{\kappa^2 N^4}\right)} \end{aligned} \tag{5.43}$$

from which the level splitting may be evaluated analytically in the WKB approximation.

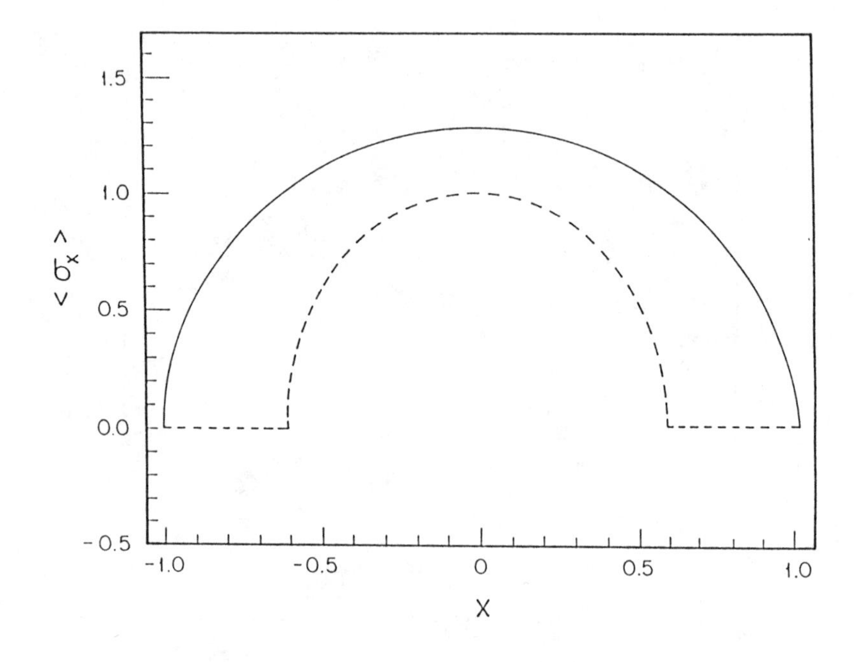

FIGURE 5.3
Collective path for the model in constrained mean-field theory (dashed line) and in imaginary-time mean-field theory (solid line).

The energy splitting in the imaginary-time mean-field theory is given by

$$\Delta E = \frac{\omega}{\pi} e^{-\frac{N}{2}\int_{-T/2}^{T/2} d\tau \langle\phi(-\tau)|\frac{\partial}{\partial\tau}|\phi(\tau)\rangle}, \tag{5.44}$$

where the wave function ϕ is the solution to

$$\left[\frac{\partial}{\partial\tau} - \frac{1}{2}\frac{\partial^2}{\partial z^2} + \frac{1}{2}z^2 + \kappa\langle\sigma_z\rangle_\tau z + \kappa\langle z\rangle_\tau \sigma_z + 2\lambda\langle\sigma_x\rangle_\tau \sigma_x\right]\phi(z,\tau) = \epsilon\phi(z,\tau) \tag{5.45}$$

with periodic boundary conditions and $\langle\mathcal{O}\rangle_\tau \equiv \langle\phi(-\tau)|\mathcal{O}|\phi(\tau)\rangle$.

Whereas the penetrability is well-defined and straightforwardly calculable for any energy, the pre-multiplying factor in Eq. (5.44) is more

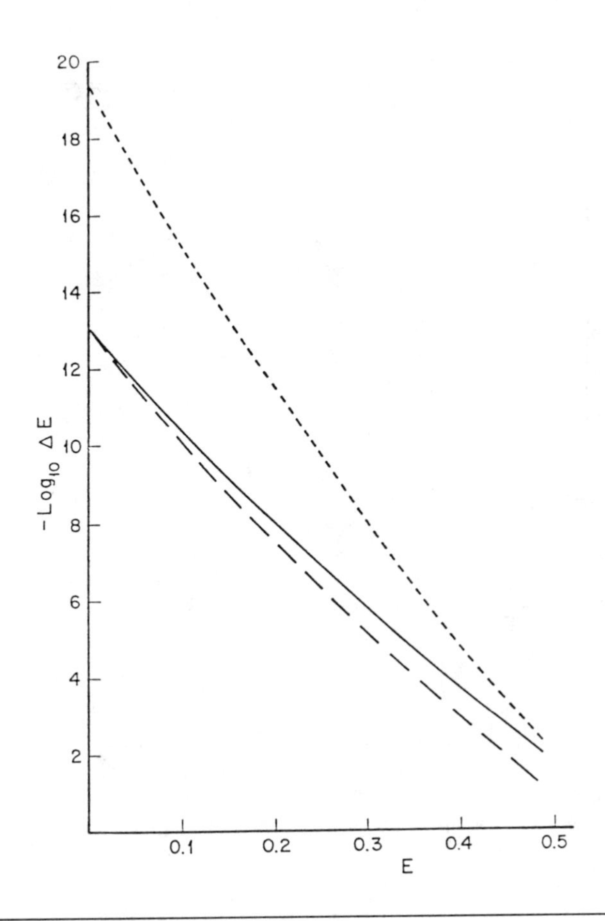

FIGURE 5.4
Comparison of exact splittings as a function of energy (solid curve) with the imaginary-time mean-field theory (dashed lines) and constrained mean-field theory (dotted).

difficult. For the ground state, it is unambiguously defined as the ratio of two determinants in the dilute instanton gas approximation [9] and should be well-approximated by the collective frequency ω in the outer well. We have thus used this oscillator frequency for all energies. As the energy is increased, it is clear physically that other degrees of freedom play increasingly important roles, so this prescription will systematically overestimate the splitting.

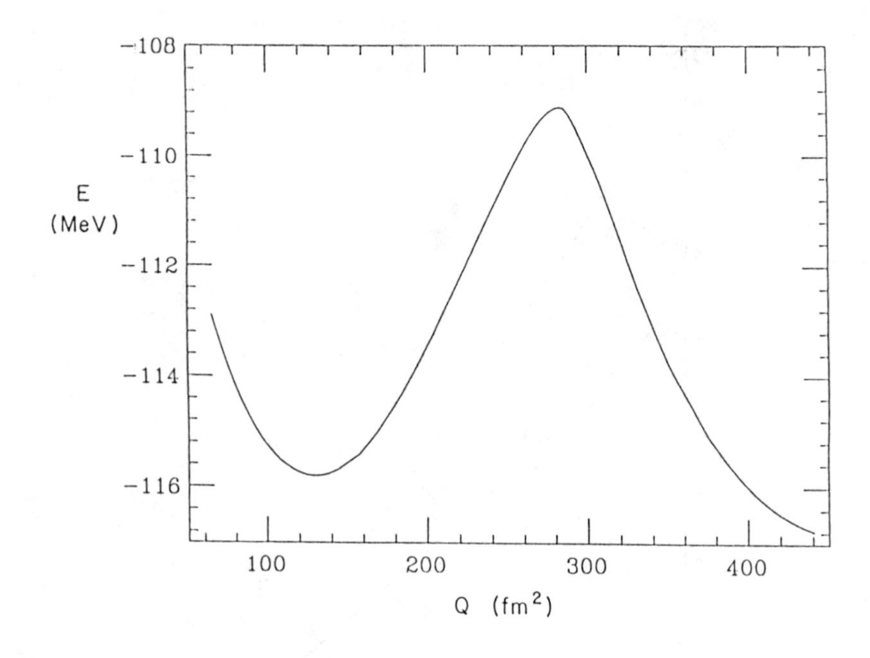

FIGURE 5.5
Hartree-Fock energy as a function of deformation.

The self-consistent path defined by the solution to Eq. (5.45) is shown by the solid line in Fig. 5.3. Note that in contrast to the constrained case, $\langle \sigma_x \rangle$ is non-zero for the entire trajectory, so that the symmetry is broken for the whole optimal trajectory. In this model, we can improve the static constrained theory appreciably [14] by constraining $\langle \sigma_z \rangle$ instead of $\langle z \rangle$, which generates a path much closer to the optimal path. This constraint effectively grabs hold of the relevant level-crossing degree of freedom from the start. In a realistic fission calculation, however, there is no way to constrain the corresponding single-particle degrees of freedom, so one must necessarily constrain a collective deformation variable analogous to z

The difference between the optimal path and the static constrained path is dramatically demonstrated by the approximations obtained for level splittings below the barrier. For the ground state splitting, $\Delta E = e^{-12.96}$, the imaginary-time mean field yields $e^{-13.02}$ whereas the constrained mean

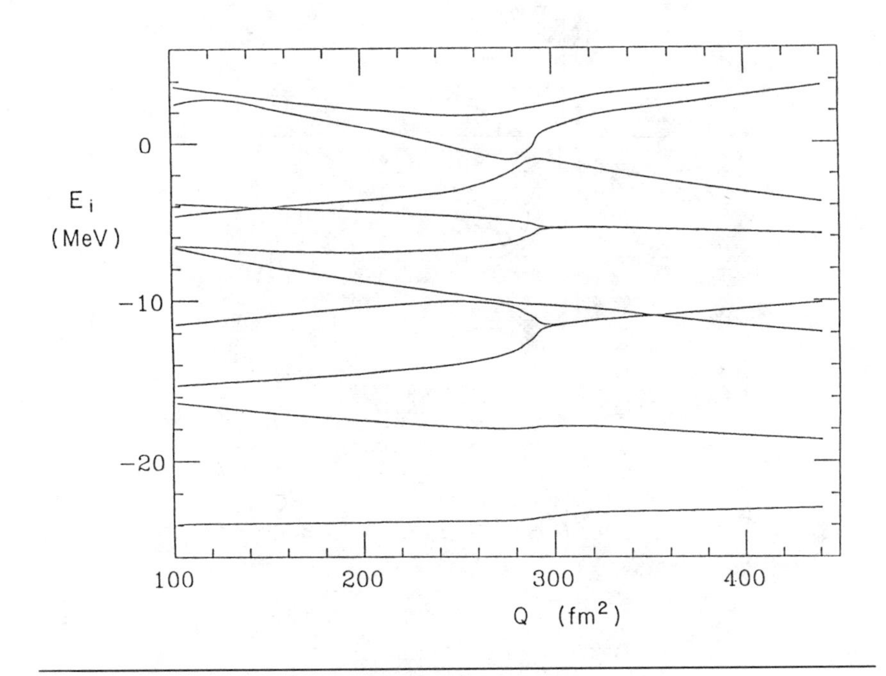

FIGURE 5.6
Single-particle energies as a function of deformation $\langle Q \rangle$. The lowest eight orbitals are occupied.

field yields $e^{-19.5}$. The corresponding results for excited states are graphed in Fig. 5.4.

5.4 Fission of ^{32}S in Three Dimensions

Finally, I will show the results of a calculation [16] which displays symmetry breaking associated with orbital rearrangement in a nucleus in three dimensions. We calculate the symmetric fission of ^{32}S using a density-dependent interaction with the strength of the Coulomb force increased to produce fission. The collective variable analogous to $\langle z \rangle$ in the previous model is the quadrupole moment, and the total energy and single-particle energies in a constrained Hartree Fock calculation are shown in Figs. 5.5 and 5.6. The eighth orbital is the last occupied state and changes from having two nodes

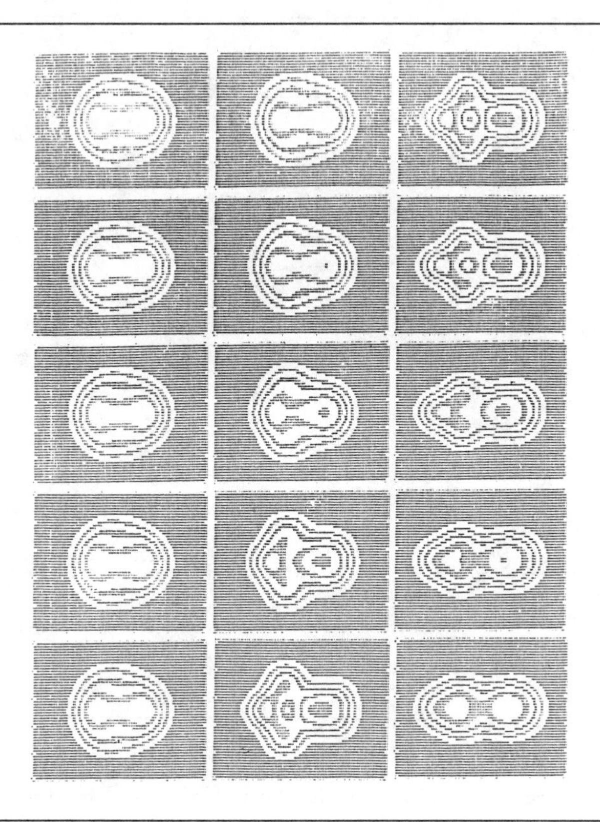

FIGURE 5.7
Contour plots at sequential times of the density integrated over z.

in the transverse direction at small deformation to having three nodes in the longitudinal direction at large deformation.

The time-dependent mean field equations are solved as follows [17]. We begin by solving the imaginary-time analog of the RPA equations at the top of the barrier shown in Fig. 5.5, with a constraint on the time average of the quadrupole moment to render the calculation stable. The period is then successively increased, and for each period the eigenvalue problem of Eq. (5.39) is solved iteratively. Since the largest eigenvalue of the evolution operator $e^{-\int_{-T/2}^{T/2} h_{sp}(\tau)d\tau}$ corresponds to the lowest occupied state $\phi_1(-T/2)$, successive application of the evolution operator to an arbitrary state converges to ϕ_1. Higher states are calculated by application of the evolution operator and orthogonalization to all lower states. In this way, the

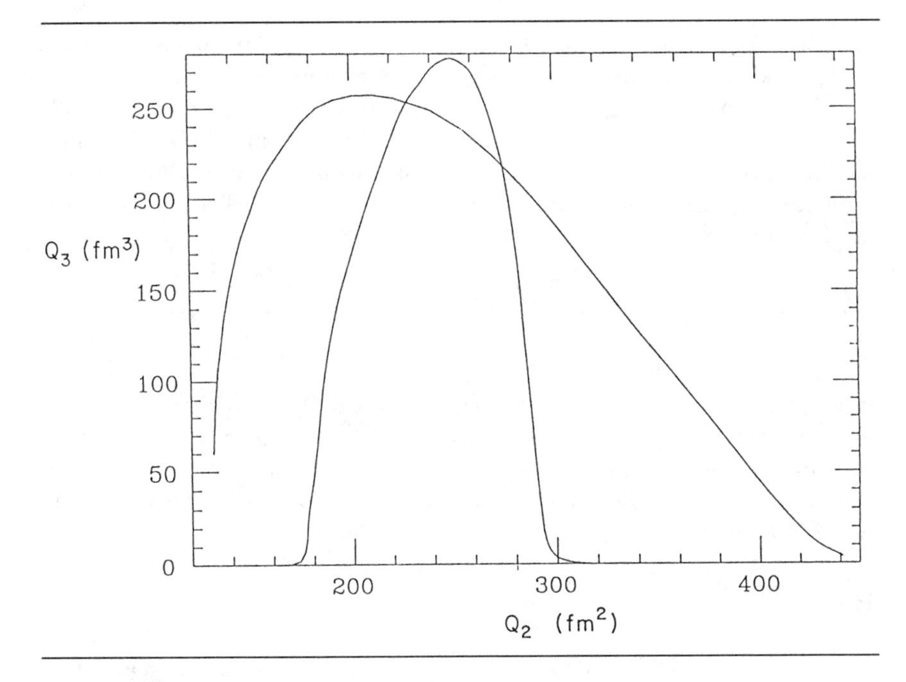

FIGURE 5.8
Collective motion path for the fission of ^{32}S in constrained mean-field theory and in imaginary-time mean-field theory.

eigenstates $\left\{\phi_i^{(n+1)}\right\}$ in the potential generated by $\left\{\phi_i^{(n)}\right\}$ are calculated and iterated to self-consistency. As T is increased, the solution approaches the bounce solution to any desired accuracy.

The sequence of shapes through which the nucleus evolves in shown in Fig. 5.7. The salient feature is the significant symmetry breaking throughout the entire evolution through the barrier, and it is clear that nodal rearrangement is playing a significant role in the dynamics. The similarity between this calculation and the pedagogical model of the last section may be seen by expressing the trajectory by the quadrupole moment Q_2, which characterizes the collective deformation and corresponds to $\langle z^2 \rangle$, and the octupole moment, Q_3, which characterizes the symmetry breaking and corresponds to $\langle \sigma_x \rangle$. The paths for the solution of the imaginary-time mean-field theory and for static mean-field theory with a quadrupole constraint are shown in Fig. 5.8. The similarity to the paths in Fig. 5.3 for the spin model is striking. Again the symmetry is broken along the entire optimal path, whereas the symmetry breaking is restricted to a small domain in the immediate vicinity of the level crossing in the constrained theory. Hence, as in the case of the spin model, solving the equations for the optimal path is expected to provide a major quantitative improvement in the theory.

In summary, the quantum mean-field theory which arises naturally from path integrals, self-consistently specifies the optimal collective path, and incorporates the basic physics of nodal rearrangement. It thus provides a microscopic understanding of many key elements of collective dynamics in many-Fermion systems.

Acknowledgements

It is a pleasure to acknowledge the essential contributions of George Bertsch, Shimon Levit, Henri Orland, Giovanni Puddu and Rüdiger Wolff to the developments discussed in this talk. This work is supported in part by funds provided by the U. S. Department of Energy (D.O.E.) under contract #DE-AC02-76ER03069.

Bibliography

[1] J.W. Negele and H. Orland, *Quantum Many-Particle Systems* (Addison–Wesley, Reading, MA, 1987).

[2] J.W. Negele, Rev. Mod. Phys. **54** (1982) 947.

[3] S. Levit, J. Negele and Z. Paltiel, Phys. Rev. **C21** (1980) 1603; **C22** (1980) 1979.

[4] H. Reinhardt, Nucl. Phys. **A367** (1981) 269.

[5] H. Kleinert, Phys. Lett. **B69** (1977) 9.

[6] J. Langer, Ann. Phys. (NY) **54** (1969) 258.

[7] A.M. Polyakov, Nucl. Phys. **B121** (1977) 429.

[8] G. 't Hooft, Phys. Rev. Lett. **37** (1976) 8.

[9] S. Coleman, Phys. Rev. **D15** (1977) 2929.

[10] J.P. Blaizot and H. Orland, Phys. Rev. **C24** (1981) 1740.

[11] A.K. Kerman and S.E. Koonin, Ann. Phys. (NY) **100** (1976) 332.

[12] H. Tang and J.W. Negele, Nucl. Phys. **A406** (1983) 205.

[13] J.W. Negele, *Nuclear Structure and Heavy Ion Collisions* (Varenna Summer School LXXVII, Soc. Italiana di Fisica, Bologna, Italy, 1979).

[14] P. Arve, G.F. Bertsch, J.W. Negele and G. Puddu, Phys. Rev. **C36** (1987) 2018.

[15] H. Lipkin, N. Meshkov and A. Glick, Nucl. Phys. **62** (1965) 188.

[16] R. Wolff, G. Puddu and J.W. Negele, to be pub.

[17] G. Puddu and J. Negele, Phys. Rev. **C35** (1987) 1007.

6

Quantum Mechanics and Macroscopic Realism

A. J. Leggett
Department of Physics
University of Illinois at Urbana-Champaign
1110 West Green Street, Urbana, IL 61801

If one wishes to provoke a normally phlegmatic group of physicists into a state of high animation, few tactics are better assured of success than to drop into the conversation the topic of the foundations of quantum mechanics, and specifically of the quantum measurement problem. To get an idea of the range of opinion on this subject, let me show you two quotations. The first is by the late Leon Rosenfeld, commmenting in 1967 on a paper published in that year by Daneri, Loinger and Prosperi on the quantum theory of measurement: "[Daneri *et al.*] have conclusively established the full consistency of [the quantum-mechanical] algorithm, leaving no loophole for extravagant speculation." Rosenfeld clearly felt that there was simply no problem of quantum measurement, and that anyone who even spent time thinking about it would be wasting his or her time. For a different view we turn to E. T. Jaynes, writing in 1981: "Quantum theory not only does not use— it does not even dare to mention— the notion of a 'real physical situation.' Defenders of the theory say that this notion is philosophically naive, and that recognition of this constitutes deep new wisdom about the nature of human knowledge. I say that it constitutes a violent irrationality, that somewhere in this theory the distinction between reality and our knowledge of reality has become lost, and the result has more the character of medieval necromancy than of science."

So views are certainly both widely varying and strongly held. To see why this should be so, let us go back for a moment to a very well-worn thought-experiment that we all of us meet when we first encounter quantum mechanics in our undergraduate courses, namely the Young's slits experiment. The experiment can be done with either electrons or photons, but somehow it always seems a bit more spectacular when done with electrons, so let us consider for definiteness that case. As you know, the basic ingredients are a source of electrons which can be made arbitrarily weak, some kind of collimating and monochromating device which we need not specify in detail, an opaque screen containing two slits, reasonably far apart, which can be opened and closed at will, and a final screen covered with some kind of scintillating material. Let us imagine that we turn the source down very low, leave both slits open and watch the final screen. Our first observation is that we get apparently random individual flashes on the screen, not some kind of continuous diffused glow; at this stage it seems natural to conclude that whatever is coming through the apparatus from the source to the screen is coming in discrete units, *i.e.*, particles. However, if we keep a record of the flashes at each point of the screen and make a plot in which the "brightness" of any point is proportional to the number of flashes we have detected there, we see that as time goes on we build up a pattern of light and dark bands very reminiscent of that produced by a classical light source in the same type of apparatus. At this stage it might reasonably occur to us to do the following experiment: first, we close off slit 2, say, and measure the number of flashes per minute that we get on the screen at a particular point: call this N_1. Then we repeat the operation with slit 1 closed and slit 2 opened, getting a number N_2. Finally, we measure the number of flashes per minute at the same point on the screen with *both* slits open: call this N_{12}. Needless to say, if whatever is coming through the apparatus is "particles" in the classical sense, and if the slits are sufficiently far apart that we may reasonably assume that opening or closing one does not physically affect propagation through the other, then we should expect that $N_{12} = N_1 + N_2$. As we all know, this is not in general what is seen experimentally: indeed, by a judicious choice of measurement point we can ensure that N_1 and N_2 are individually finite, but N_{12} is zero ("total destructive interference").

Of course you are all familiar with the standard method of accounting for these results in quantum mechanics. We assign to the ensemble of electrons a *wave function* Ψ, and assume that the intensity I (or number N) of electrons arriving at a given point on the screen is proportional to $|\Psi|^2$. Note that what is crucial here about the notion of a "wave" is not that it undergoes any obviously undulatory motion, but rather that it is associated with an *amplitude*, that is a quantity (here the wave function Ψ itself) which can be either positive or negative (or more generally complex). If now we assume that it is the *amplitude* of the "waves" propagating through the two slits which is additive, then we find (apart from an irrelevant overall constant)

$$N_{12} = |\Psi_{12}|^2 = |\Psi_1 + \Psi_2|^2 \neq |\Psi_1|^2 + |\Psi_2|^2 = N_1 + N_2$$

and, of course, the predictions of this description are more generally quantitatively confirmed.

Actually, this familiar "Young's-slits" thought-experiment is just a special case of a much more general phenomenon in quantum mechanics. Suppose we have an ensemble of systems– for the moment let them be microscopic, such as electrons, kaons or neutrons– which can start, as in the diagram, from state A, propagate from there to either of the different states B or C, and from there to D, E or F. (Actually, the diagram as shown does not preserve unitarity, but that is irrelevant in the present context.)

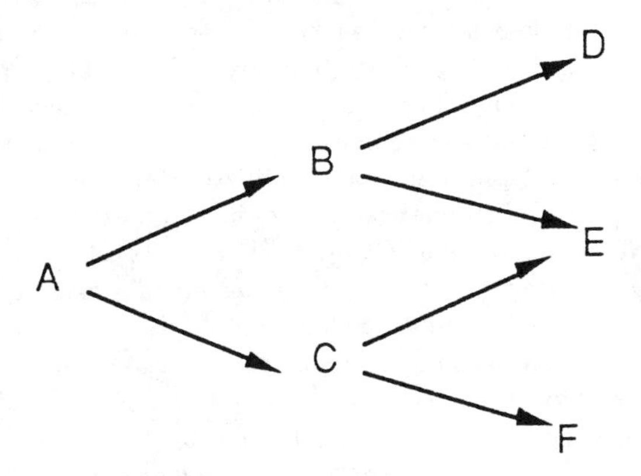

Suppose we first block off channel C, say (in any concrete case experimentalists know how to do this) so that the only way the system can get from A to E is via state B; we can then measure the number of systems per unit time passing from A through B to E, $P_{A\to B\to E}$. Similarly, by shutting off possibility B we measure $P_{A\to C\to E}$. Finally, we leave both channels open and measure the total probability of passing from A to E through "either" B or C, $P_{A\to E}$. The remarkable observation, of course, is that in general

$$P_{A\to E} \neq P_{A\to B\to E} + P_{A\to C\to E} \tag{6.1}$$

and, needless to say, the formalism of quantum mechanics predicts just this. From an intuitive point of view, the inequality of Eq. (6.1) would seem to indicate rather strongly that when both intermediate possibilities are left open, there is what Einstein would have called an "element of reality" associated with each; any particular system in some sense does not select a unique intermediate state but passes through both. On the other hand, if we set up a measuring device to ascertain which of the intermediate states any particular system of the ensemble passed through, we always get a definite result— each system does, it seems, pass through *either* B *or* C, not both. Needless to say, under these conditions the inequality Eq. (6.1) is replaced by an equality.

This situation usually seems very strange when we first meet it as undergraduates: microscopic systems seem to be behaving quite differently depending on whether or not they are observed! Those of us who are too persistent in our questions are usually sent by our professors to read the classic writings of Bohr and Heisenberg on the subject. What do we find? Let us consider Bohr's views for definiteness. These are very clear: there is really no paradox here, because to detect which of the intermediate states any given system went through we should have to make it interact with a macroscopic device which acts as a measuring apparatus. But, Bohr emphasizes, microscopic objects such as electrons or neutrons are simply *not the kinds of things* which we can even conceive of as having "properties" in the absence of a coupling to a macroscopic apparatus; thus, the electron in interaction with the measuring device is simply not the same thing as the electron left alone. A philosophically sophisticated defense of this point of view was given by the empiricist philosopher Hans Reichenbach in his book *Philosophic Foundations of Quantum Mechanics.*

So far, so good. However, to make this argument work one needs, at least implicitly, to be able to make a sharp distinction between the micro-objects under investigation and the macroscopic device which we use to make measurements on them (which, Bohr always emphasized, must be described in classical terms). Is it actually possible to do this? To examine this question, let us examine the measurement process itself in a little more detail. Specifically, let us consider what is often viewed in the literature as the paradigm of a measurement process, namely a Stern-Gerlach experiment. (Actually, I would claim that in a modern context the Stern-Gerlach experiment is not only not at all paradigmatic, but actually rather atypical, but that is a different story: in the present context it will do.) The principal ingredients are, first, an oven which serves as the source of a beam of atoms, let us say of spin $\frac{1}{2}$: second, a "black box" selector device which enables us, by appropriate setting of the knobs, to guarantee that all atoms emerging have a spin projection of $+\frac{1}{2}\hbar$ on any chosen axis: and, third, a separating magnet whose magnetic field gradient is, let us say, in the z-direction. The purpose of the magnet is to separate the atoms into distinct beams depending on their spin projections on the z-axis: an atom with spin projection $+\frac{1}{2}\hbar$ will be deflected upwards, one with projection $-\frac{1}{2}\hbar$ downwards. This can, if need be, be directly checked by setting the selector knobs appropriately.

It is tempting to think that this spatial separation of the atoms according to their spin components already constitutes a measurement. However, such a view would rapidly lead to contradictions. For, according to the axioms of quantum measurement theory as set out in most textbooks, once a measurement has been made it is impossible to see the effect of interference between different possible values of the measured quantity. Now in principle there is nothing to stop us recombining the separated beams and exhibiting the effect not only of spatial interference but also of the interference of the two spin components; and while in practice this has not been done in a Stern-Gerlach experiment, a conceptually very similar effect has been demonstrated in a beautiful series of experiments using a neutron interferometer. So, we had better not have made a "measurement" at this stage! But, come to think of it, why did we ever think we had? All we have done is to ensure that *if* a particular atom of the beam has spin up, it will be deflected upward, and vice versa: but since we do not at this stage know whether any given atom has in fact been deflected upward or downward, we have obtained no information about its spin; and no information, no measurement.

Thus, to complete our "measurement" it is necessary to install some device which will tell us that a particular atom has indeed been deflected, say, upwards, *i.e.*, a *counter*. Such a counter is, typically, a macroscopic device which starts in a metastable state and is triggered, by the passage of the particle in question, into an irreversible process, at the end of which it has lost a macroscopic amount of free energy. It is often argued that the property of macroscopic irreversibility is essential in order to stabilize and preserve the results of the measurement against thermodynamic fluctuations, and it does indeed seem to be a feature of most if not all real-life measuring devices. (For example, in a proportional counter the voltage across the counter has, initially, a nonequilibrium value, which is metastable because the gas is neutral and hence insulating. The passage of a fast charged particle produces ion pairs, which conduct a pulse of current across the counter and cause an irreversible drop in the voltage.) This *macroscopic* and irreversible event can be used to produce other macroscopic events such as the (directly observable) flash of a lamp or the printing-out of a message on a computer tape. Alternatively, as in Schrödinger's famous thought-experiment, we can connect the counter up to a closed box containing a cat, in such a way that if the counter is triggered the cat is executed, while so long as it is untriggered the cat lives. This famous example makes it dramatically clear that it is possible to use a *microscopic* difference— the difference between a single atom having spin up or spin down— to trigger *macroscopically* different final states, in this case the states of life or death for the cat.

Now, where in this process, exactly, has the "measurement" taken place? After all, not only the magnet but the counters and the electronics, and presumably even the cat, are nothing but collections of atoms and molecules: and since we all believe that quantum mechanics describe atoms and molecules, it presumably describes equally well these macroscopic collections of them. So, while it may not be *necessary*, it should at least be *legitimate* to apply the quantum mechanical description right through the whole process, at least up to the point where we open the box to inspect the state of the cat. And now we come to a somewhat disturbing state of affairs. First, let us consider the quantum-mechanical description of the cat before the experiment starts. Actually, we should not really describe the cat by a wave function— at the very least we should do it the minimal honor of attributing to it a density matrix— but purely to make the argument concise and intuitive at this point, let us temporarily attribute to the cat a wave function corresponding to "being alive," say Ψ_{alive}. Consider now the final state of the cat following the emergence of the first atom from the oven. If this atom is drawn from an ensemble with spin $+\frac{1}{2}$ along the z-axis, it will go harmlessly into the upper beam, the lethal counter will remain untriggered and the cat will live, and will therefore continue to be described by the wave function Ψ_{alive}. If on the other hand the atom was drawn from an ensemble with z-component of spin $-\frac{1}{2}$, it will be deflected downwards, the lower counter will be triggered and the cat will be executed; in which case its final state will be some other wave function, say Ψ_{dead}, which is clearly not only orthogonal to Ψ_{alive} but corresponds to a *macroscopically different* state of affairs.

So far, so good. But what if the atom in question is drawn from an ensemble corresponding to an eigenstate of spin, say $+\frac{1}{2}$, along the *x-axis*? We all know that according to the formalism of quantum mechanics, such a state is a linear superposition of the eigenstates of spin along the z-axis: in an obvious notation $\Psi_{+}^{(x)} = \frac{1}{\sqrt{2}}(\Psi_{+}^{(z)} + \Psi_{-}^{(z)})$. But it is a very fundamental and non-negotiable property of the formalism of quantum mechanics as currently understood that it is a *linear* theory: if initial state A at the microlevel leads to final state A′ at the macrolevel, and state B correspondingly to B′, then the intitial state $2^{-\frac{1}{2}}(\text{A}+\text{B})$ leads inexorably to the final state $2^{-\frac{1}{2}}(\text{A}' + \text{B}')$. Thus, in this case the final state of the cat is

$$\Psi_{\text{cat}} = 2^{-\frac{1}{2}}(\Psi_{\text{alive}} + \Psi_{\text{dead}}) , \tag{6.2}$$

that is, a *linear superposition of macroscopically distinct states.* It is a question whether this peculiar-looking final state is more uncomfortable for the cat or for the theoretical physicist who has to contemplate it.

Needless to say, this schematic presentation of the Schrödinger's Cat paradox is far too naive; in particular, to make it technically correct one must take careful account of the fact, noted above, that even the initial state of the cat is not in practice describable by a pure-state density matrix. Nevertheless, when these necessary technical caveats are taken into account, we are still left with the fundamental paradox: under certain circumstances, such as the ones described, the quantum formalism apparently requires us to describe even a macroscopic object such as a cat (or, if you prefer, the "universe" containing the atom, counters, electronics, cat, *etc.*) as not being in a definite macroscopic state: neither of the possible macroscopically distinct outcomes has been selected. Nevertheless, while considerations of humanity may have prevented the large-scale carrying out of this particular experiment, we all of us believe that were it to be done, and were we at this point to inspect the state of any particular cat of the ensemble, we should most certainly find it to be either definitely alive or as definitely dead— indeed, it is clear we would not even know how to recognize the occurrence of the "anomalous" state described by Eq. (6.2). In a nutshell: under certain circumstances quantum mechanics apparently says that even at the macroscopic level no definite outcome is selected, whereas we all know that in real life it is.

This famous paradox has now been around for more than fifty years, and there must by now be literally thousands of papers in the literature which claim to have resolved it; it would certainly take me an hour's lecture in itself to review them adequately. Let me however just remind you briefly of the nature of some of the most popular ones. It is convenient to divide them roughly into "exotic" solutions, by which I mean those which invoke considerations from outside what we would normally think of as the domain of physics, and non-exotic ones. In the first class I would put, first, what I call the "mentalistic solution", by which I mean the solution which says: we should take the final state Eq. (6.2) of the cat seriously— indeed no definite outcome has been selected. However, when a human being opens the box and inspects the state of the cat, something new comes on the scene, namely Consciousness (as it were with a capital C), and this is not describable by the laws of physics but requires something quite different. My reaction to that proposal, in a nutshell, is that badly as we may understand the foundations of quantum mechanics, we understand the phenomenon of human consciousness even worse; so that intellectually this is a leap from the frying-pan into the fire. A second "exotic" solution is the so-called relative-state, alias Many-Worlds or Everett-Wheeler, interpretation, which goes even further: not only is Eq. (6.2) the correct description of the final state of the cat *before* we look, but it (or something similar) is still the correct description *after* we have looked and believe we have found a definite result; the "non-observed" possibility represented in the wave function of Eq. (6.2) continues to "exist" (or so it is alleged) as a "parallel world" of "of equal reality" to the observed one. (Various formal theorems of quantum mechanics are invoked to show that we cannot obtain from this interpretation any experimentally inconsistent result.) To this proposal my reaction is quite literally unprintable, since in the most exact and literal sense I simply do not know what the words used by its advocates in formulating it, apparently English, are supposed to mean.

Of the "non-exotic" solutions, probably the most conservative is the "statistical" one. This says, in effect: Wait, you have been taking the quantum-mechanical formalism much too seriously. All it ever was, or was intended to be, was a formal calculus for predicting the probabilities of various directly observable events— and I really mean observable, with the naked eye if necessary. Thus, the *only* meaning of the description given by Eq. (6.2) is that we have a 50% probability of finding any particular cat alive and 50% of finding it dead: no other significance whatever should be attributed to it. This interpretation, while apparently internally self-consistent, has the very unsettling feature that if one adopts it, the apparently legitimate question "What was the state of this particular cat before I chose to inspect it?" simply has no answer, even in principle, within the framework of the theory and indeed is probably meaningless. A second possibility is to try to find some characteristic feature of the measurement process which can be used to demarcate where quantum mechanics stops and some other mode of description was to take over. Most commonly, the phenomenon of dissipation or irreversibility is taken to play this role, and one sometimes finds it argued or implied that the quantum description is valid only to the extent that there is no, or negligible, dissipation. The problem with this view is that there really seems no good reason, within the quantum-mechanical framework itself, to mark out "dissipation" as something special— it is simply the macroscopic manifestation of a phenomenon apparently well understood at the microscopic level, namely the transfer, irreversible in practice for well-known reasons, of energy from a single macroscopic degree of freedom to a myriad of microscopic ones. Indeed, as we shall see later, the description of dissipation within the framework of quantum mechanics is now a major, and apparently reasonably successful, industry, and appears to lead to no obviously undesirable consequences.

Probably, the "solution" of the Schrödinger's Cat paradox which is most popular among practising physicists is what I shall call for want of a better name the "orthodox" one— which may or may not be regarded as effectively a variant of the Copenhagen interpretation by those who believe they know what the latter is. This "solution" comes in several dozen different varieties, but in essence all of them say: OK, if you insist, then Eq. (6.2) is indeed a legitimate description of the final state of Schrödinger's Cat. However, no paradox is thereby implied, because there is no conceivable experiment which will discriminate between this description and the familiar classical "mixture" description, according to which each particular cat is by this stage either alive or dead, but we don't know which and are therefore forced to give a description in terms of probabilities. This is equivalent to saying that no conceivable experiment can detect the effect of the interference between the states Ψ_{alive} and Ψ_{dead}. Why not? Because the cat is a macroscopic body and moreover— and this is crucial in the context of this lecture— the states Ψ_{alive} and Ψ_{dead} are *macroscopically distinct.* The principle which is either stated or implied in all such arguments is that *it is impossible to see the effects of quantum interference of macroscopically distinct states.* The reasons given differ in detail, but all rest in the last result on the fact that macroscopic bodies, and the variables which describe them, are coupled strongly and usually irreversibly to their "environment" (which may be taken in the present context as including these degrees of freedom of the body itself which are not of primary interest) and that this interaction always destroys the effects of interference. Numerous model calculations supporting this conclusion may be found in the literature.

Once this point is granted, the "orthodox" argument then proceeds by saying in effect: Since the two descriptions, by Eq. (6.2) and by a classical mixture, predict identical results for all conceivable experiments, they are equivalent, and therefore it is perfectly legitimate to say that by this stage each individual cat of the ensemble is indeed definitely alive or dead. Thus there is no paradox.

Is this last step in the argument legitimate? That depends crucially on one's philosophy of science, and mine says that it is not. At the microscopic level, we are (most of us) happy to believe that a quantum superposition of states can*not* legitimately be interpreted as simply a classical probabilistic statement about an event which has definitely occurred. Why? Because we have, or think we have, direct evidence to the contrary in the form of the observability of interference effects (compare the argument given above). As we go from microscopic to macroscopic, nothing in the formalism changes: all that changes is that it is no longer possible to produce evidence, in the form of interference effects, against the "mixture" interpretation. So: is it legitimate to keep a particular interpretation of the formalism so long as the experimental evidence forces it on us, but to drop it as soon as the evidence disappears, even though the formalism itself is of whole cloth and unchanged throughout? I think not, but as I said this is a matter which cannot be decided by the considerations of physics alone.

Let us at this point step back and ask: How did we get into this mess? We got there because we implicitly assumed that quantum mechanics applies, not just to electrons and neutrons, not just to atoms and small molecules, but to large molecules, microbes, counters, bits of electronics and even cats— in other words, to anything composed of atoms and molecules, however macroscopic, complex or organized. An inexorable consequence of this assumption was that under certain circumstances a macroscopic body may have a description which does not obviously correspond to a definite macroscopic state. Let us contrast this with an alternative hypothesis, which we may characterize as "macrorealism", namely: A macroscopic body is at all times in a definite macrostate, whether it is observed or not. In other words, the "potentiality" apparently described by quantum mechanics at the microlevel is converted into "actuality" at the macrolevel.

It should be strongly emphasized that although almost identical words to the above are to be found in the writings of Heisenberg, the meaning I am giving to them is entirely different from his. Heisenberg appears to have believed that, in some mysterious way which he never found it necessary to specify, the *formalism of quantum mechanics itself* would produce the desired transition from microscopic "potentiality" to macroscopic "actuality". The hypothesis I am proposing is far more radical: namely that *quantum mechanics fails* by the macroscopic level, and that other physical laws, so far unknown, come into play and effect the conversion for us. And now let us ask: How do we know that's not right?

A number of answers can be dismissed at once. Direct evidence? Apart from the experiments I will discuss below, which are in any case at best circumstantial, we have none. Neutron interferometry, *etc*? This only appears to be relevant because of a verbal ambiguity. It is often said that the beautiful experiments carried out using neutron interferometers confirm that "quantum mechanics still works at the macroscopic level". In *one* sense of the word this is true, but it is clearly not the one that is relevant to our question. The same applies to the so-called "macroscopic quantum phenomena" seen in superconductors, superfluid helium, lasers, *etc.*; none of these demonstrates the occurrence of superpositions of macroscopically distinct states. Agreement of theory and experiment for the properties of complex, macroscopic bodies? Again, totally irrelevant. (If you doubt this, I suggest that those of you who are engaged in standard condensed-matter theory should ask yourselves: When, in my everyday work, did I last have to deal with a superposition of macroscopically distanct states? I would take a large bet that the answer is "never".) Prejudice? Well, prejudice is not always a bad thing, and some "prejudices" have served us very well in physics in the past, among them reductionism, in the sense of the belief that since complex bodies are composed of electrons and nuclei, their properties must "in principle" be entirely derivable from those of electrons and nuclei— which in the present context would most naturally imply that "in principle" they can be completely obtained from the solution of an appropriate giant Schrödinger equation. Indeed, were it not for the quantum measurement paradox, I would be perfectly happy with this belief. But a prejudice it is nonetheless, and physics, we are always told, is an experimental subject.

One frequently encountered answer deserves more detailed discussion, both because the subject is interesting in its own right and because the formal techniques used to explore it have close analogs later in my talk. This is the claim that macroscopic realism has been definitely refuted by the experiments done in connection with Bell's theorem. Let me remind you of the nature of these experiments and what is inferred from them.

The source is an ensemble of atoms which are excited by a discharge into an excited state from which they cascade down to the ground state, in the process emitting two photons. We focus on the case in which the photons (call them 1 and 2) are emitted in opposite directions. Some distance from the source, photon 1 is subjected to a switch which directs it into one (and only one) of two "yes-no" devices, that is, devices which are guaranteed, for each photon, to give one of two answers. We do not need to know in detail the nature of the devices at this stage; in practice they would consist of a polarization analyzer followed by a photodetector (so that in the usual interpretation, "yes" and "no" correspond to the photon's being transmitted and rejected by the analyzer respectively). Call these devices a and b; a similar arrangement is made for photon 2, with devices c and d. A crucial feature of some of the most recent experiments using this setup is that the space-time points at which photon 1 is switched into one of a or b, and at which photon 2 passes the analyzers of c and d, are separated by a spacelike interval, so that according to the postulates of special relativity the act of switching 1 cannot causally effect the behavior of 2.

To analyze this experiment we introduce variables A, B, C, D in a *purely operational* way as follows: If for a particular pair of photons the photon 1 is switched into device "a" and this device gives the answer "yes", then A is defined, for that photon pair, to be +1; if photon 1 is switched into device a and the latter gives the answer "no" then A is by definition −1 for that pair. In the present context it is crucial to note that if photon 1 is switched into device "b", then *A is (so far) not defined.* We define variables B, C, D in the obviously analogous way. Experiment can measure $\langle AC \rangle$ the expectation value of the correlation of A and C, on one set of pairs (namely, those of which photon 1 was switched into "a" and 2 into "c"), $\langle AD \rangle$ on a different set, *etc.* Thus, we can measure experimentally the quantity

$$K_{\text{exp}} \equiv \langle AC \rangle_{AC} + \langle AD \rangle_{AD} + \langle BC \rangle_{BC} - \langle BD \rangle_{BD} \,, \tag{6.3}$$

the subscripts being added to emphasize that the ensembles on which the expectation values are measured are strictly speaking different. Given a particular physical implementation of the devices a,b,c,d (*e.g.*, as the analyzer-photodetector combination described above with particular settings of the analyzers) quantum mechanics makes quite unambiguous and rigorous predictions about the value of K_{exp}.

Why is this of interest? To see this, let us consider the consequences of applying to the above experiment a particular conjunction of apparently common-sense assumptions about the world which nowadays is conventionally taken to define an "objective local" theory. These defining hypotheses are:

1. local causality.

2. induction

3. macroscopic counterfactual definiteness.

1 simply says that events which are spacelike separated cannot exert any causal influence on one another. **2** says that if a particular subensemble of systems drawn from a given initial ensemble is subjected to a given set of measurements, then the expectation values measured on the subensemble are characteristic of the original ensemble. **3** is the most subtle and, in the present context, most interesting assumption; in most formulations of Bell's theorem it is substituted by the assumption of microscopic realism ("photons have definite properties, whether the latter are measured or not") but in the present context a macroscopic formulation is more relevant. To explain **3**, imagine that of a particular photon pair photon no. 1 was actually switched into device b. Then, as we have seen, the quantity A is not defined. However, we now argue as follows: Had I so wished, I could perfectly well have switched this photon into device a. Had I done so, then device a would have given either the answer "yes" or the answer "no": *i.e.*, A *would have* had either the value +1 or the value −1. Therefore, I can treat the "value of A which would have occurred" in this possible but not actually performed experiment as an objective characteristic of the *actual* physical situation. Note that this formulation (originally due, I believe, to H. Stapp) works entirely in terms of macroscopic events and avoids any attriution of "reality" to micro-objects such as photons.

Out of the apparently innocent conjunction of hypotheses **1-3** Bell and subsequent workers have obtained an important inequality. The argument goes as follows: for each photon pair, from **3**, the quantity A exists and has the value ± 1, irrespective of whether the photon 1 in question was switched into device a. Similarly for B, C, D. By assumption **2**, the value of A must be independent of whether photon 2 of the pair was switched into c or d, *etc.* Thus, for each pair the products AC, AD, BC, BD are defined and take the values ± 1. Elementary arithmetic then establishes the inequality

$$\mathrm{AC} + \mathrm{AD} + \mathrm{BC} - \mathrm{BD} \leq 2 \,. \tag{6.4}$$

This inequality clearly remains true when the left-hand side is averaged over the complete ensemble of pairs; call the relevant averages $\langle \mathrm{AC} \rangle_{\mathrm{ens}}$, *etc.*

Finally, since by assumption **2** we have $\langle AC\rangle_{ens} = \langle AC\rangle_{AC}$, *etc.*, we have from Eq. (6.3) the inequality

$$K_{exp} \leq 2 \, , \tag{6.5}$$

which is a prediction about an *experimentally measurable* quantity.

The punch-line is, of course, that for certain easily specified settings of the analyzers, *etc.*, a quantum-mechanical treatment of the problem predicts that $K_{exp} > 2$. Thus, we can force nature to choose between quantum mechanics and the whole class of objective local theories. As you all know, the experiments of the last fifteen years or so have come out overwhelmingly in favor of quantum mechanics; and while it is true that there are a few "loopholes" in the application of the Bell-type inequality (6.5) to these real-life experiments (connected with the limited efficiency of the detectors and, with the formulation of assumption **3** used here, also with the fact that the separation between the relevant *macroscopic* events is not necessarily spacelike), only a small (though vocal!) minority of physicists currently believes that exploitation of these loopholes will enable us to save local objectivity. While this is, of course, a matter for experiment, for the purposes of the present argument I will assume that the class of objective local theories has effectively been refuted by experiment.

This being so, which of the assumptions **1-3** is the least painful to give up? If we sacrifice **1**, the whole structure of special relativity theory appears to crumble, and very few physicists (as opposed to popular writers on the subject!) choose this way out. Similarly, were we to doubt assumption **2** (induction) most of experimental physics (not to speak of the insurance industry!) would grind to a halt. Thus most practising physicists, placed between a rock and a hard place, find it least painful to sacrifice assumption **3** and to agree that, in the words of one pithily titled paper on the subject, "unperformed experiments have no results". It is often claimed that this conclusion is exactly in the spirit of Bohr's formulation of the Copenhagen interpretation, according to which "properties" of a physical system in a given situation are a function only of the *actually employed* measuring setup (and hence by implication not of a setup which could have been used but was not).

Returning now to the main theme of this talk, I want to stress that the above conclusion— that macroscopic counterfactual definiteness must be rejected— while interesting in its own right, is not directly relevant to the question of macroscopic *realism.* "Counterfactual definiteness" is a statement about the existential status of the results of an experiment which could have been, but was not, carried out, *i.e.*, about a *hypothetical* situation: realism, by contrast, is a statement about the behavior of the system in an *actually occurring* situation. While there are subtle connections between the two concepts, the falsity of macroscopic counterfactual definiteness certainly does not imply the falsity of macroscopic realism.

How, then, are we to test quantum mechanics vis-á-vis macrorealism? Clearly, this will only be possible if, contrary to the established wisdom in the quantum measurement literature, there exist cases where we cannot only ensure that a macroscopic body (or ensemble of such bodies) is described by a linear superposition of macroscopically distinct states, *but also see the effects of interference between these states*, *i.e.*, distinguish the superposition experimentally from a classical mixture.

Do such cases in fact exist? There are two essential conditions that must be fulfilled (and, in practice a host of subsidiary ones which I will not go into). First, we need a *macroscopic* body whose dynamics is nevertheless controlled by a *microscopic* energy. The reason goes back to an old argument of Bohr: We expect, crudely speaking, that the predictions of quantum mechanics will approximate to those of classical mechanics as soon as the relevant action, S, is large compared to $\hbar$. But S is $\sim Et \sim E/\omega_0$, where ω_0 is a characteristic order of magnitude of the frequency of classical motion, so we need the energy E to be $\lesssim \hbar\omega_0$, *i.e.*, microscopic. A second essential condition is that the dissipation in the system is very low: this requirement is connected with the standard argument, quoted earlier from the quantum measurement literature, that the effects of superposition of macroscopically distinct states are always washed out by the inevitable interaction of a macroscopic body with its environment. Detailed calculations in recent years have confirmed that the only circumstances in which we have a reasonable hope of evading this argument are when the *dissipative* effects of the interaction are very small. (That is another whole hour's lecture.)

Of the various systems which have been proposed as candidates, probably the most promising is a class of superconducting devices based on the Josephson effect. The Josephson effect is the paradigm of a phenomenon where the dynamics of a macroscopic variable such as the current circulating in a superconducting ring (see below) is controlled by a microscopic energy (typically, of the order of the thermal energy of a single atom at room temperature!); moreover, at low temperatures, where the normal quasiparticles are mostly frozen out, the dissipation in such devices can be made extremely small. A particular example I shall discuss for concreteness (though it is not the only system which has been used in the experiments I shall describe) is a so-called rf SQUID ring, that is a bulk superconducting ring closed by a Josephson junction (which in the present context may be regarded as simply a sort of gate, through which Cooper pairs of electrons can pass, but with more difficulty than in the bulk), and through which an external magnetic flux, here regarded as a c-number controlled by the experimenter, can be applied. The "macroscopic variable" relevant in this case is the electric current circulating in the ring, or equivalently the *total* flux trapped through it (which has a contribution from this current superimposed on the externally applied contribution). Since the various states we shall be discussing differ in the value of electric current by something of the order of 1 μA, we may reasonably regard the latter as a "macroscopic" variable.

What kinds of experiments can we do with such a system to test quantum mechanics, either in its own right or vis-á-vis macrorealism? To explore this question, we note that the potential energy of the system as a function of circulating current may have, depending on the parameters, one or more metastable minima separated by classically impenetrable barriers. Thus, the obvious thing to do is to look for various effects connected with quantum mechanical tunneling through the barrier, an effect which has no classical analog. In the existing literature one finds, broadly speaking, three kinds of experiment. In the first, carried out by groups at Sussex and Kharkov, one chooses the parameters to be such that the system has many metastable minima, couples the SQUID ring to a tank circuit, measures the current-voltage characteristics of the latter and interprets the results in terms of quantum motion of the system between the minima. The principal drawback of this type of experiment is that very few of the relevant parameters can be measured independently, so that the interpretation given requires a large number of assumptions; thus, if one is interested in establishing the applicability of quantum mechanics to such systems in the face of possible doubt, they are not particularly well suited to the task. A second type of experiment, carried out at Leiden, consists of adjusting the parameters so that there are two nearly degenerate potential wells only, and looking for tunneling between them. While this kind of experiment has a much more direct interpretation, and there seems little doubt that the results constitute strong evidence for the reality of the tunneling phenomenon in such systems, in this case also it has not been possible to measure all the relevant parameters independently, so that no quantitative comparison of theory and experiment has so far been possible.

The most informative type of experiment in the context of the present discussion, which has been done at various laboratories throughout the world over the last ten years or so, consists in trapping the macroscopic variable in a metastable well and measuring the rate of escape through the barrier into what is effectively a continuum. The important point to emphasize here is that the values of current corresponding to the points of entry into and exit from the barrier are such that they can reasonably be said to correspond to macroscopically distinct states. Thus, to the extent that (quantum-mechanical) theory and experiment agree, this can be taken as at least circumstantial evidence that the quantum formalism does indeed apply to the motion of a "macroscopic" variable. A second important point is that in some of these experiments, in distinction to those mentioned above, it has proved possible to measure all the relevant parameters *in situ* in independent, purely classical experiments. To formulate the theoretical predictions it has proved essential to take careful account of the dissipation which, though possibly small, is always inherent in these devices, and much of the theoretical work of the last ten years has been devoted to this problem. The upshot is that the comparison of theory and experiment, though not perfect, is satisfactory and indeed in one or two recent experiments excellent, and in particular suggests strongly that the general prescriptions developed to handle the effects of dissipation are not qualitatively in error. This is an essential precondition for the further round of experiments to be discussed below.

These existing experiments, then, offer convincing circumstantial evidence that the formalism of quantum mechanics indeed still applies at the macroscopic level, in the sense relevant to the Schrödinger Cat paradox. If that is so, then it would imply that macrorealism is excluded (just as the validity of the quantum description of two separated particles which have interacted in the past implies, as in effect originally pointed out by Einstein, Podolsky and Rosen, that local realism must fail). However, it would be very much preferable to draw this conclusion— that nature does not believe in macrorealism— directly from experiment. Thus, we are led to seek a way of forcing Nature to choose between quantum mechanics and the class of macrorealistic theories, in the same way as Bell's theorem and the related experiments have forced her to choose between quantum mechanics and the class of objective local theories. One possible such experiment is what is usually known in the literature as the "macroscopic quantum coherence" experiment, and I now sketch its nature, stressing by implication the formal analogy with the Bell's theorem work.

We take a SQUID ring and apply through it an external flux such that the potential energy has two degenerate potential wells separated by a barrier, as in the Leiden experiments mentioned above; the difference is that we now deliberately make the dissipation as low as possible so as to enhance the possibility of *coherent* tunneling backwards and forwards between the wells, as in the case of the ammonia molecule. In a practical realization of the experiment, the two wells might correspond to a current of the order of 0.1 μA circulating clockwise in the one case and anticlockwise in the other, so by most people's criteria the two states in question would qualify as macroscopically distinct. We suppose that whenever actually measured, the current corresponds to one or other of these two values (perhaps with some small deviation due to fluctuations which can in principle be taken into account); this expectation has already been checked in the Leiden experiments, albeit under somewhat different conditions, and can of course be re-checked in a preliminary experiment. Given this, we define a variable $Q(t)$ operationally as follows: If the current is measured at time t and found to be (say) in the clockwise direction, then $Q(t) = +1$; if it is measured at time t and found to be anticlockwise, then $Q(t) = -1$. If the current is not measured at time t, then $Q(t)$ *is for the moment undefined.*

We now imagine that we perform a series of experimental runs in each of which we start with some definite initial conditions at $t = 0$ and measure the current at two times, t_1 and t_2, only. By doing this we can establish the experimental value of the correlation

$$\langle Q(t_1)Q(t_2)\rangle_{t_1,t_2}$$

where the subscripts emphasize that the sub-ensemble on which the correlation is measured is that defined by measurements at times t_1 and t_2. (This may seem tautologous, but see below.) In a similar way we can establish the correlations $\langle Q(t_1)Q(t_3)\rangle_{t_1,t_3} + \langle Q(t_2)Q(t_4)\rangle_{t_2,t_4}$, *etc.* Note that we never measure more than twice on any run. In this way we can establish the experimental value of the quantity

$$\begin{aligned} K_{\text{exp}} \equiv\ & \langle Q(t_1)Q(t_2)\rangle_{t_1,t_2} + \langle Q(t_2)Q(t_3)\rangle_{t_2,t_3} \\ & + \langle Q(t_3)Q(t_4)\rangle_{t_3,t_4} - \langle Q(t_1)Q(t_4)\rangle_{t_1,t_4} \quad . \end{aligned} \tag{6.6}$$

Quantum mechanics makes definite predictions about the value of K_{exp}, although the calculations needed to establish it are less trivial than in the Bell's theorem case: see below.

We now want to contrast the predictions of quantum mechanics with those of the class of theories which embody "macroscopic realism." To do so we must of course have a precise definition of the latter. We choose to define macrorealistic theories as those embodying the conjunction of the following two postulates:

1. A macroscopic body which has available to it two or more macroscopically distinct states is at all times in a particular one of these states, whether or not it is observed.

2. It is possible in principle to determine the macroscopic state with arbitrarily small effect on that state or on the subsequent dynamics of the system.

It will become clear in the course of the argument that postulate **2** is essential to enable nontrivial predictions to be made; postulate **1** alone does not permit a meaningful experimental test. At first sight one might be skeptical about **2**, remembering Heisenberg's argument about the "γ-ray microscope" thought-experiment. However, Heisenberg rejected **2** (at the *micro*level) only because he had in effect already rejected **1**. To see that **2** is, at least at the macrolevel, barely distinguishable from **1**, imagine that I do an "ideal-negative-result" type of measurment: that is, given that the system is known to be either in state A or the macroscopically different state B, I devise an experiment in which I interact physically with it only if it is in state B. I then keep only these trials in which I did *not* get a response (*e.g.*, if A and B represent different spatial positions, I shine a flashlight on B and keep only these cases in which I see nothing). If it is now argued that, despite the fact that I have not interacted with the system, (which "is" in state A) I have nevertheless affected its subsequent dynamics, then I can see no meaning in the statement that it "was" uniquely in state A in the first place. Thus, I would claim that **2** is so natural a corollary of **1** at the macrolevel that it makes little sense to treat them as separate.

We now derive an inequality which must hold in any theory embodying postulates **1** and **2**. First, it follows from **1** that $Q(t)$ actually exists for all t, whether or not the current is measured, and has one of the values ± 1. It then follows immediately, as in the Bell's theorem work, that the quantity

$$\begin{aligned}&\langle Q(t_1)Q(t_2)\rangle_{\text{ens}} + \langle Q(t_2)Q(t_3)\rangle_{\text{ens}} \\ &+\langle Q(t_3)Q(t_4)\rangle_{\text{ens}} - \langle Q(t_1)Q(t_4)\rangle_{\text{ens}}\end{aligned}$$

is less then or equal to 2, where the subscript emphasizes that all expectation values are taken on *the same* ensemble. Unfortunately, it is experimentally impossible to measure all the terms on the same ensemble, since by the mere act of measuring at (*e.g.*) t_1 we may affect the value at t_2. This is where postulate **2** comes in: it assures us that in principle (*e.g.*, by a suitable set of ideal-negative-result measurements) we can in fact treat all measurements as made on the same ensemble, namely the one specified by the initial conditions alone. (Thus, postulate **2** in effect stands in for the *conjunction* of local causality and induction in the Bell's theorem argument: though it may perhaps be argued that we should really add induction as a separate postulate in the present context). Once given this, we can immediately derive an inequality, which must be satisfied by all macrorealistic theories, for the experimentally measurable quantity K_{exp} (for arbitrary values of the times t_1, t_2, t_3, t_4):

$$(K_{\text{exp}})_{\text{MR}} \leq 2 \; . \tag{6.7}$$

What, now, does quantum mechanics predict about this quantity? If we treat the system as a simple "two-state" system similar to the NH_3 molecule, it is very straightforward to work out the correlation $\langle Q(t_i)Q(t_j)\rangle$; this is independent of the ensemble (and of temperature) and given by

$$\langle Q(t_i)Q(t_j)\rangle = \cos \Delta(t_j - t_i) \; , \tag{6.8}$$

where Δ is the characteristic oscillation frequency (the energy splitting of the even-parity and odd-parity eigenstates, divided by $\hbar$). If we choose $t_2 - t_1 = t_3 - t_2 = t_4 - t_3 = \pi/4\Delta$, we find that this implies $K_{\text{exp}} = 2\sqrt{2}$, in conflict with Eq. (6.7).

If, however, I were now to rush off to the laboratory, set up and perform the experiment, find a value of $K_{\rm exp}$ less than 2, and come back and announce to a startled world that I had found definitive evidence of the breakdown of quantum mechanics at the macrolevel, then I would be justifiably jeered out of town. Why so? Because it is well known that no macroscopic system is likely to behave anything like an ideal two-state system such as the NH_3 molecule; as has repeatedly been argued in the quantum measurement literature, any macroscopic body, almost by its very nature, will be strongly and, usually, dissipatively coupled to its environment, and this may well change the nature of the dynamics not only quantitatively but even qualitatively. Thus, it is essential to do a proper quantum-mechanical calculation for the specific system used in the experiment, making realistic assumptions about the coupling to the environment and taking careful account of all possible sources of noise, dissipation, *etc.* At first sight it might be thought that this is in practice impossible, since we would never be in a position to be sure *a priori* that we had thought of everything; however, very fortunately, it turns out that we can relate the expected effects of dissipation in quantum mechanics not to *a priori* microscopic models, but rather to its *experimentally observable* effects in *purely classical* experiments on the same system; thus, if we can measure the relevant parameters in a preliminary classical experiment, we are in a position (or so we hope!) to make reliable quantum-mechanical calculations independently of any detailed microscopic model. Various calculations along these lines have been done over the last few years, and while there is still some dispute among theorists, our own calculations, which I believe to be rigorous as far as they go, indicate firmly that with not obviously unattainable parameter values we can reach a situation where the quantum mechanical prediction is

$$(K_{\rm exp})_{\rm QM} > 2 \,, \tag{6.9}$$

so that there is still a conflict with the macrorealistic inequality (6.7). If this is so, then it is indeed possible in principle to force Nature to choose between quantum mechanics and macrorealism.

This experiment is currently being set up by Claudia Tesche at IBM Yorktown, and we may hope for the first results within a year or two. At least three possible outcomes may be envisaged: First, it may well turn out that when the experiment is actually run, we shall discover that there are various previously unsuspected sources of noise and dissipation in the relevant frequency regime, which results in the impossibility of ever in practice attaining a situation where the quantum-mechanical prediction satisfies the inequality (6.9). This would certainly be disappointing, but the program I have outlined above has had a considerable amount of spin-off on both the theoretical and experimental fronts, so I don't think we need be ashamed of having pushed it, even in this case.

The second possibility is that the experiment is done under the "right" conditions, and comes out in favor of quantum mechanics, and hence automatically against macrorealism. Whereupon, of course, 99% of the physics community will say, in effect "What a waste of time and money— I could have told you that for nothing!" The implication is that since we all know that quantum mechanics is the whole truth about the world, it was obvious ahead of time that the experiment would turn out the way it did. Well, perhaps: but at the very least, we should have brought the experimental demonstration of quantum interference effects up from the level of, say, the plutonium nucleus— probably the most complex system for which until recently there was even circumstantial evidence for such effects— to the level of a SQUID ring, which on a logarithmic scale is about 3/4 of the way to Schrödinger's Cat! Moreover, we should have provided a direct experimental demonstration that one major and apparently "common sense" prejudice about the world, macrorealism, under certain circumstances fails. If the physics community wishes to claim that that result is trivial, it is of course free to do so; I am, equally, free to disagree.

Finally, the third possibility is that the experiment is done under conditions such that the predictions of quantum mechanics are incompatible with those of macrorealism, and comes out in favor of macrorealism (or at least against quantum mechanics). Whereupon 99.999...% of the physics community says: It must be a bad experiment. (We all know that quantum mechanics is the whole truth about the world, so obviously any experiment which claims to say otherwise must be wrong.) Let's suppose, for the sake of argument, that the experiment is redone, in six different laboratories throughout the world, on six different latitudes, under six different political systems and supported by six different funding agencies, ...and still comes out against the quantum-mechanical predictions. At that point, of course, the roof falls in on the theory: it will be claimed that we must have forgotten some essential noise source, that the relation claimed between the classical and quantum behavisor is not rigorously demonstrated, *etc., etc.* We are trying to plug the obvious loopholes, of course, but certainly it will be impossible to plug every conceivable one ahead of time. Let's suppose, again for the sake of argument, that eventually they are all plugged to everyone's satisfaction, and that the foolproofed quantum-mechanical predictions are still in disagreement with experiment. What should then be our reaction?

At that point, I think, we might have to consider very seriously the possibility that the reductionist prejudice which has served us so well in physics, and indeed in most of science, for the last three hundred years— the prejudice that the behavior of complex bodies is in some sense no more than a consequence of the behavior of their components, that "the whole is no more than the sum of the parts"— is actually false; and, in particular, that the fact that quantum mechanics describes brilliantly the behavior of electrons and nuclei, and that complex, macroscopic bodies are composed of electrons and nuclei, does *not* necessarily imply that quantum mechanics even in principle describes the behavior of complex bodies. If that were to turn out to be the case, it would obviously be of interest not just to professional physicists but also to philosophers of science and many others.

So, listening to my peroration, I think you can guess which outcome of the experiment I am secretly hoping for! But, please, don't ask me— at least when I am sober— to place bets on the outcome.

7

Quantum Gravity: Whence, Whither?

S. Deser
Department of Physics
Brandeis University
Waltham, MA 02254

A review, for extreme non-specialists, is given of some of the properties and problems of quantum gravity. We first obtain classical Einstein theory from the point of view of well-understood Lorentz-invariant ideas, and explore its role as a "classical regulator". The ultraviolet problems of perturbatively quantized local geometrical models are presented, and their implications drawn; in this connection, we also consider supergravity. Finally, we discuss some new work on a possibly consistent quantum system, topologically massive gravity.

7.1 Introduction

The purpose of this lecture is to give a coherent (if necessarily superficial) account of quantum gravity and of its problems to an audience with no background in the field. The subject is vast, and we will necessarily have to omit large and active areas of research, as well as any technical questions. Our main point will be to show that general relativity, as well as its various generalizations necessarily suffers at the quantum level from its classical (geometrical) virtues: the perturbative nonrenormalizability of the theory will be a major theme.

Our survey will begin, after some historical and motivational remarks, with a "derivation" of classical Einstein theory from a non-geometric point of view, starting only from the basic ideas of Lorentz invariant classical and quantum physics and some qualitative observed properties of gravitation. As a cautionary note, we will then consider the self-energy problem in classical gravity and show that the full non-perturbative result and its perturbative expansion have diametrically opposite properties. This should be borne in mind when we then come to the corresponding ultraviolet problems of the quantized theory, which can only be expressed in perturbative language at present. In particular, we will see that the equivalence principle, which is one of the glories of the classical theory, leads unavoidably to intractable problems at the quantum level . We will see that in 4 or more dimensions, naive power counting of radiative corrections (closed loops) implies that Einstein or any other local geometrical theory necessarily loses its predictive power (is nonrenormalizable) if it is otherwise physically acceptable *i.e.*, has excitations none of which are ghost or tachyonic in character. We will then discuss supergravity as an example of a generalization of the theory which has a wider gauge invariance, unifies matter with gravity and thereby provides a hope for compensation of the infinities among the different fields. We will see why, although very beautiful, it is unlikely to succeed as a fundamental unification (but may still be a correct effective low energy theory). We will then turn to the currently active field of $D = 3$ gravity, particularly to a recent candidate for quantum consistency. This is a model, called topologically massive gravity which involves both the Einstein action and the (by now universally fashionable) Chern-Simons invariant. In addition to its intrinsic interest, we will see that this is the only local gravity theory which could *a priori* be viable in the quantum regime, *i.e.*, renormalizable. Although we cannot yet be sure of its success or failure, it does provide a good illustration of the quantum questions we have raised. Finally, we will present some conclusions and make connection with the nonlocal generalization offered by string theory.

7.2 A Brief History

The Planck length, which supplies the scale of quantum gravity, was indeed first obtained by Planck in his series of papers around 1900; he defined the quantity $\kappa \equiv (G\hbar c^{-3})^{\frac{1}{2}} \sim 10^{-32}$ cm, which, together with c and $\hbar$, provides natural units of length, time, and mass ($\hbar\kappa^{-1}c^{-1} \sim 10^{-5}$ gm or 10^{19} GeV is the corresponding mass). That κ as the basic coupling constant in gravity, would lead to ultraviolet problems was already noted in the early thirties by Heisenberg and others. Actual quantization of the gravitational field also began in the thirties through the work of Rosenfeld and of Pauli and Fierz, who studied the massless spin 2 field action which is the weak field limit of the full Einstein theory. I believe it was realized quite early on that the gravitational field, being universally coupled to all matter, could not be left classical (except as a – frequently very good – approximation), any more than could the electromagnetic field; in either case, it makes no fundamental sense to couple the field only to some expectation value of the matter sources. What seems to have caused some confusion (and sometimes still does) was what "quantized geometry" would mean, *i.e.*, what a "metric operator" would keep of geometrical significance. The answer is, roughly speaking, that classical geometry is only effectively valid as a limiting low energy (and $\hbar \rightarrow 0$) concept, whether from quantized local Einstein theory or (more likely) from some nonlocal generalization such as string theory.

Modern work on quantization began in the fifties, where the rather peculiar dynamics of the theory (such as the absence of any extrinsic notion of time, and the fact that its conjugate, the "Hamiltonian," vanishes) associated with its gauge (= coordinate) invariance was disentangled. Although the canonical and covariant quantization of the classical theory has been better and better understood in different ways ever since, there still remain problems of regularization in the operator transcription far worse than in the relatively mildly nonlinear analog of gravity, the Yang-Mills field. The covariant, diagrammatic approach to closed loops – radiative corrections – was well-understood by the early seventies when many explicit one-loop calculations were performed (with negative results); corresponding tree-level effects were also calculated, despite the horrible proliferation of indices in the graviton-graviton vertices. These results were all within the standard perturbative framework in which the gravitational field was developed about some background (usually flat space) and nonlinearities were expanded in a series of powers of κ. There have also been parallel ongoing attempts at summing parts of the series or inferring non-perturbative consequences of the theory, in the hope that these would in fact lead to gravity as a universal regulator which would make everything finite, essentially because of the old hope that κ would act as a natural small-distance cutoff for all quantized fields. Unfortunately, none have been very conclusive, since we have no well-defined closed form approach to these problems. There are also a number of conceptual questions involving the path-integral formulation and the configurations to be summed over (*e.g.*, should different topologies be included, and if so with what weight), how to continue to Euclidean signature *etc.* as well as the meaning of the "wave-function of the universe" that are actually the basis of much current activity, particularly in connection with the cosmological constant problem and, more generally, with quantum cosmology. I cannot cover these equally important ideas, nor those regarding the quantum mechanical effects of black holes first studied systematically by Hawking.

A completely different aspect of quantum gravity was uncovered in the then very distant context of (old-fashioned) string theory by Scherk and

Schwarz and by Yonea around 1974; they noted that among the infinity of excitations described by these nonlocal systems, there appeared, for closed strings, a massless spin 2 particle , which by the uniqueness and universality I will discuss in the next section, must be identified with the graviton. The notion of supersymmetry, which also evolved from string theory, led in 1976 to a profound generalization, supergravity, which is a local theory unifying spacetime and matter into a single multiplet. This was in fact a much more radical unification than in the Kaluza-Klein geometrization of electromagnetism of the twenties, because fermions, and not just bosonic gauge fields were associated with gravity. However, strings, Kaluza-Klein extensions and supergravity did have the common theme that the "true" spacetime dimensionality need not be 4, but could be higher. This is likely to remain a fruitful notion, especially if we ever find a compelling way to account for compactification into just the observed 4 macroscopic dimensions together with the curling up of the others.

More recently, we have seen the 1984 string revolution which emphasized the role of "anomalies" associated with gauge theories, the required avoidance of which would leave (what was then thought to be) an essentially unique model whose low energy excitations correspond to the observed particles. This chapter is of course far from complete, but the deep changes in our notions of what occurs at "small distances" is likely to remain important; I am sure all this will be covered in the article by Gross. However, one should caution that nonlocality also has its price, which is still to be completely understood.

Another interesting development of this decade is that of gravity in 2 and 3 dimensions. The former is in fact closely related to strings through the work of Polyakov, while we will see that $D = 3$ gravity theories show us an entirely different world than we might expect from our $D = 4$ experience.

7.3 Classical Gravity

Although Einstein arrived at classical general relativity by profound geometrical intuitions which found their setting in Riemannian geometry, it is in fact possible to arrive at it by entirely complementary means, based on the more familiar (and experimentally incontrovertible) ideas of special relativistic dynamics, particularly special relativistic quantum theory (in the tree level sense).

In the special relativistic and Galilean macroscopic world, the forces of gravity are found to have certain simple qualitative properties which require an essentially unique field for their description. These properties are that there are attractive macroscopic forces between static (as well as moving) masses, that the forces fall off as $\frac{1}{r^2}$, and that light is also "bent." On the other hand, all forces in special relativity are due to exchange of particles – which are characterized by two invariants – their mass (which can also be zero) and spin. All $\frac{1}{2}$ integer spin particles are immediately excluded by the fact that the forces are macroscopic (they must then involve exchange of at least 2 fermions) and $\frac{1}{r^2}$ (such exchange implies faster falloff). Likewise, all particles of spin greater than 2 and long range (= zero rest mass) can only couple to matter "currents" which have zero static limit, and are also immediately excluded. This leaves spins (0,1,2). Spin 1 is excluded by the fact that like "charges" attract. (Actually, implicit in the attraction/repulsion induced by even/odd spin exchange is the requirement that the intermediate field enter with positive energy, which is in turn based on the observed stability of matter.) A scalar (spin 0) field cannot be the main gravitational intermediary, because the only local generalization of Newtonian mass density is of course the stress tensor $T^{\mu\nu}(x) = T^{\nu\mu}(x)$ of a system, and its scalar part T^{μ}_{μ}, which would couple to spin 0, vanishes for the electromagnetic field, so light would not be bent. (The fine print is that a scalar could couple to some other quantity such as $F_{\mu\nu}F^{\mu\nu}$, but then the "bending" would be radically different from that observed.)

We have almost run out of candidates: there remains only the spin 2 field, either massive or massless; being described by a symmetric tensor field $h_{\mu\nu}$, it is ideally suited for coupling to $T^{\mu\nu}$. One totally unexpected dividend here is that one can show that this field must be strictly massless (*i.e.*, of infinite range) even though there is clearly no observable solar system difference between a Yukawa potential with range the size of the universe, say, and the strictly Coulomb one . Nevertheless, it turns out that there is a discontinuity here, and that there is a 25% difference in the light bending predicted by the $m = 0$ and $m \neq 0$ choices, the observed bending leading of course to $m = 0$. Now there is only one way to describe a (positive energy) infinite range spin 2 field (just as the Maxwell action is unique under similar assumptions). In both cases, these fields are governed by actions and field equations of the gauge type, with currents/stress tensors as sources. Being gauge fields, these systems have identically conserved field operators: for Maxwell, they are $\Box A_\mu$ as leading term, minus the term $\partial_\mu \partial^\nu A_\nu$, *i.e.*, $\partial_\nu F^{\nu\mu}$. For a symmetric tensor $h_{\mu\nu}$, one similarly has $\Box h_{\mu\nu}$ minus analogous terms, the identically conserved combination being called $G_L^{\mu\nu}(h)$, the linearized Einstein tensor.

The respective sources $(J^\mu, \kappa^2 T^{\mu\nu})$, where κ has the dimensions of length (we use $\hbar = 1 = c$) and $h_{\mu\nu}$ is dimensionless, must therefore be conserved (not identically, but by virtue of the matter field equations) not only for free sources but precisely also when their coupling to the gauge fields is included. This is indeed the case for the electric current: charge remains conserved because photons are neutral. However, the stress-tensor is a different story: it represents the energy of a system and is only conserved

($\partial_\mu T^{\mu\nu} = 0$) when that system is isolated. But it is now in interaction with the h-field and so no longer conserved; only the stress tensor of the total matter plus h-field system is. We must therefore add the term $T^{\mu\nu}(h)$ – quadratic in h – to the source side of the field equations. But this requires adding a cubic term in h in the action, which in turn changes the $T^{\mu\nu}$ of the h-field (since $T^{\mu\nu}$ is derived from the action), and so on – indefinitely. While there is no *a priori* guarantee that this series converges, *i.e.*, that a consistent self-coupling exists, it, of course, does. (Indeed, by a suitable choice of variables, the whole process stops at the first nonlinear (cubic) order in the h-field action.) The net result is two-fold: first, the original gauge invariance of the $h_{\mu\nu}$ field, which was the obvious generalization ($\delta h_\mu = \partial_\mu \xi_\nu + \partial_\nu \xi_\nu$) of the Maxwell $\delta A_\mu = \partial_\mu \Lambda$ becomes nonlinear, and depends on h; second, the total h-field action depends on $h_{\mu\nu}$ only in the combination ($\eta_{\mu\nu} + h_{\mu\nu}$) and does not involve the Minkowski $\eta_{\mu\nu}$ alone. This universal dependence is also true of the matter action in its coupling to $h_{\mu\nu}$ – it also only depends on the combination ($\eta_{\mu\nu} + h_{\mu\nu}$). Furthermore, the gravitational Lagrangian has a very specific form, homogeneous in second derivatives at every power of $h_{\mu\nu}$; "miraculously" this form is a purely geometric quantity – the scalar curvature density of a Riemann space with metric $g_{\mu\nu} \equiv \eta_{\mu\nu} + h_{\mu\nu}$. The underlying symmetry is just general coordinate invariance. Furthermore the result is unique, up to a possible additional part, the so-called cosmological term, and can be extended to provide a derivation of supergravity as well.

We have just sketched how the qualitative observed properties of gravitation at the macroscopic level, together with the kinematic framework of special relativity which is valid there, lead to the geometrical picture of matter propagating in a Riemann manifold whose curvature is determined by the matter through the nonlinear Einstein equations, *i.e.*, "as if" Minkowski space $\eta_{\mu\nu}$ had turned in to $\eta_{\mu\nu} + h_{\mu\nu}$. All this describes the effective low energy classical Einstein theory, and does not imply that it is the "true" underlying Lagrangian to be quantized, anymore than one must take photons as "true" fundamental particles.

7.4 Classical Gravity as Regulator?

This interlude will illustrate both the nonperturbative function of classical general relativity as a universal "regulator," *i.e.*, as providing a physical cutoff for infinities that would otherwise be present in systems coupled to it, and the pitfalls of the perturbative expansion in such strong-field problems. We emphasize that the discussion here is purely classical, so that the lessons are not immediately applicable in the quantum domain, but it is one worth bearing in mind when we come to the latter in the next section.

Consider first the self-energy of a simple distribution of mass and charge at the Newton-Maxwell level. We are interested in the behavior of the self-energy as the size ϵ of the particle tends to zero, *i.e.*, in the point limit. The total mass (or energy) of this system is given by

$$m = m_0 - \frac{Gm_0^2}{2\epsilon} + \frac{e^2}{2\epsilon} \tag{7.1}$$

as a function of its bare mass m_0 and charge e, and the usual $\frac{1}{2}$ factor for the self-energy has been inserted. Bare mass here means that value the mass would have in the absence of interactions (the charge is of course unaffected by gravity).

In this naive picture, we see that the point limit is catastrophic: the usual infinite repulsive Coulomb self-energy is now joined by an infinite attractive Newtonian self-attraction (except for the special case $Gm_0^2 = e^2$). Now let us make an equally naive "equivalence principle" argument, namely that since *all* energy gravitates (that is the standard "elevator" argument), Eq. (7.1) should be replaced by

$$m = m_0 - \frac{Gm^2}{2\epsilon} + \frac{e^2}{2\epsilon} \ . \tag{7.2}$$

The only change is the apparently minor one of replacing m_0 by m in the Newtonian self-energy term, but the consequence is dramatic. For simplicity, consider first the neutral system ($e = 0$). We can see that the total mass m can never turn negative since, as it decreases to zero, the self-attraction in fact vanishes. Indeed, we can solve the quadratic Eq. (7.2) explicitly for m, with the result

$$m = \frac{\epsilon}{G}\left\{(1 + 2m_0 G\epsilon^{-1})^{\frac{1}{2}} - 1\right\} = m_0 - \frac{Gm_0^2}{2\epsilon} + O\left(\frac{G^2 m_0^3}{\epsilon^2}\right) . \tag{7.3}$$

The last equality indicates the perturbative expansion of m in terms of G, and shows that each new term is more and more divergent as $\epsilon \rightarrow 0$; things seem even worse than in Newtonian theory. However, the correct result for m is of course

$$m \rightarrow (2m_0\, \epsilon/G)^{\frac{1}{2}} \rightarrow 0 \ . \tag{7.4}$$

As predicted, zero is a lower bound – point particles of finite bare mass having zero total mass. Had we included the Coulomb term, we would have obviously found $m \rightarrow \mid e \mid G^{-\frac{1}{2}}$, a finite (though numerically totally unrealistic) limit. Thus, the nonperturbative result is perfectly finite – the infinities of naive Newtonian theory have been regulated away by the more complete model – but a perturbative analysis would only have made the divergence problem appear (infinitely) worse.

The full general relativistic treatment of this system can be given in closed form and leads to precisely the same result. The only refinement is that the "size" ϵ of the particle is a coordinate artifact, and its true (invariant) size remains finite. Also, while the neutral particle has no visible exterior mass, space is curved in its interior, as we would expect since there is after all a difference between this system and one with no source at all. Although some thought has been given to a quantum translation of this result, we do not know how to carry out the infinitely more complicated analog there. I should, however, mention that classical Einstein theory (even without sources) is subject to the well-known singularity theorems of Penrose and Hawking, which essentially state that all solutions will in general break down somewhere, given some reasonable (?) assumptions. To date, there is also no clear connection between these results and the quantum problems.

7.5 Quantum Gravity

We now come to the full quantum gravity theory. Although we are primarily interested in general relativity in $n = 4$ spacetime dimensions, we will work in a more general framework in which n is unspecified and the action to be quantized is not necessarily Einstein's, but can be any geometrical one, involving other terms than the scalar curvature. We will be rather summary about the formal aspects of the quantization, being more interested in the divergence problem itself.

Associated with the geometric nature of gravity is an underlying gauge invariance of its description under arbitrary coordinate charges. This restricts the form of the allowed Lagrangian, just as electromagnetic gauge invariance requires the Lagrangian to depend only on the field strength $F_{\mu\nu}$, rather than on the vector potential A_μ. Here the equivalent of $F_{\mu\nu}$ is the curvature $R^{\mu}_{\nu\alpha\beta}$ of space-time: the vanishing of either is the necessary and sufficient condition for no excitations (A_μ is a pure gauge, spacetime is flat), and they are both defined in terms of the commutator of two covariant derivatives acting on a "charged" field, $iF_{\mu\nu} = [\partial_\mu + i\,A_\mu,\ \partial_\nu + i\,A_\nu]$, and a similar expression for R. Thus any gauge invariant Lagrangian density must be a scalar built from curvatures and their (covariant) derivatives, just as the vector action could be $F^2_{\mu\nu}$ or $F_{\mu\nu}\Box F_{\mu\nu}$ *etc.* Since the curvature is of second derivative order in the tensor potentials (= metric), field equations of second derivative order will be obtained only if the Lagrangian is linear in curvature – this is the Einstein term, which is the scalar $R^{\mu\nu}_{\mu\nu}$.

However, it is conceivable that there are also higher terms present, such as $R^2_{\mu\nu\alpha\beta}$, which introduce 4th derivatives, without contradicting low energy observations, since these terms are damped to $O(\kappa^2\partial^2)$ compared to R. The basic property of relevance to us is that whatever the leading part of the action, whether just R, or R^2 *etc.*, each term is homogeneous of a given derivative order in its expansion about any background, say flat spacetime. That is, curvature is a nonlinear function of the metric $g_{\mu\nu}$ since it involves its matrix inverse $g^{\mu\nu}(g^{\mu\alpha}g_{\alpha\nu} = \delta^\mu_\nu)$, and unlike standard matter fields, which we normally expand about zero, it must be divided into $g_{\mu\nu} = \eta_{\mu\nu} + \kappa\, h_{\mu\nu}$ where $h_{\mu\nu}$ is the field variable defining the expansion. (The fascinating possibility of using a "zero vacuum" is currently being explored in lower dimensional gravity.) The Einstein action then has the schematic expansion

$$I_E = \int d^n x \left[(\partial h)^2 + \kappa h \partial h \partial h \ldots + (\kappa h)^n\, \partial h\, \partial h + \ldots\right] \quad . \tag{7.5}$$

This means there is the usual kinetic term $(\partial h)^2$ plus a series of vertices involving three or more h's, each with two derivatives; we have introduced κ in the expansion (and so made h dimensional) in order that the leading term $(\partial h)^2$ have the proper dimension, just as the Maxwell action is $\sim (\partial A)^2$. The higher terms just represent the universal self-coupling of the stress tensor $T^{\mu\nu}(h)$ to the h-field, as we discussed in our derivation of the theory. This is just the equivalence principle in another guise.

In quantum language, the h-field action, Eq. (7.5), tells us that we have an h-propagator (from the $(\partial h)^2$ part) $D \sim k^{-2}$ in momentum space (just as does any normal particle), and that the interaction vertices (stress tensor coupling) which we collectively call V, represented by the higher terms, behave as k^{+2}, unlike electrodynamics, where the Feynman rules tell us the electron-photon vertex (current) is γ^μ, *i.e.*, momentum-independent. In particular we deduce that $(DV) \sim k^0 \sim 1$. Now consider the loops we can build from our action. A moment's thought shows that at lowest, 1-loop, order we have an equal number p of D's and Γ's, together with an integration over virtual states, *i.e.*, $\int d^n k$ for dimension n. The net result for one loop is thus

$$\int^\Lambda d^n k (DV)^p \sim \int^\Lambda d^n k \sim \Lambda^n \quad . \tag{7.6}$$

Here Λ is a high frequency momentum cutoff, and we see that these corrections diverge more and more with increasing dimension, being quartic for $n = 4$. Worse than this however, is that at each higher loop, which is obtained by attaching a new internal line connecting 2 new vertices in the lower loop diagram and integrating over the new set of virtual states, we obtain the new factor:

$$\int^\Lambda d^n k' D (DV)^2 \sim \int^\Lambda d^n k' D \sim \Lambda^{n-2} \tag{7.7}$$

times the Λ^n of the original one. So each higher loop correction merely worsens the situation: the degree of divergence Δ grows with each loop order L, being in general $\Delta \sim \Lambda^{L(n-2)+2}$. To remove these infinities, we must absorb them in newly defined counterterms, *i.e.*, new terms of the Lagrangian whose coefficients must be defined experimentally. This is what is meant by a non-renormalizable theory, one whose predictive power is lost because of this ever-increasing set of infinities. The cause is indeed the positive dimension of κ which means that $(\kappa\Lambda)$ is dimensionless, and can appear to arbitrary powers.

For $n = 4$ Einstein theory, there happens to be (in absence of matter) an accidental divergence which allows the 1 loop counterterms to be removed; however, already at 2 loops there is a truly dangerous counterterm and it has been found to be present by explicit calculations. In the Einstein plus matter system, one-loop order is already generically non-renormalizable; we will return to this point in our discussion of supergravity.

Is there any remedy to the above problem? Looking back at Eq. (7.6), we see that the 1-loop counting is independent of the model, but depends only on dimension. However, from Eq. (7.7), we see that if $n = 4$, *and* we let the propagator behave as k^{-4} (even though V will then $\sim k^{+4}$), then higher loops will not make things worse, since they will now contribute $\sim \Lambda^{4-4}$. This seems a small enough price to pay for a viable theory, since adding the appropriate $R^2_{\mu\nu\alpha\beta}$ terms in the action (remember that $R \sim \partial^2 h$, so that $R^2 \sim h\partial^4 h$ to leading order in h) will have little macroscopic observational effect, as we have seen. These theories ($\sim R^2 + R$) can indeed be shown, by careful calculation, to be renormalizable. Unfortunately the price – loss of unitarity – is even worse than the disease. This fact is of course the reason we have avoided, ever since Newton, theories in which higher derivatives entered at the fundamental level. What happens is that systems whose kinetic terms are $\sim h(\Box^2 - \lambda\,\Box)h$ have propagators of the form

$$D(h) \sim (k^4 + \lambda\, k^2)^{-1} \sim \lambda^{-1}[k^{-2} - (k^2 + \lambda)^{-1}] \tag{7.8}$$

and it is the residues at the poles of a propagator which define the character of the associated particle. In particular, a negative sign (unavoidable here) means negative probability quantum-mechanically (and instability classically through runaway solutions), and of course the sign of λ in Eq. (7.8) cannot simultaneously save both residues (in addition, one could risk tachyonic behavior as well with the wrong sign of λ!). The above loss of unitarity persists in higher dimensions; barring some unexpected mechanism for its restoration, we must abandon these models. We will return to lower dimension later.

One may argue that the above power-counting estimates do not take into account the possibility that there might be cancellations between different infinite contributions, such as those from different matter sources in "the" correct theory. This is a perfectly reasonable idea, but is hardly likely to hold accidentally, *i.e.*, unless the correct theory has a new invariance which orchestrates the cancellations. For example, it is known that no combination of scalars, vectors and spinors coupled to Einstein gravity can conspire to provide even a 1-loop cancellation. Furthermore, the putative new symmetry must include fermions, since we know that they tend to generate factors of (-1) when they run around a closed loop. This motivates us to find such a symmetry – and indeed there is one, supersymmetry, whose generalization to supergravity precisely involves new fermions, and is therefore purely quantum in nature.

7.6 Supergravity

The idea of unifying gravity with matter is of course an ancient one. As we have noted, Kaluza-Klein theory was originally designed to integrate the then two known gauge fields, gravity and Maxwell. It uses the fact that a field in a higher number of dimensions looks, in lower ones, like a collection of fields of several spins, and that the higher dimensions can be "curled up," and be invisible at sufficiently large scales. However, this is fundamentally a bosonic unification, and we want to include fermions. The latter notion appears crazy at first, given the spin-statistics connection. However, let us persevere, and consider the much milder idea of global supersymmetry in the absence of gravity. This is an invariance which is of purely quantum origin since fermions are involved, and differs from internal symmetries because it is intimately involved with the Poincaré group itself – which makes it sure to have gravitational repercussions. (Its generators do not commute with the Poincaré generators and, more important, their own commutators are proportional to the translations.) Consider for simplicity two free fields of adjoining spin, say a scalar and real (Majorana) spinor field, or the Maxwell field plus this spinor. It is clear that if the two fields are to transform into each other, the gauge parameter must itself be fermionic. For example, it is easy to check explicitly that the action

$$I = -\frac{1}{4}\int F_{\mu\nu}^2\, d^4x - \frac{i}{2}\int \bar{\psi}\partial\!\!\!/\psi\, d^4x \tag{7.9}$$

is invariant under the transformation

$$\delta\, A_\mu = \bar{\alpha}\,\gamma_\mu\,\psi, \qquad \delta\,\psi = iF^{\mu\nu}\,\sigma_{\mu\nu}\,\alpha \tag{7.10}$$

where α is an anticommuting (Grassmann) constant parameter. Establishing this symmetry involves a bit of γ-algebra, and use of the cyclic identities $\partial_{[\mu}F_{\alpha\beta]} \equiv 0$. There is a corresponding conserved (spinorial) current $\mathcal{J}^\mu \sim F\cdot\sigma\gamma^\mu\psi$, and a corresponding conserved charge $Q = \int d^3x\mathcal{J}^0$, which has the amazing property that (the trace of) its square is in fact the energy of the system Eq. (7.9). For us the important point is that there is an extension of this global invariance to a local gauge symmetry where α becomes a spacetime function $\alpha(x)$, and that the corresponding gauge field is the doublet of massless spin $(\frac{3}{2}, 2)$. The latter is of course gravity, the former is represented by a vector-spinor field ψ_μ, which in the absence of gravity obeys essentially the Dirac equation *modulo* some gauge terms (it is clear that for ψ_μ to represent a massless particle it must have just the 2 helicity $\pm\frac{3}{2}$ components, and so not all 4 spinors ψ_μ, $\mu = 0 \cdots 3$, can be dynamical) with action

$$I_{\frac{3}{2}} = i \int \epsilon^{\mu\nu\alpha\beta}\, \bar{\psi}_\mu \gamma_5 \gamma_\nu \partial_\alpha \psi_\beta \, d^4x \, . \tag{7.11}$$

The action, Eq. (7.11), is easily seen to be invariant under $\delta\psi_\beta = \partial_\beta\, \eta(x)$ and $\delta\bar{\psi}_\mu = \partial_\mu\, \bar{\eta}(x)$, due to the antisymmetric ϵ-symbol. This local abelian spinorial invariance only involves ψ_μ and is very different from the global invariance under rotation into the spin 2 field analogous to Eq. (7.10). The two are unified upon considering the Einstein action together with the generally covariantized version (essentially $\partial_\alpha \rightarrow D_\alpha$) of Eq. (7.11). This system is consistent (a very difficult thing to achieve for spins > 1 coupled to gauge fields), a property which can be shown to imply invariance under the transformation

$$\delta\, g_{\mu\nu} = \bar{\alpha}(x)(\gamma_\mu \psi_\nu + \gamma_\nu \psi_\mu), \qquad \delta\, \psi_\mu = D_\mu \alpha(x) \tag{7.12}$$

which combines the local abelian $\eta(x)$ invariance of Eq. (7.11) with that of global supersymmetry. So spacetime is no longer invariant, since metric and spin $\frac{3}{2}$ field rotate into each other here (in the D_μ of $\delta\psi_\mu$ there lurk derivatives of $g_{\mu\nu}$). This simple supergravity can be extended in various ways, both to include other matter (of lower spins) in addition to ψ_μ and to higher dimension (one of the mysteries is the significance of the unique highest-dimensional version permitted, namely $D = 11$; at least $D = 10$ is a local limit of superstring theories). For us, however, the question here is not so much the elegance or uniqueness properties of this "Dirac square root" of gravity, but rather whether the fermions do indeed provide a less unpleasant ultraviolet behavior. The answer is yes, but not enough. For example, one finds for the simplest version given above that there cannot be any one– or two–loop infinities (because there are no corresponding invariant counterterms) in contrast to gravity plus generic matter (as mentioned earlier). However, from three loops on there are candidate infinities which cannot be excluded on invariance grounds, although no one is likely to verify this by explicit calculation soon. So, in absence of any gauge reason to cancel the divergences, we must give up the idea that any finite combination of matter fields will provide sufficient compensation; there are no other symmetries possible uniting gravity with lower spin matter in finite numbers.

7.7 Topologically Massive Gravity

We noted, in our discussion of the divergences of quantum gravity models, that no renormalizable ones (involving only a finite number of counter-terms), could exist, generically, in 4 or more dimensions unless they contained ghosts, *i.e.*, violated unitarity. At this stage, *any* viable dynamical model would be welcome, if only to teach us what sort of miracle it benefits from. I want to discuss briefly here a theory – topologically massive gravity – in 3 spacetime dimensions which almost makes it, and recent work on its quantum properties (with Z. Yang) that is not yet fully understood. Nevertheless I present it as the only possible candidate that is local, has dynamical excitations, no ghosts, and is power-counting renormalizable. It is also very interesting in its own right, involving a new type of geometrical quantity, the Chern-Simons form which is currently of great interest in the study of gauge fields, and which appears in very different contexts, from string theory to models of high T_c superconductivity. Going back to our power-counting Eqs. (7.6) and (7.7), we see that for $n = 3$, a propagator which behaved asymptotically as k^{-3} would give rise to no new divergences beyond those at 1-loop. This would require a kinetic term with 3 derivatives, something that cannot be accomplished in an even number of dimensions, but which is possible here, due to the presence of the antisymmetric $\epsilon^{\mu\nu\alpha}$ symbol. A simple example of odd derivative order is provided by the vector field, for which $\int d^3x \epsilon^{\mu\nu\alpha} A_\mu \partial_\nu A_\alpha$ is an invariant under $\partial A_\mu = \partial_\mu \Lambda(x)$, even though it is not constructed from $F_{\mu\nu}$ alone; the density $\epsilon A \partial A$ is not invariant, but its integral obviously is. Now one can immediately construct a gravitational analog, $\int d^3x\; \epsilon^{\mu\nu\alpha}\; h_{\mu\lambda} \Box \partial_\nu\, h^\lambda_\alpha$ at the linearized level. The question is whether it has a fully non-linear geometrical generalization, and indeed it does – the (unique) Chern-Simons form, which is again only coordinate invariant as an integral rather than as a density because it does not depend on curvatures only. I cannot do more here than mention some tantalizing aspects of the theory, even at the classical level, but let me at least describe it in words. In 3 dimensions the Einstein action alone does not generate any dynamics (like electromagnetism in 2 dimensions) because there the full curvature tensor $R_{\mu\nu\alpha\beta}$ and its contraction, the Einstein tensor $G^{\lambda\sigma}$, have the same number of components; they are essentially each other's double duals (using ϵ to dualize, as usual). Thus spacetime is flat except where the matter sources are located, so there are no gravitational interactions in the usual sense and no excitations of the metric field (no gravitons). This theory can still be formally quantized and, with use of the proper variables, is in fact finite; its coupling to quantized matter also has some unusual properties. The Chern-Simons term alone is also non-dynamical because its contribution to the field equations is just the conformal tensor, *i.e.*, a quantity whose vanishing implies that space-time is conformally flat ($g_{\mu\nu} = \phi(x)\eta_{\mu\nu}$) without any dynamics for $\phi(x)$. However, the sum of the two actions is more dynamical than its parts: it describes a gauge particle that propagates with finite mass, has spin 2 with a single P-odd helicity and no ghost problems, despite the 3rd derivative order of the Chern-Simons part. (There is also a supersymmetric extension involving a massive spin $\frac{3}{2}$ field obeying second derivative equations.)

When we quantize and go through the power counting, we find that the graviton propagator has two parts: one, describing the conformal part of the metric, still decreases as k^{-2}, while the remainder is dominated by the Chern-Simons term and so falls off as k^{-3}. The structure of the conformal factor part is, however, so special that it does not lead to any catastrophes in power-counting, while the k^{-3} sector is manifestly acceptable. Furthermore, at one loop, one may show that everything is explicitly finite. So far, then, all is going miraculously well and the reason is that the Einstein constant κ, while still present, is never in the numerator, but only appears as κ^{-1}, which is "super-renormalizable"; the true coupling constant is the dimensionless coefficient of the Chern-Simons term (dimensionless because the term itself is dimensionless). Thus both the dimensional coupling constant problem and the corresponding power counting one are avoided by the model. What stops the proof from being complete is the technical requirement of defining a well-behaved covariant higher loop regulator, *i.e.*, a prescription for giving meaning to the *finite* number of divergences which can appear. Here we are currently stymied, because any covariant regulator brings in κ to positive powers. Appeal to other regularization methods shows that there could appear something akin to an "anomaly," which is a problem known to other gauge theories. If present, this counterterm would upset the renormalizability. To decide whether these anomalies are really there would require an impossible higher-loop calculation. So there is one (and only one) model which may be the exception to our negative conclusions, but then again the "technical" problem may be unavoidable; this is a novel situation, where our usual intuition is not applicable, and requires further work.

7.8 Summary

We have attempted to follow a few of the problems and ideas in quantum gravity theories, starting from the classical origin of general relativity as a "normal" Lorentz invariant field theory. We have seen that Einstein theory emerged as the unique model in 4 dimensions capable of fitting even the qualitative observed properties of gravitation within this framework. It was also noted that classically, at least, the divergence problem was in fact solved by gravity due to its attractive nature and to the equivalence principle requirement that all energy gravitate. It was only when we made a perturbative expansion that pathologies arose. We then turned to the – necessarily perturbative – quantum version of geometrical theories. Here we were careful to point out that while classical Einstein theory was indeed the correct effective low energy model, this does not mean that it is the theory to be quantized, nor even that it is part of that theory, and that in any case we were stuck with the (suspect?) perturbative approach. As expected from simple dimensional arguments, we indeed found that all local theories of gravity (with the possible exception of the topological massive model in 3 dimensions) were rejected by simple power-counting unless they involved (unacceptable) violation of unitarity through ghost excitations. We also noted that while the possibility of cancellation amongst possible infinities due to some hidden new symmetry motivated supergravity, that theory was really not "crazy enough" either, at least on *a priori* grounds.

It seems that there are two ways out. Either the classical Einstein theory (or some local generalization) is the proper start for quantization, but requires as yet nonexistent nonperturbative techniques to vindicate it, or more likely, the correct finite quantum theory is not so closely tied to classical spacetime roots but is something like string theory, which is nonlocal (and has no classical spacetime in it at all!) and whose local limit involves an infinite number of local excitations. Although finiteness has not yet quite been proven for any superstrings, it is undeniable that they reduce to (super)gravity in the appropriate limit. Whether their nonlocality will be physically acceptable and will yield a correct particle spectrum is not yet clear. In any case they certainly broaden our horizons in seeking a deeper unification and raise our expectations as to what a successful theory should predict, including the dimensionality of spacetime and the observed internal symmetries of matter, while unifying all the forces through some higher invariance principle.

Acknowledgements

I thank Z. Yang, who collaborated on the research described in Sec. 7.7, for discussions. This work was supported in part by NSF grant PHY88-04561.

Bibliography

The present list is only meant to provide a small sample of reviews on the topics covered here; some of the original papers directly used in the text are also included. I have not cited the many equally important topics in quantum gravity which time did not permit me to mention. However, between these reviews , the standard texts *(e.g.*, Misner, Thorne and Wheeler; Weinberg; Green, Schwarz and Witten), and the appropriate journals (including Classical and Quantum Gravity, besides those of wider coverage) it should be possible to delve successfully in the literature.

Classical Relativity

R.P. Feynman, Caltech Lectures (1962).
S. Weinberg, Phys. Rev. **B135** (1964) 1049; **B138** (1965) 988.
S. Deser, GRG **1** (1970) 9; D. Boulware and S. Deser, Ann. Phys., **89** (1975) 193.
H. Van Dam and M. Veltman, Nucl. Phys. **B22** (1970) 397; D. Boulware and S. Deser, Phys. Rev. **D6** (1972) 3368.
S.W. Hawking and G.F.R. Ellis, *The Large-scale Structure of Space-time* (Cambridge, 1973).
C. Will, *Theory and Experiment in Gravitational Physics* (Cambridge, 1985).

Classical Gravity as Regulator?

R. Arnowitt, S. Deser and C.W. Misner in *Gravity, an Introduction to Current Research*, ed. L. Witten (Wiley, NY, 1962).

Quantum Gravity

B.S. DeWitt, Phys. Rep. **19** (1975) No.6.
S. Deser in *Gauge Theories and Modern Field Theory*, ed. R. Arnowitt and P. Nath (MIT, 1976).
C.J. Isham in *Quantum Gravity - an Oxford Symposium*, ed. C.J. Isham, R. Penrose and D. Sciama (Oxford University Press, Oxford, 1975); in *Proceedings of the Eighth Texas Symposium on Relativistic Astrophysics*, ed. M. Papagiannis (New York Academy of Sciences, New York, 1977); in *Quantum Gravity II* , ed. C.J. Isham, R. Penrose and D. Sciama (Oxford University Press, Oxford, 1980); in *Superstrings and Supergravity*, ed. A.T. Davies and D.G. Sutherland (SUSSP Publications, Edinburgh, 1986); in *General Relativity and Gravitation*, ed. M. MacCallum (Cambridge University Press, Cambridge, 1987).
T. Appelquist, A. Chodos and P.G.O. Freund, *Modern Kaluza-Klein Theories* (Addison-Wesley, Reading, MA, 1987).
K.S. Stelle, Phys. Rev. **D16** (1977) 953.
M.H. Goroff and A. Sagnotti, Nucl. Phys. **B266** (1986) 709.

Supergravity

P. Van Nieuwenhuizen, Phys. Rep. **68** (1981) No. 4.
S. Deser, J.H. Kay and K.S. Stelle, Phys. Rev. Lett. **38** (1977) 527.
D. Boulware, S. Deser and J.H. Kay, Physica **96A** (1979) 141.

Strings

M.B. Green, J.H. Schwarz and E. Witten, *Superstring Theory* (Cambridge, 1987).
L.D. Eliezer and R.P. Woodard, Nucl. Phys. **B** (in press).

Three-Dimensional Gravity Theories

S. Deser, R. Jackiw and S. Templeton, Ann. Phys. **140** (1982) 372.

S. Deser and J.H. Kay, Phys. Lett. **120B** (1983) 97.

S. Deser, R. Jackiw and G. 't Hooft, Ann. Phys. **152** (1984) 220.

G. 't Hooft, Comm. in Math. Phys. **117** (1988) 685.

S. Deser and R. Jackiw, Comm. in Math. Phys. **118** (1988) 495.

E. Witten, Nucl. Phys. **B311** (1988) 46.

J.H. Horne and E. Witten, Phys. Rev. Lett. **62** (1989) 501.

S. Deser, J. McCarthy and Z. Yang, Phys. Lett. **B** (in press).

S. Deser and Z. Yang, in preparation.

8

String Theory: Current Status and Future Prospects

David J. Gross
Department of Physics
Princeton University
Princeton, NJ 08540

8.1 Introduction

In physics we make progress by looking deeper and deeper into the structure of things, by observing nature at shorter and shorter distances. This unfortunately requires larger and larger energies and therefore larger and larger budgets. Until now, we have explored nature to distances of 10^{-17} centimeters, almost a millionth millionth of a centimeter, with the aid of accelerators with the energy of a trillion electron volts. This energy marks the border between where we have explored nature with our large accelerators, which serve as microscopes, and where we have not yet explored the structure of nature. Now, at the moment, theoretical elementary particle physics is in the very unusual situation of possessing a complete theory of everything that has so far been measured. Within the last twenty years we have constructed theories of all of the traditional forces of nature; and most of the traditional questions of particle physics have been answered. We now believe that we possess adequate and complete theories of the strong nuclear forces and the electromagnetic and weak forces which are three of the four basic forces of nature. The reason, of course, that we are able to do this is that the distance scale, or the energy scale, characteristic of these forces, is between 1 GeV and a hundred GeV, the region which we have been exploring with the big accelerators for the last decade or two. We have learned that all of these forces are very similar. They are all consequences, we believe, of a local symmetry of nature which is both responsible for the strong interactions, the symmetry group being SU(3), and for the electroweak interactions, with the symmetry group being SU(2)×U(1).

The carriers of these forces are the gluon (the glue that holds the nucleus together) which has been observed indirectly at the DESY accelerator in Hamburg, the photon or light ray that was discovered theoretically by Maxwell; as well as the W and Z particles which were discovered at CERN a few years ago. These forces act on particles of matter, so to complete the story we must understand what matter is made out of. This too has been achieved. The nucleus of each atom is made of quarks which carry a *color* charge as well electric charge and experience both strong and electroweak forces. In addition there exist particles, like the electrons and the neutrinos, which feel only the electroweak interactions.

So, we have photons, W's, Z's and gluons which provide the forces between matter which is made out of quarks and leptons: these are the ingredients of what is called the standard theory. This theory, over the last fifteen years has passed all the tests posed by experiment. Time and time again experiment has confirmed the predictions of the standard theory and, more importantly, has not discovered any new phenomenon that cannot be encompassed within the theory.

This success does not leave us totally happy, for many reasons. The first reason is that there are still many mysteries that are not explained by the standard theory. There are many features that have to be put into the theory by hand, and there are many numbers that we cannot calculate. There are many phenomena, such as the reason that these forces arise in the way they do, that is as consequences of local gauge symmetries, which we would like to understand. Also we would like to unify the strong and electroweak interactions, joining them together as one force, instead of having two or three separate forces.

In physics there are two stages of understanding: you first ask the question, "How? How does it work?" After you have understood how it works you begin to ask, "Why? Why is it so?" Now that we understand how it works, we are beginning to ask why are there quarks and leptons, why is the pattern of matter replicated in three generations of quarks and leptons, why are all forces due to local gauge symmetries? Why, why, why?

The standard, conventional way of answering such why questions and of achieving greater unification is to study physics at smaller distances or greater energies. Since we have a good theory of low energy physics, we have a platform from which we can try to extrapolate to higher energies. This effort has been going on for fifteen years or so. The most important thing that emerged early in these investigations was the realization that if we are going to unify all the interactions together, the natural distance or energy scale is very far from present day experiments. It is probably at an energy of 10^{16} to 10^{19} GeV, which is close to the energy where gravity becomes a strong force. In other words, the simple extrapolation of what we know seems to imply that nothing fundamentally new will happen until we get way above present energies, until we go to energies so large or distances so small that gravity, otherwise ignorable as a very weak force, becomes important.

This poses a serious problem for the theorists. We face the question of how to discover the truly new physics if the energy at which the new physics shows up is 17 orders of magnitude bigger than present experimental investigation. Usually, we theorists have had the luxury of being presented by our experimental friends with new discoveries, new paradoxes and new phenomena, which made it easy for us to discover new theories and new explanations.

Indeed, some of my colleagues believe that an attempt to guess or conceive of a new theory without experimental input is wrong, immoral and dangerous and therefore that we should wait for experiment to catch up. I agree that such an attempt is dangerous, but I cannot wait that long. However, if we are to proceed in the absence of experiment it must be in a different style than before. Largely it must be a style that is based on mathematical ideas and on the search for new beauty in the fundamental structure of nature. One lesson that we can extract from the successes of the last ten or twenty years is that Nature, for some reason, is fundamentally based on principles of symmetry. Certainly the secret of unification, *i.e.*, of bringing together different forces, is symmetry. If we are to achieve more unification we must invent new symmetries.

It is not easy to invent new symmetries. It requires discovering new degrees of freedom, as well as new dynamical mechanisms for hiding the symmetry, otherwise the symmetry would not be new, it would be obvious to everyone, an old symmetry. For example, to understand the nuclear force we had to discover the hidden constituents of protons and neutrons, namely the quarks. In addition, we had to discover a new dynamical mechanism called confinement which explained why we had never seen the quarks before. In the last decade two truly new symmetries have been explored at great length in the attempt to achieve greater unification: one is based on the idea that there exist more than three spatial dimensions in the world and therefore one has extra space-time symmetries associated with the extra dimensions, and the other is supersymmetry.

The idea that there are more than 3 dimensions (right-left, forward-backward and up-down) is an old idea. It goes back to 1921, shortly after Einstein's theory of relativity, when it was invented by a Polish mathematician called Kaluza. In modern language, Kaluza said the following:

Einstein tells us that space-time is a dynamical property of the world, so let us imagine that there are actually five space-time dimensions, one more spacial dimension. Imagine that the dynamics of space-time are such that this dimension is not a straight infinite line, but rather a small, little circle of size, perhaps 10^{-33} centimeters, the characteristic dimension of gravity. At every point, we could move right-left, forward-backward or up-down, or around the little circle as well. However, if the circle was very small we would never notice that there was an extra dimension. The reason we would not notice the little circle is that, as I said before, it takes enormous accelerators or big microscopes to see very short distances. What does it mean to say that we couldn't see the little circle? It means that, each atom and every particle is smeared out around this extra dimension, so that there is no structure in that extra dimension. So a low energy physicist, who cannot do experiments at these very high energies, would never see these extra dimensions.

You might then ask, "If you can't see them why hypothesize them?" What Kaluza discovered is that there is some effect of gravity in five dimensions that persists even if one of the dimensions is a very small little circle. He discovered that the momentum of particles in the fifth dimension, which is conserved and quantized in integer units, can be thought of as the electric charge and the remnant of the gravitational forces in the fifth dimension appear to a low energy physicist as electromagnetic interactions between these charges. This, in fact, was the first attempt to unify electromagnetism and gravity. Kaluza's theory explained that both the gravitational and electromagnetic forces that we experience in four dimensions can be though of as arising from pure gravity in five dimensions.

This idea has been generalized and extended in the last decade with the hope of extending it to unify all of the forces of nature, all of which are gauge interactions similar to electromagnetism, together with gravity. Even more, this program offers the hope of explaining all other forces as consequences of gravity. To do so, of course, one has to imagine more than five dimensions, perhaps ten. But one can imagine with equal ease a world of ten dimensions, with nine spacial dimensions in which six of these are curled up into little circles so that they are unobservable, except for the remnants of gravity which would appear to us as the nuclear, weak and electromagnetic interactions.

The other new symmetry that has been much discussed is called supersymmetry. It is a marvelous extension of ordinary space-time symmetry which offers the possibility of giving a reason for the existence of matter, a reason, that is, for the existence of quarks and leptons. Photons and gluons can be said to exist as a consequence of gauge symmetry. In a similar sense one might hope to explain quarks and leptons as a consequence of supersymmetry. These symmetries, as a basis for unifying all interactions together with gravity, have been explored for the last fifteen years in the context of the relativistic, quantum-mechanical, field theories that we have used to construct the standard theory. This program, however, does not seem to work. Most importantly, the standard relativistic quantum mechanical framework that we have cannot incorporate gravity. We do not know how to join together Einstein's theory of gravity and quantum mechanics in a consistent field theory. This brings us to string theory, which does offer the possibility of a consistent unified theory of gravity.

8.2 String Theory

There are three significant achievements of string theory. First, string theory is a consistent logical extension of the conceptual structure of physics. Second, it produces a finite and consistent theory of quantum gravity. Finally, it might describe the real world.

There have been a very few times in the history of physics where consistent, logical and nontrivial extensions of the framework of physics have been successfully made. The best examples are the theories of relativity, quantum mechanics, and now perhaps string theory. Relativity is an extension of classical physics to the realm where the velocity of light, c, must be regarded as finite; quantum mechanics is an extension to the realm where Planck's constant, $\hbar$, is not zero. Perhaps string theory completes the trio of fundamental dimensional constants by extending classical physics to the regime where the Planck mass cannot be taken to be infinite.

String theory is a conservative extension of the logical framework of physics. It changes nothing save the attempt to base physics on point particles. It suggests that the fundamental constituents of matter, and indeed of spacetime, are not pointlike objects but rather extended stringlike objects. From then on string theories have been developed in a totally conservative fashion, without relinquishing any of the traditional structure of relativistic quantum mechanics. Nonetheless, proceeding in this conservative fashion, one has found structures which hint at the need for a serious reevaluation of the conceptual foundations of physics.

8.2.1 The construction of string theory

The basic idea of string theory is that everything is made out of strings. We used to think that the proton was an elementary point like particle and then, we learned that at distances of a Fermi (or 10^{-13} centimeters), it has structure. In fact, it is made out of quarks. At present energies we can only explore these distances, and the quarks look pointlike. Many people have wondered whether when we look at shorter distances, we will not see that each quark is made out of three preons or subquarks. But history does not always repeat itself. String theory says that if we look at a quark with a good microscope, that can see distances of 10^{-33} centimeters, we will not see smaller constituents, but rather the quarks will look to us like little closed strings.

To say that matter is made out of extended objects like strings represents an enormous increase in the complexity of the world. In traditional physics we deal with fields, or wave functions that describe particles as localized at space-time points. In string theory, instead, we have functionals—functions that depend not just on a point, but on a whole curve. There are many, many more curves than there are points. Point particle theories, field theories (which includes all theories of physics until recently) discuss one particle at a time; string theories, on the other hand, automatically discuss an infinite number of particles together.

A point particle has no structure, but a string can do many things and each vibration of a string (like the harmonics of a violin string) corresponds to a separate elementary particle. So, you might say that this is absurd. We only observe in nature a few dozen particles, why introduce a structure that contains from the very beginning an infinite number of particles, where are they all? One answer is that all, except for a very small number, of the particles described by the string, are very, very heavy and it would require enormous accelerators, that we do not have, to make such particles. That is why we have never seen them. But still, is not this machinery somewhat cumbersome? The answer is that we need an enormous increase in the degrees of freedom of the world if we want to have an enormous increase in the possible symmetries of the world.

String theory certainly possesses an incredibly large symmetry. What is this marvelous symmetry of string theory? All I can say is that I wish I knew the answer. It is one of the main concerns of current research to discover the full symmetry of string theory. The full symmetry is incredibly large, but it is hard for us to see because it is largely broken or hidden from us. What we do know is that string theory contains automatically, without our arranging for it ahead of time or adjusting anything, the largest symmetry that has ever been conceived by point particle physicists. It contains automatically the symmetries that are responsible for the emergence of gravity and the other gauge interactions of nature.

I cannot explain here how string theory works in detail, but I would like to emphasize that the way we have constructed string theory is a natural generalization of the way we construct theories of particles. For example, in classical physics (in other words, forget about quantum mechanics), particles move, as time evolves, along trajectories that are such so as to have minimal length. In other words, of all possible motions the actual motion is the one for which the path traversed has the smallest possible length. In flat space a particle, if there are no other particles around, will therefore move in a straight line. To quantize such a system is also straightforward, following Feynman's path history description of quantum mechanics. One calculates the probability amplitude for a particle to propagate from one point to another by summing over all paths connecting the two points, weighting the sum with a phase factor which is simply $\exp[\frac{i}{\hbar}S_{classical}]$, where $S_{classical}$ is the classical action (= invariant length) of the path.

The dynamics of strings is constructed by generalizing this same principle to extended objects. We say that strings as well, as they evolve in time, move along a trajectory in such a way that the area of the tube they span is as small as possible. Based on that principle one can construct both the classical and the quantum mechanical description of the propagation of strings.

When this was originally done people first studied the modes of vibration of both closed strings and open strings. They calculated their properties, *i.e.*, the masses and quantum numbers of the natural vibrations of these strings. The remarkable thing that they discovered was that closed strings always contained a particle that could be identified with the graviton, the quantum of gravity, and that open strings always contained a particle that could be identified with light rays, the quanta of gauge theories. This came out of the theory without having to be put in by hand. In fact, it was very embarrassing because originally string theory was constructed as a theory of nuclear force. As such there was no room for gravity or electromagnetism. It is only with the revival of string theory in the 1980's, as a unifying theory of everything, that this feature is very welcome. The other remarkable, and originally embarrassing, feature of string theory was that these theories were only consistent if one imagined that space time was 26 dimensional (Later, for the superstring, the dimension of spacetime was determined to be ten). Again, as a theory of the nuclear force this is absurd, but it is quite tolerable in the contest of a unified theory of gravity.

There is even a bigger difference between particles and strings when we come to think about interactions between particles in terms of the trajectories that describe their motion by saying that two particles, when they meet at the same point, have some probability of turning into a third particle, which then has some probability of splitting into two particles. This can produce a scattering process. The interaction is all concentrated at the point where the trajectories meet, a singular point of the diagram that describes the space-time evolution. The introduction of such an interaction at a point is an *ad hoc* and highly non-unique procedure, which is one of the reasons there are so many particle theories in the world. The situation is much more appealing in the case of strings.

How do strings interact? We would like to let strings interact locally as well by having two strings come together and when they touch at a point become a third string. We clearly can describe this by the so-called pants diagram. Think of horizontal slices through your pants and you will see that this describes the time history of two strings, coming together and forming a third string. However there is no particular point, no singular point, where the strings join together. Thus if one considers time histories in which the moving strings map out two dimensional surfaces which contain handles (to be precise one must analytically continue these surfaces from their original Lorentzian signature to Euclidean space to get a smooth manifold) they describe strings coming together forming a third string, which can then break into two strings. But unlike the particle picture there is no point that you can pick out and say "this is where the interaction took place." The surface is completely smooth. It is essentially because of this natural geometrical and unique way of introducing interactions that strings are so symmetrical and unique.

8.2.2 String theory of gravity

Not surprisingly these interactions contain gravity. This is to be expected once the graviton appears in the spectrum of the string, since according to general principles the only consistent interactions of a massless spin two meson must be those of general relativity, at least at low energies. Since strings contain both gravitons and gauge particles they yield a theory which contains and reduces to ordinary Einsteinian gravity and Yang-Mills theory at low energies. More than that, strings, for the first time, provide us with a consistent and well behaved quantum mechanical theory of gravity, thus providing us, at the very least, with a viable model of quantum gravity. It also naturally contains quantum gauge theories of precisely the type that we need to describe low energy physics. The two come together; in string theory you cannot turn off gravity and there is no real distinction between gravity and Yang-Mills interactions.

The final, quite remarkable, feature of string theory, which is appropriate for a unified theory of everything, is its incredible degree of uniqueness. In principle, there are no adjustable parameters, no numbers that you can change by hand. In physics, we always need three units (standards of length, of time, and of mass) in terms of which we express physical quantities. We can choose them to be Planck's constant, the velocity of light and the Planck mass that characterizes gravity. In string theory, in principle, one can calculate everything in terms of these dimensional units. For example, the so-called fine structure constant, which is a dimensionless number that characterizes the strength of electrical forces and is approximately equal to $\frac{1}{137.03}$, should be calculable in string theory.

Unfortunately, at the present time, this potential predictive power is not realized. The reason is that the fine structure constant, for example, always appears in the theory multiplying a dynamical variable, which so far is not determined by the dynamics that we know. Thus we are allowed to choose the value of this variable to be whatever we want, thus adjusting by hand the effective value of the fine structure constant. We expect and hope that we will get enough control over the dynamics of the theory to fix uniquely the ground state of string theory. If so then all the dynamical variable will be dynamically determined and then all dimensionless parameters will be calculable. In addition string theory can determine things that previously were undeterminable, such as the dimension of space-time and the size of the symmetry group of nature.

8.2.3 A realistic theory of the world

Finally, the reason string theories are so exciting is that they offer attractive, realistic theories of the world. Of course, the potential of string theory for providing a unique description of the real world is far from being realized. But there is no aspect of the real world that we have so far observed that is not contained in some sense in string theory. In particular, the heterotic string seems very promising.

The heterotic string is a strange object which combines two separate halves of two other strings to form a new kind of object. That is why it is called the heterotic string, heterotic means a vital hybrid. It is this combination that gives rise to a closed string that lives in ten dimensions and produces a gauge symmetry of an especially beautiful kind. According to the heterotic string theory the unbroken symmetry of the world is a symmetry group called $E_8 \times E_8$. It is based on the group E_8 which is a very exceptional group. Eli Cartan classified all continuous groups in the early part of this century. He discovered, in addition to infinite families of groups such as the rotation group in any dimension, five exceptional groups. The largest and in many ways the most beautiful of these is E_8.

I said that this was an attractive theory of the real world— but the real world is four dimensional and has a much smaller symmetry group. The heterotic theory has 496 gauge bosons, whereas we have only observed 12 so far. So if we are to make contact with reality we have the opposite problem to the one that I started with. We know, perhaps, the physics at 10^{19} GeV, and we have to make our way down to very low energies to understand what we see around us. In other words, if the heterotic string theory were correct, I could easily make predictions about what experiments would see at 10^{19} GeV, but to make predictions of what will be seen at 100 GeV might be as hard or harder than guessing this theory from low energy experiments.

What has been done so far has been to ask the simplest and broadest questions. The first one is "Are there any solutions of the theory that are not ten dimensional, but look more like our world of four dimensions?"

This means can we find a solution in which the six extra dimensions that we have never seen are curled up into some little space which is very small and the rest of space-time is four dimensional and flat? Since string theory is a theory of gravity this is a dynamical question. Gravity describes the dynamics of space-time and this question can be approached by looking at the equations of string theory. The answer that has been discovered is "Yes, there are solutions of this type." In fact, there is an embarrassment of riches, there are too many solutions and we don't understand why one is preferred over the others. But many do exist with four flat dimensions and with the other six curled into a small little space.

The second question is, "Are there solutions with the known forces, in which the symmetry that generates the forces is not the big $E_8 \times E_8$, which might be present at high energies, but a smaller one that could give rise to the strong, electromagnetic and weak forces that we observe at low energy?" The answer is very interesting, "The compactification of space-time from ten dimensions to four dimensions, by curving six of the dimensions into a little space, forces the symmetry to be broken down to a smaller symmetry. There exist many solutions with essentially the observed low energy symmetry."

Next, "Are there solutions with the matter content that we see about us, with the observed quarks and leptons?" The answer is "Yes." In fact very interestingly one can count the number of generations of leptons and quarks (electrons, neutrino, up and down quarks, muons, ...) simply by counting the number of holes in this six dimensional compact space.

This is also very pleasing. We can find solutions with the observed structure of space-time, with the observed forces of nature and with the observed content of matter.

The final question is, "Are there any problems?" And the answer unfortunately is again, "Yes." The main problem, in my opinion, is that there are simply too many solutions and we do not know enough about the dynamics of the theory to pick one out over the others. Probably, they are all wrong solutions, since all of them have symmetries that are so far unbroken.

When all of this was discovered, in the beginning of 1985, it was very heady. Many of us hoped that we could get away, as theorists sometimes have been able to do, with fudging the solutions to the remaining hard problems. The hope was that one could find a special solution among the large list that one could generate, and that that solution would look like the real world, and then one could put in supersymmetry breaking by hand. Then one would be in a position to see if low energy physics emerges, make predictions and do calculations.

Well, it is not that simple. There are millions and millions of solutions, none of them is particularly special and putting in supersymmetry by hand just does not work.

8.3 The Problems of String Theory

The real problem in string theory is not to find more and more solutions but to find the dynamics that picks the unique vacuum. It is not unusual to have more than one classical vacuum in a quantum mechanical system, especially in supersymmetric theories where there are often many "flat directions" in which one can vary the expectation value of some field without changing the energy of the vacuum. In such a situation one can develop a perturbation theory about each vacuum separately. However, unless perturbative instabilities arise, the correct vacuum involves non-perturbative physics— *e.g.*, tunneling. The trouble is that in string theory we have a similar situation but we do not know the Lagrangian, we do not even know what the coordinates are. All we have are the classical solutions and perturbation theory about them. All of our understanding of string theory is based on the following two ingredients: a rule for finding the classical vacua ($\equiv$ classical solutions) of some theory whose Lagrangian we do not know, the rules for calculating S-matrix elements in perturbation theory. This is not good enough.

One of the reasons it is not good enough is that perturbation theory diverges very badly. This is a recent result of Periwal and myself (our proof is for the bosonic string theory, mainly for technical reasons, but I believe it is true for all string theories). That is, if you construct string scattering amplitudes perturbatively in powers of the coupling g, (*i.e.*, $A(g) = \sum_n a_n g^n$), a parameter which in principle is determined by the dynamics but in practice, namely in perturbation theory, is a free parameter, then you discover that this perturbation expansion has zero radius of convergence. The coefficients a_n grow faster than $n!$, so not only is it divergent it is also not Borel summable. This is bad for perturbation theory, but very good for physics. If it were not the case that perturbation theory diverged then you could sum it, ending up with billions and billions of classical vacua, and the full quantum mechanical perturbations about them. Since these are all consistent theories, order by order in the coupling, you would end up with billions of consistent theories. All of them would contradict observation since all of them have unbroken supersymmetry and all of them have a massless dilaton. So it is good that we are not allowed to sum the series.

The meaning of a perturbation theory that diverges in this way is that the vacuum is unstable. One indication of this is that if you try to sum such a perturbative expansion (say by the Borel technique) you would find that the vacuum has complex energy, which indicates vacuum instability. Such series occur often in quantum mechanics, for example in the case of the double well potential. They also occur in totally consistent physical field theories, such as QCD. In both of these cases the non-Borel summable divergences are related to the existence of instantons, which are signals of quantum tunneling which mixes different vacua. Tunneling effects behave as $\exp[-\frac{1}{g^2}]$, which cannot be reproduced by a perturbative expansion. This divergence we regard as an indication that all of the string solutions that people have found are unstable at the quantum level due to nonperturbative effects. We must develop a formalism that allow us to go beyond perturbation theory, find the nonperturbative effects that will pick out a unique vacuum, break supersymmetry, generate a mass for the dilaton and all other good things.

The main trouble in going beyond perturbation theory is that we do not have the analog of the Lagrangian, or a second quantized Hamiltonian. We do not even know what the natural variables of string theory are nor what the natural description of configuration space is. All we have are these rules for constructing S-matrix amplitudes in perturbation theory. There have been many attempts to go beyond perturbation theory and to find the correct basic formulation of the theory. I think it is honest to say that none of these suggestions has gotten very far.

Perhaps the simplest set of open problems in string theory, where early progress is most likely, is in the area of classical string theory. String theory is a generalization of general relativity at both the classical and quantum level. Surely the simplest problem is to understand the structure of the classical limit of the theory. So far little is known. We have a prescription for finding stationary classical solutions of the theory— these correspond to two dimensional, conformally invariant, field theories— but beyond that we have no control over the classical theory outside of weak field, perturbative expansions.

Are all classical solutions of the theory in correspondence with two dimensional conformal field theories? What is the appropriate description of the classical configuration space, or the phase space of string theory? What is the topology of finite energy, or finite action solutions? How does one set up an initial value problem in string theory; how does one impose boundary conditions? What is a useful formulation of the equation of motion of classical string theory? What is the Lagrangian or Hamiltonian of the theory? Much formalism remains to be developed, many questions remain to be answered.

There are, of course, many many problems and an enormous amount of work is going on to try to understand the theory and to apply it to the real world. In fact, some people think that it might take decades, if not longer, to fully explore the structure of this theory. That we are far from a full understanding is clear if we note that one main problem is to arrive at an elucidation of the logical structure of string theory. We still lack a unifying principle that can guide us. Most of the advances that have taken place so far have occurred almost by accident. Einstein developed relativity by having an idea— the principle of equivalence— and then he constructed his equations. String theory has largely developed in the opposite direction, by discovering mathematical structures and then groping towards the physical concepts. Presumably, enormous advances will be required in order to obtain greater dynamical control and in order to make calculations and testable predictions. Nonetheless, one can speculate.

Let us return to a discussion of the problems that we face in string theory. One of the main problems is that our knowledge of string theory is limited to weak coupling and low energy. String theory has two relevant parameters, the Planck mass which divides low and high energies, and the coupling constant which controls the loop expansion. We might therefore distinguish four separate regions of string theory. The first regime is the regime of weak coupling and low energy. It is the only regime of string theory that we know how to deal with. Indeed, the only thing we know how to do in string theory is to calculate scattering amplitudes in weak-coupling perturbation theory and all of our understanding of physics is based on concepts that we have learned at very low energies compared to the Planck mass. In this regime strings look like particles and their dynamics can be represented by an effective local field theory.

This domain is very familiar, but it might not be the simplest domain for strings. Indeed it appears that the simplest regime is the regime of weak coupling and very high energy. Here strings behave as strings, and cannot be represented by particles. Their dynamics cannot be reproduced by an effective local field theory. Nonetheless, this might be the regime where strings reveal their deepest secrets. I argue that this is the case, that in this regime the dynamics of strings is very simple— albeit very nonfamiliar.

Finally, we have the regimes where the coupling is large, namely where we can no longer trust perturbation theory. These are the hardest to understand. Here we must come to grips with strong quantum mechanical effects as well as strong gravity. In the regime of low energies we know something from field theory about the interesting dynamical effects that can occur for strong coupling, for example confinement in gauge theories. We know very little about the interesting strong coupling effects that can occur in quantum gravity, although there are beginning to be recent ideas in this direction (for example topological field theory as a model of strongly coupled gravity). Since the only mass scale of the theory is the Planck scale it seems likely that the wave function of the vacuum will be dominated by physics on this scale. To understand the string vacuum *it will be necessary to have at least a qualitative understanding of strings in all four regions, for both weak and strong coupling and at both high and low energy.* Then we might proceed to construct, or guess, the structure of the vacuum, which is the first stage making contact with low energy phenomenology.

8.4 Strings at High Energies

What can we say about some of the other regions? To this end I have been exploring string scattering at very high energies. In string theory this means high compared to the natural scale— the Planck mass. Although this region is experimentally inaccessible nothing prevents you from doing Gedanken experiments, since in a Lorentz invariant theory one can boost particles to an arbitrarily high energy. The idea is to use the theory to see what happens.

The reason that strings are simple at high energies is because in that limit the sum over surfaces is dominated by a single saddle point. In fact you can think of the zero Planck mass limit as the semiclassical limit of the first quantized theory. What you find in this limit is universal, exactly calculable and very string behavior. The elastic scattering amplitude, at center of mass squared energy s and scattering angle θ, behaves (in N loop order) as

$$A(s,\theta) \approx f(s,\theta) \exp \frac{-s}{(N+1)m^2_{Planck}} \times \left[\sin^2\left(\frac{\theta}{2}\right) \ln \sin^{-2}\left(\frac{\theta}{2}\right) + \cos^2\left(\frac{\theta}{2}\right) \ln \cos^{-2}\left(\frac{\theta}{2}\right)\right],$$

falling very rapidly at high energy. The exponential term is universal, the prefactor depends on the details of the theory and the particles. The prefactor has been calculated for the bosonic string and for the heterotic string through two loops. Because of the fact that the path integral is dominated by a saddle point a very simple spacetime picture of the string scattering emerges. The sum over Riemann surfaces is dominated by a particular set of moduli and the space-time trajectory of the string is given by a particular classical solution. This trajectory is (for N loop scattering) $X^\mu_N(z) = \frac{i}{m^2_{Planck}(N+1)} \sum_k P^\mu_k \ln |z - a_k|$, where P_k is the momentum of the kth particle, and z runs over the two dimensional worldsheet with punctures at the points a_k. The cross ratio of the a_k's is given in terms of the scattering angle.

One interesting feature of this formula is that X^μ is imaginary. What does this mean? We are familiar with imaginary trajectories, they appear in the quantum mechanical evaluation of the semiclassical limit of classically forbidden processes. Since we can think of m_{Planck} as a first quantized version of Planck's constant, and since in the limit as $m_{Planck} \to 0$ the amplitude behaves as $\exp\left(-\frac{1}{m^2_{Planck}}\right)$, we also obtain the exponential suppression characteristic of classically forbidden processes. We might interpret this result as indicating that the scattering of strings is an exponentially suppressed process that can only be calculated by analytically continuing to imaginary time. This is not a tunneling process since the scattering is certainly allowed as a real process. (For example the velocity flow $\frac{dX^i(\sigma,\tau)}{dX^0(\sigma,\tau)}$ is real and physical). It is analogous to a particle that backscatters, with an exponentially small amplitude, at energies way above a potential barrier.

I have speculated that this result might imply that the high energy limit of string theory, which can be thought of as a first quantized classical limit, is such that no scattering is allowed. What does it mean to say that the high energy phase of the theory has a trivial S-Matrix? It clearly implies the existence of an enormous symmetry of the theory, since the individual momenta of each string mode is separately conserved. Perhaps at high energies the maximal symmetry of the string is restored and the maximal symmetry of the theory is so powerful as to render the theory trivial. *The nontrivial dynamics of the string is then a consequence of symmetry breaking.*

Another remarkable feature of the above saddlepoint is its large degree of universality. $X^\mu_{classical}$ is independent of which particle is being scattered, of the particular string theory one is considering, of the nature of the compactified dimensions and (except for the overall scale), of the order of perturbation theory (N). This implies the existence of certain relations between the scattering amplitudes of different string states.

These relations are a strong hint that something very simple is going on. This appears to be consistent with the previous speculation, that as $m_{Planck} \rightarrow 0$ all interactions disappear. Perhaps, when m_{Planck} turns on, only a few degrees of freedom are involved, leading to an infinite set of relations between the high energy scattering amplitudes.

In trying to interpret the simplicities that seem to emerge at high energies, particularly in terms of the restoration of a large symmetry, one is faced with a difficult problem. The feeling of many string theorists is that there is a much larger symmetry in the theory than that which is visible at low energies. Such a high energy symmetry is not going to be easy to discover. It is hard to find a new symmetry without being able to visualize the situation of unbroken symmetry. In this case the unbroken phase is probably not describable in terms of our normal concepts of spacetime. Perhaps space-time itself is only a crude and approximate concept, useful at distances large compared to the Planck length, but that when we probe the structure of space-time at distances of order the Planck length, space-time itself will melt away to be replaced by something else. There are many other speculations that one can make, but this is certainly the most interesting to explore; namely, what replaces space-time?

I shall present a handwaving argument that the behavior we see when we scatter strings at very high energies is an indication that spacetime loses its meaning at the Planck length. This is not happening because of strong gravitational phenomena, as many have speculated in the past (*i.e.*, space-time foam, wormholes, *etc.*), but rather because of stringy effects, which occur since when you use strings for probing spacetime you have to worry about the fact that the strings themselves are extended nonlocal objects.

Let us use the semiclassical description of string scattering at high energies to examine whether strings can be used to construct a microscope that would allow us to probe arbitrarily short distances. We have always used high energy collisions as microscopes. We have no direct way of probing the structure of space. We only do so indirectly through measurements of momenta and we thereby use high energy accelerators as powerful microscopes. The high energy scattering of the strings was described by a specific surface $X^\mu(\xi)$. If we examine, in a particular gauge, the nature of this space-time trajectory, we can conclude that the size of the strings at the time of collision, which defines the size of the interaction region or the size of the distances that can be explored using these probes, grows with increasing energy as $X_{\text{coll}} \approx \alpha' \frac{E}{N}$, where E is a characteristic energy of the scattering. This is unfortunate, given that we are trying to use strings as local probes, since we find that as we increase the energy in order to probe shorter distances, the effective size of the strings increases.

There is one problem with this conclusion, namely the saddlepoint trajectory is imaginary (*i.e.*, lies in complex Minkowski space). Nonetheless, even if the trajectory described above is not the real Minkowski trajectory, it should describe averages of the real trajectories and thus the average size of the interaction region.

At low energies (compared to the Planck mass) we gain in spatial resolution as we increase the energy of our microscope. However, it appears that when we get to the Planck length things change. The fact that our probes themselves are nonlocal strings becomes relevant. In fact the strings themselves expand with increasing energy. This contributes to the fuzziness with which we can resolve distances, so that the total fuzziness is

$$\Delta X \approx \frac{\hbar c}{E} + \frac{G_N E}{g^2 c^5} \; .$$

Here we have used the fact that (in the heterotic string theory) the string scale (set by α') and the Planck scale are related by g^2 . (We can think of g^2 as the fine structure constant and G_N is Newton's constant of gravity.) This is fortunate, since it means that with our assumption of small g^2, the strings never get within their Schwarzchild radius. Thus we do not have to come to grips with strong gravitational effects. The above smearing is not due to string gravity, rather it is due to the nonlocal nature of string probes. It implies that the minimal length that can be probed with strings is of order $\frac{1}{m_{Planck}}$! If this argument is valid then space-time could have no physical meaning below the Planck length.

There is another indication in string theory that the Planck mass is the minimal length. That is the amusing fact that you get the same theory if you compactify a string on a torus or alternatively on the dual torus, whose radii are related by $R \to \frac{1}{m^2_{Planck} R}$. (You must also interchange states with internal momenta, which is quantized in units of $\frac{1}{R}$, with states that wind about the torus, whose length is quantized in units of R.)

I expect that, in the final formulation of string theory, space and time will emerge only as an approximate concept, which is valid or useful only for certain approximations to the theory. Thus, for example, many people believe that the proper description of the configuration space of string theory is something like the space of all (cutoff) two-dimensional field theories. Conformal field theories are just the classical solutions of this theory. For these we have a space-time interpretation of the perturbative expansion of string scattering amplitudes. However, for a general two dimensional field theory there is no such interpretation. In such a formalism the space-time manifold would not be a primary concept, rather it would emerge as a useful description of string physics at low energies and weak coupling (where, presumably, perturbation theory is justified). If such a thing is true, or something even crazier is true, then we are in deep trouble, because it will be very hard to guess the conceptual framework that replaces our conventional notions of spacetime.

8.5 The Prospects for String Theory

Fundamental theories in elementary particle physics invariably have important applications in other areas of physics. These are welcome spinoffs from the exploration of a domain of nature which is increasingly removed from everyday experience and which is unlikely to have direct technological applications.

Quantum field theory, the basis of our theories of the electro-weak and strong interactions, has many such applications. Renormalization theory plays a crucial role in the modern understanding of critical behavior; the phenomenon of spontaneous symmetry breaking is important for superconductivity and other many-body phenomena as well as for particle physics; and recently gauge theories are proving of great value in many condensed matter problems. In addition, the structure of non-Abelian gauge theories turns out to be of enormous mathematical interest. Physical ideas from gauge theory have already had a great impact on the exploration of the geometry of three and four manifolds and more is promised.

Will string theory have such interesting and important applications and spinoffs? It is of course too early to tell, but we might hope that this will be the case. These might occur in other areas of physics, such as in surface physics, which studies the behavior of real surfaces and phase boundaries, embedded in three dimensions; or in QCD, where one believes that the low energy structure of hadrons might be described in terms of an effective string theory. To date, however, the most important spinoffs have been in mathematics. String theory has excited the mathematical imagination of many, since it appears to have essential connections to many, hitherto disconnected, areas of mathematics. This is very exciting but also sobering. If string theory raises deep, fundamental problems in modern mathematics then we are likely to be able to enroll the best mathematicians in the exploration of this theory. On the other hand it means that much of the mathematics that we need for developing the theory does not yet exist.

What are the prospects for the future? Clearly the times are hard for both experimentalists and theorists. They are hard because the standard model works very well. We are paying the price of our success with the lack of experimental surprises and the absence of clues as to where to go from here. Theorists are pretty good at extrapolating from what we already know to guess where new physics might arise, where problems will appear, where new thresholds will show up, even if they are very bad at guessing the correct physics at those new thresholds. Thus, after the Fermi theory of the weak interaction people could guess that there had to be new physics at 100 GeV, even though they had no idea what that new physics would be. As far as we can tell, if we use the standard model to extrapolate the known forces, we find that new physics— fundamentally new thresholds— will only appear at extraordinarily high energies, 17 orders of magnitude removed from where we presently do experiments. It is only true of the forces that the extrapolation is straightforward. The matter content is harder to extrapolate. Heavy particle thresholds and new symmetries that are broken at low energies turn on and off as powers of the energy. However, if we regard the nature of the forces as more fundamental than the existence of another family or two of quarks and leptons, or even the existence of another gauge group, then truly new physics will only appear at the Planck mass.

Of course there are likely going to be many new experimental discoveries in between the TeV region and the Planck energy. All unified theories, string theory especially, predict that there will be lots of new stuff in this region. In addition to the standard stuff, there might be even stringy stuff. But it will not be the fundamental modification of the laws of physics that we believe will show up at the unification scale or at the Planck mass. As one who used to hunger after every bit of data from deep inelastic scattering, which I was convinced would yield the secret of the strong interaction, as it did, I desperately long for relevant experimental data that would give us hints about the unification of the forces. I also know that one new surprising and reliable experimental result is worth a ton of theoretical speculation. Where should we look? Of course, there are the standard places to look, where one is bound to find new physics— such as the Higgs sector. More interesting is the discovery of supersymmetry, which hopefully will occur with the SSC and which would be one of the fundamental discoveries of the twentieth century. These results would be of crucial importance to string theorists. In addition maybe there will be surprises. String theory suggests that there might be extra gauge interactions, which might be visible at low energies. There might also be massive stringy remnants of Planck mass physics with strange properties. There may be all sorts of strange things in the scalar sector of the theory, with gravitational strength couplings. After all the dilaton remains one of the biggest mysteries in string theory. If it develops a small mass and couples coherently to matter it could be detectable as an attractive correction to gravity. The indications for this would be similar to those that have been advanced for a "fifth force", however this dilaton induced force would necessarily be attractive. A fifth force that is attractive, would be very attractive; whereas a fifth force that is repulsive is very repulsive. A repulsive fifth force implies the existence of some light vector meson with a dimensionless coupling of order 10^{-40}. This would be a strong blow against string theory as well as any other possibility of unification (since at the unification scale all dimensionless couplings should be of the same order of magnitude.)

Can we succeed in making this extrapolation? You can easily give arguments both pro and con? The arguments against success are easy— history teaches us that without direct experimental clues and tests theorists tend to go wrong. In favor we have the fact that we are lucky to possess a good starting point for extrapolation. We have a very comprehensive theory of low energy physics that seems to work very well. It is not easy to extend such a theory without contradiction, so consistency is a guide. We also have the incredible luck of knowing an important bit of Planck mass physics— namely gravity, which turns off like a power of the energy at low energies but fortunately couples to large objects. Indeed, much of the motivation to construct unified theories is based on the desire to combine gravity with the other things that we know from low energies. Gravity is our only direct handle on the Planck mass physics. Finally, we can be lucky.

Is there any chance of direct experimental verification of string theory? I certainly do not know, but I do not think it is impossible. For example some people are very disturbed by theorists who imagine extra dimensions that cannot be seen directly. I did some historical research and discovered that the first person to object to more than three dimensions was Ernst Mach, who wrote in 1883 that "Spaces of more than three dimensions may be used, but it is not necessary to regard these as anything more than mental artifices." What was Mach talking about? He certainly was not aware in 1883 of Kaluza-Klein nor of string theory. What he was referring to was the interest in higher dimensional theories which was provoked by the mathematical work of Riemann. He noted that "the use of the fourth dimension was a very opportune discovery for the spiritualists and for the theologians who were in a quandary about the location of hell". I certainly hope the spiritualists don't find out about ten dimensions!

Mach's criticism should provide no solace for the opponents of string theory because on the same page he noted that atoms, which cannot be perceived by the senses, are "these mental expediences have nothing to do with the phenomenon itself." Mach did not believe in atoms because he thought that *one* could not observe them. Little did he know that only 22 years later Einstein and Smolokowski would realize that the observation long before of the botanist Brown would provide us with indirect evidence of the existence of atoms, that couldn't be perceived by the senses directly but only via their effect on small particles suspended in a solution. Maybe this is how strings, or extra spatial dimensions, will eventually show up, as effects that might be lying around today and that we cannot yet recognize.

A more interesting and practical question is: even if we were to succeed, how long will it take? This is hard to predict. Let me give a mountain climbing analogy. It used to be that as we were climbing the mountain of nature the experimentalists would lead the way. We lazy theorists would lag behind. Every once in a while they would kick down an experimental stone which would bounce off our heads. Eventually we would get the idea and we would follow the path that was broken by the experimentalists. Once we joined our friends we would explain to them what the view was and how they got there. That was the old and easy way (at least for theorists) to climb the mountain. We all long for the return of those days. But now we theorists might have to take the lead. This is a much more lonely enterprise. In the past we always knew where the experiments were and thus what we should aim for. Now we have no idea how large the mountain is, nor where the summit is. Thus it is very hard to predict how long it will take to make substantial progress.

The progress in string theory is good and healthy. Those of us working in the field feel that it is an exciting area in which one is making serious progress. But at the same time the mountain looks like it is getting bigger and bigger. There are no contradictions in the theory; there are no insoluble problems. It is just that as we learn more the theory acquires more of a grandiose structure. So it is very difficult to predict how long it will take to make contact with experiment.

9

Gamma Ray Stars

M. Ruderman
Department of Physics
Columbia University
New York, NY 10027

Various techniques and results of searches for γ-rays from stars are surveyed briefly. Most of the γ-ray sources are otherwise unidentified and where the source is a known one the mechanism for γ- ray production is not yet agreed upon. A model is considered for the origin of γ-rays from solitary, rapidly spinning, strongly magnetized neutron stars such as the Crab and Vela radiopulsars. The expected evolution of such γ-ray sources suggests that they may evolve into stars which will closely resemble the strongest unidentified sources and may ultimately evolve into the observed family of transient γ-ray burst sources.

9.1 Introduction

Separating us from the rest of our Universe are 10^3 g cm^{-2} of atmosphere. Visible light and high frequency radio waves pass through it relatively easily but it is opaque to most of the rest of the electromagnetic spectrum. As a result our view of the universe has been extremely restricted until the past several decades when platforms with appropriate detectors have been carried above the atmosphere. The last electromagnetic window to be opened (and it is still only slightly open) is the γ-ray one.

A low energy cosmic γ-ray is Compton scattered before penetrating through much less than 10^{-2} of our atmosphere. An ultrahigh energy γ-

ray will initiate an electron-positron shower when passing through less than 10^{-1} of it. To observe cosmic γ-ray sources either a detector must be lifted above the atmosphere or the atmosphere itself must be incorporated into the detection technique. This is being accomplished in a variety of ways:

1. **Balloons, Rockets, and Satellites** [1]. The limited detector areas and weights have, so far, limited observations to relatively strong sources whose γ-ray luminosity (L_γ) exceeds the total optical luminosity of the sun ($L_\odot$) for sources at typical Galactic distances and to γ-rays of energy less than several GeV. The angular resolution achieved is typically about 1° at a GeV and of order 10° at an MeV.

2. **Atmospheric Cerenkov Technique** [2,3,4]. Electron-positron showers from γ-rays with energies $\gtrsim 5\cdot 10^{11}$ eV give enough optical Cerenkov radiation to allow ground based detection under special conditions. These include dark clear nights, an intense already identified source with $L_\gamma \gg L_\odot$ and, usually, a known source periodicity to allow discrimination in the presence of a very much greater background of γ-rays from the decay of π^0-mesons created by cosmic rays near the top of the atmosphere. In principle, these background γ-rays from $\pi^0 \to \gamma + \gamma$ result in a Cerenkov light pattern at sea level sufficiently different from that caused by a single incident cosmic γ-ray that up to 98% of the background can be eliminated. The angular resolution for detecting cosmic γ-ray sources from the Cerenkov light they make in the atmosphere is about 1°.

3. **Extensive Air Showers** [4]. For incident γ-ray energies $\gtrsim 10^{15}$ eV the resulting $e^\pm$ shower can reach the ground and be detected directly. As in the Atmospheric Cerenkov Technique, the angular resolution $\sim 1^\circ$, and a source in our Galaxy must have $L_\gamma \gg L_\odot$ and, generally, an already well-determined periodicity. The very large background of cosmic-ray produced γ-rays in our atmosphere will be accompanied by μ-mesons from the decay of the π-mesons which are coproduced with the π^0-mesons. It has, therefore, been expected that $e^\pm$ showers from atmospheric π^0- production would be accompanied by an order of magnitude more μ-mesons than are present in $e^\pm$ showers initiated by incident cosmic γ-rays. The diminishing γ-ray fluxes with increasing γ-ray energy have limited the air shower detection technique to γ-ray energies $\lesssim 10^{16}$ eV.

4. **Induced Atmospheric Fluorescence** [4]. For γ-rays of energy exceeding 10^{17} eV the fluxes from possible sources are so weak that very large regions of the sky must be simultaneously monitored for the fluoresence caused by these γ-rays. Each of the $10^8 - 10^{10}$ $e^\pm$ of the shower initiated by such a γ-ray gives of order 10 N_2 fluoresence optical pho-

tons per meter which may, on dark cloudless nights, be detected by a "fly's eye" detector which sees the whole sky.

All of the above methods for detecting localized γ-ray sources have yielded of order 10^7 photons (about as many as an optical telescope collects from a typical star in 10^{-1} s) at a cost of $10 per photon. Such expensive but relatively meager γ-ray photon data have, of course, been pushed very hard to yield information about possible sources. We turn next to a brief summary of the results.

9.2 "Point Sources" of Cosmic Gamma-rays

Two point-like (for $\gtrsim 1^\circ$ resolution) γ-ray source families were anticipated before any were detected. Cosmic ray protons within dense interstellar clouds produce π^0-mesons which decay into pairs of γ-rays. Several such sources have been observed. A second anticipated source was the Crab supernova nebula. Its optical light is synchrotron radiation from 10^{12} eV electrons (positrons). These same very energetic $e^-(e^+)$ will collide with the optical photons and exchange energy with them (inverse Compton scattering) to produce 10^{12} eV γ-rays. The Crab nebula has indeed been confirmed as such a γ-ray source [5].

Other published reported sources include the following:

1. **Active galaxies including M 31, Cen A, and the quasi-stellar object 3C 273**. Claimed observations have been in the range 10^2 MeV–1 GeV (HE), at 10^{12} eV (VHE), and at 10^{15} eV (UHE), but all have been rather marginal and unconfirmed [2].
2. **A variable source of 0.51 MeV e^+–e^- annihilation γ-rays from the Galactic Center Region**. This has been attributed to processes around a massive black hole or to an accreting neutron star in the close binary system GX1+4 [6,7].
3. **Radiopulsars**. Of the 400 observed radiopulsars four have been reported as γ-ray sources. The Crab pulsar, the 935 year old neutron star which powers the Crab supernova remnant nebula, is a source of energetic quanta from optical frequencies up to 10^{12} eV [1,2,8]. It is most powerful in the hard x-ray regime. The 10^4 year old Vela pulsar has not been observed as a pulsed x-ray source but resembles the Crab pulsar in its emission above a few MeV. Its reported VHE emission at 10^{12} eV has not yet been confirmed. VHE emission has also been reported but unconfirmed for the radiopulsars PSR 1953+29 and PSR 1801-23 [2]. All four of the claimed γ-ray radiopulsars have very short spin periods in the range 10^{-2}–10^{-1} s.

4. **Unidentified Galactic Sources** [1,9]. The Cos B γ-ray satellite detected about 20 distant HE sources (10^2 MeV–1 GeV) in the Galactic disk which do not coincide with otherwise identified objects. Their γ-ray luminosities are in the range $5 \cdot 10^{36} > L_\gamma > 4 \cdot 10^{35}$ erg $s^{-1} \sim 10^2 L_\odot$, much brighter in γ-rays than the Crab or Vela radiopulsars. There are not enough neutron stars in the Galaxy much younger than the Crab pulsar to provide plausible candidates from this family for the unidentified Cos B sources.

5. **X-ray pulsars in accreting binaries**. Of the approximately 40 neutron stars in accreting binaries with measured spin and/or orbital periods a half dozen have been claimed as sources of γ-rays with energy $\gtrsim 10^{12}$ eV: Cyg X-3, Her X-1, Vela X-1, Cen X-3, LMC X-4, and 4U0115+63 [2]. None are among the established Cos B sources at lower γ-ray energies. The two sources reported most often by different groups are Cyg X-3 and Her X-1 at 10^{12} and 10^{15} eV. Each observation has been rather marginal but the total number is statistically impressive. The neutron star binary systems which have been reported as γ-ray sources do not form a special subset with respect to spin or orbital period, x-ray luminosity, or size of the secondary (*cf.*, however, Ref. [10]). The main argument for interpreting the initiator of the detected atmospheric showers as γ-rays is the preservation of source period in the received signal after flight times greatly exceeding 10^4 years. The emitted particles should, therefore, be neutral, almost massless relative to their energies, stable enough to survive the flight, and able to penetrate through more than 10^{-2} g cm^{-2} of interstellar matter without degradation. Among presently known particles only γ-rays and neutrinos are possibilities, but at 10^{12} eV (and probably even above 10^{15} eV) neutrinos would not cause the $e^\pm$ air showers in the atmosphere. Very disquieting, however, is the fact that the claimed signals are not μ-meson poor at 10^{15} eV as expected for γ-ray initiated showers. In addition, that discrimination against cosmic ray proton induced showers which worked so well in increasing the signal to background ratio in the Atmospheric Cerenkov Technique detection of 10^{12} eV γ-rays from the Crab nebula has not been similarly successful for Her X-1 observations. Thus instead of corrobration of certain x-ray pulsars in binaries as VHE and UHE γ-ray sources we have counter-indications.

6. **Gamma-ray Burst Sources (GRB's)** [11]. Several times a week transient γ-ray emission lasting 10^{-2}–10^2 s is observed from unknown sources. While they are observed these can be the strongest γ-ray sources in the sky. When they are dormant nothing is detected from them in x-rays or optically. There seem to be at least two different populations of GRB's. One appears to be almost isotropically distributed

with a hard γ-ray spectrum extending to perhaps 10^2 MeV or beyond. These do not repeat on a time scale of 10 years or more. The other have much softer spectra with not much energy above an MeV and can repeat often but not periodically. There is now considerable evidence that strongly magnetized neutron stars constitute a major source (if not the only one) of GRB's. Cyclotron absorbtion features and submillisecond time structure indicate sources of dimension $\lesssim 10^7$ cm with magnetic fields $\sim 3 \cdot 10^{12}$ G. If the GRB population is mainly in the Galactic disk the source population is $\gtrsim 10^5$. Neutron stars in close accreting binaries which are not detectable as accreting x-ray sources but in which the secondary is close enough to cause explosive γ-ray emission probably constitute too small a population to include most GRB candidates. Single neutron stars seem the most promising possibility.

We shall consider in the next sections an interpretation of that part of the γ-ray source data which seems both least fragile and least likely to be greatly extended by new observations in the next several years. It is based upon the Crab and Vela pulsar observations. Models for them suggest speculations about some of the unknown sources such as the unknown Cos B sources of **4** and the GRB's of **6** above. However, the models and speculations are and must be very subjective since there is not yet a consensus in the community about just what is happening in these γ- ray stars.

9.3 Energetic Emission from the Crab and Vela Pulsars

About 10^{-3} of the total spin-down power of the Crab pulsar is emitted in hard x-rays and γ-rays (assuming the rotating emission beams are shaped much more like fans than cones to account for the apparently high probability for observing them in both the Crab and Vela pulsars). The "light curves" at 10^2 MeV and 10^3 GeV are shown in Fig. 9.1 from Ref. [8].Here the average background counting rate is indicated by the horizontal dashed line and the arrows indicate the arrival time for the main pulse and subpulse at all lower energies. Two unambiguous pulses exist at the same phase as those at 10^2 MeV in the entire range from IR to GeV implying that all of this radiation is emitted from the same place near the neutron star. At 10^{12} eV the signal is only slightly above that from the cosmic ray background but additional support comes from a pulse phase exactly the same as that of all the lower photon energy pulses. The presence of the second pulse in these 10^{12} eV observations is much more problematic.

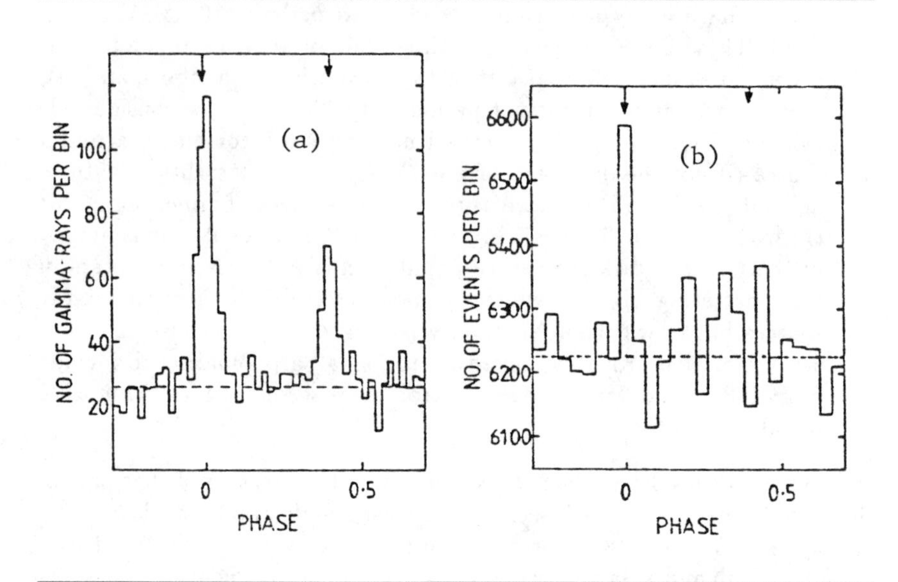

FIGURE 9.1
Light curves for the Crab pulsar at a) 10^2 MeV and b) 10^{12} eV (after Dowthwaite *et al.* [8]).

The Vela pulsar shows a similar double pulse structure with the same 0.4 phase separation as that for the Crab. This phase separation equality, however, appears to be coincidental. PSR 0540-693 is a radiopulsar similar in age, spin, and magnetic field to those of the Crab [12]. Because it is much further from us only its optical and x-ray emissions have been observed. Its optical light curve is shown in Fig. 9.2 [12]. A double pulse structure is again apparent but the pulse phase separation has narrowed to only 0.2. Thus it seems difficult to accept the common interpretation of double pulses as emission from the north and south pole of a neutron star magnetic dipole as each is rotated into view– a model which is comfortable with a phase separation near 0.5.

A final clue to the mechanism of the γ-ray emission may be the very large $e^{\pm}$ flux ($\gtrsim 10^{39} s^{-1}$) which seems to be injected by the Crab pulsar into its surrounding nebular [10,13].

From the above observations and considerations about the possible structure of the near environment around such rapidly spinning strongly magnetized (surface dipole $\sim 10^{12}$ G) neutron stars several features emerge about properties of the source of their γ-ray emission [10]:

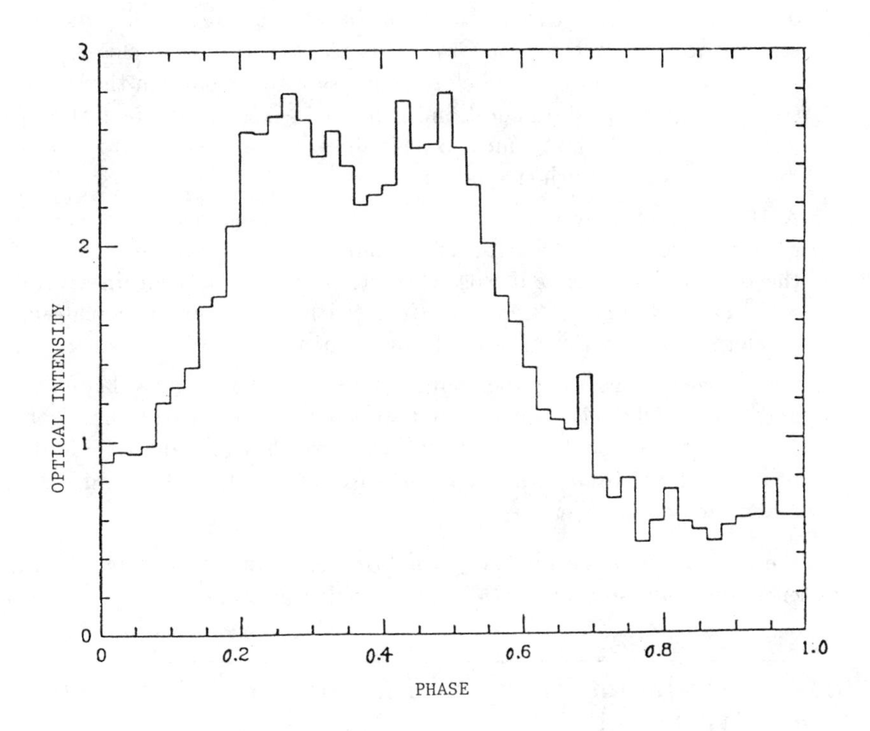

FIGURE 9.2
The optical light curve of PSR 0540-693 [12].

1. There must be a powerful accelerator near the star which can give e^-/e^+ energies exceeding 10^{12} eV. If the accelerator lies within the "light cylinder" of the star (where the corotation speed reaches c) the accelerator must be accomplished by a strong electric field ($\vec{E}$) directed *along* the magnetic field ($\vec{B}$). Ultrarelativistic acceleration perpendicular to $\vec{B}$ could be accomplished only by an $|\vec{E}| \geq |\vec{B}|$ which could not be achieved near the star by rotation.

2. The maximum net current flow through the Crab's accelerator $\sim 10^{34}$ s^{-1} of e^-, or e^+ in the opposite direction. A larger current would itself give a larger $\vec{B}$ near the light cylinder than that of the stellar dipole and quench the accelerator.

3. The potential drop of the accelerator along $\vec{B}$ must exceed 10^{14} volts

to supply the needed γ-ray power.

4. The location of the accelerator must be at least about 10^8 cm (10^2 stellar radii) away from the Crab neutron star. If it were closer, 10^{12} eV γ-rays could not escape before conversion to $e^{\pm}$ pairs in the strong stellar magnetic field. In addition the needed $\gtrsim 10^{12}$ volt potential drop would be quenched at much lower values by various $e^{\pm}$ production mechanisms in a much stronger $\vec{B}$.
5. A 10^{39} s^{-1} $e^{\pm}$ pair production could be a consequence of $\gamma + \gamma \rightarrow e^+ + e^-$ around the neutron star from the crossing γ-ray beams of the observed flux only if this interaction takes place within several $\times 10^8$ cm of the star. To be consistent with **4** this would indicate an accelerator about 10^8 cm away from the neutron star.
6. The success in observing the double pulses of the Crab and Vela pulsars and the small phase separation between those pulses in the Crab-like pulsar PSR 0540 are more consistent with a rotating double fan beam structure than with radial cones from above the north and south magnetic poles.

We turn now to consider how a pulsar model might account for the existence and suggested properties of the needed accelerator.

9.4 A Model for Energetic Emission from the Crab Pulsar [10,14,15]

Within a spinning magnetized neutron star

$$\vec{E} = \frac{(\vec{\Omega} \times \vec{r})}{c} \times \vec{B} \,, \tag{9.1}$$

where $\vec{\Omega}$ is the star's angular velocity. It follows that

$$\vec{E} \cdot \vec{B} = 0 \tag{9.2}$$

(in both the laboratory and rotating frames). Equation (9.2) fails at the stellar surface and beyond if the star is in a vacuum. However, $(\vec{E} \cdot \widehat{B})$ would then exceed 10^{12} Volts cm^{-1} outside the Crab and Vela surfaces so that charge would be pulled from those stars no matter what their surface structure. That charge flow would continue until Eq. (9.2) is satisfied outside the star as well as inside. Equation (9.1) would then also hold outside as well as inside and the outside plasma would corotate with the star ("corotating magnetosphere"). The plasma pulled outside the star would act like an extension of the star. The plasma charge density (neglecting any change in $\vec{B}$ from plasma currents)

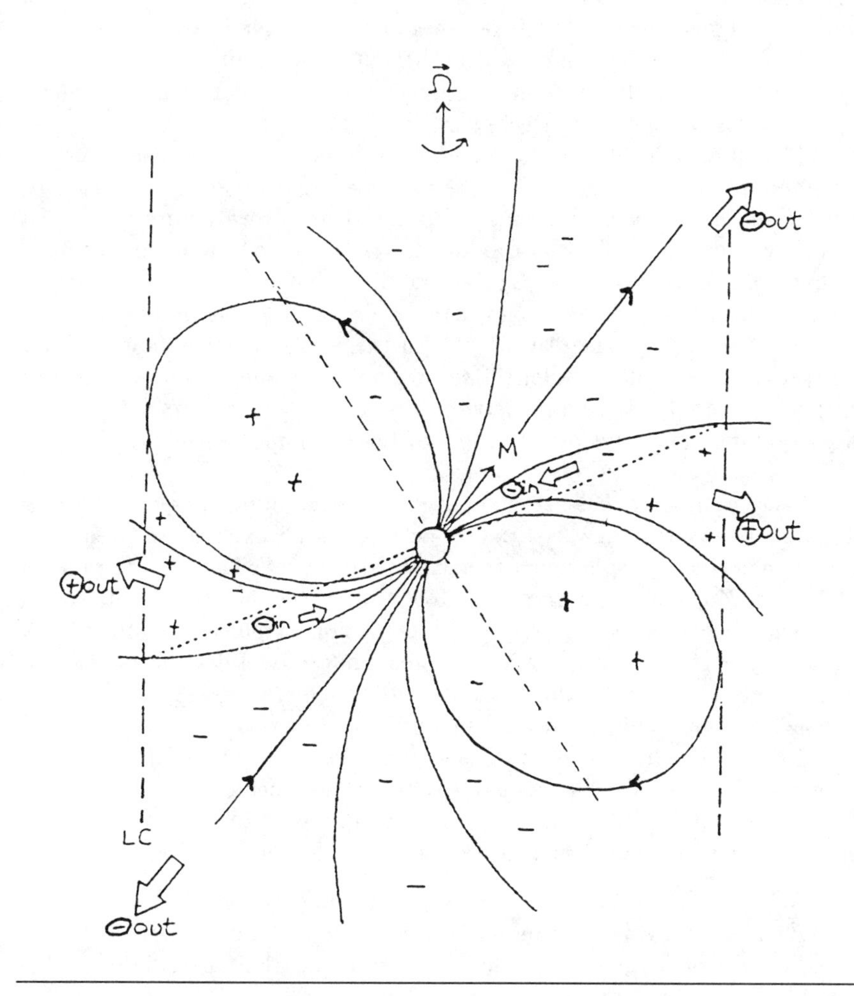

FIGURE 9.3
The corotating magnetosphere of a non-aligned neutron star with magnetic dipole moment $\vec{M}$ [10].

$$\rho = \frac{\nabla \cdot \vec{E}}{4\pi} \sim \frac{\vec{\Omega} \cdot \vec{B}}{2\pi c} \tag{9.3}$$

would be "charge separated", consisting exclusively of negative or positive charge according to the sign of $\vec{\Omega} \cdot \vec{B}$ in different regions of the magnetosphere as indicated in Fig. 9.3. Here the sign of the charge density is indicated by + or − and the dotted lines are surfaces where $\vec{\Omega} \cdot \vec{B} = 0$ and the magnetosphere charge vanishes. The arrows indicate the assumed current flow of + and − on open field lines ($\oplus$ and $\ominus$) out through the light cylinder (LC) and into the star. Strong charge depletion is expected where $\ominus$ and $\oplus$ flow in opposite directions.

The difficult problem, on which there is not yet a consensus, is understanding in detail how to describe a corotating magnetosphere near, at, and beyond the light cylinder at $r = c\Omega^{-1}$ where corotation must fail. If a cylindrical insulating sheath of radius $\ll c\Omega^{-1}$ isolated the magnetosphere from the rest of the Universe, Eq. (9.2) could be achieved everywhere within the sheath and no accelerator or γ-ray emission from the magnetosphere would be expected. When such a sheath is removed particle flow along $\vec{B}$ out through the light cylinder should begin. Charge depletion in the outer magnetosphere in the volume between the surface where $\vec{\Omega} \cdot \vec{B} = 0$ and the light cylinder can result in the failure of Eq. (9.3) and consequently of Eq. (9.2).

Such a charge deficient region ("gap") in the outer magnetosphere is a common feature of a variety of models and is an expected consequence of assumed magnetospheric current flow patterns. We shall use the model of Cheng, Ho, and Ruderman [10,13,14] for further analysis, but expect the general features of the resulting description of pulsar evolution with neutron star spin-down to be more robust than model details. Because of charge separated current flow a stable outer magnetosphere gap forms and expands in the regions indicated in Fig. 9.3. When such a gap begins to grow, the accelerating $\vec{E} \cdot \vec{B}$ in it also increases until the resulting e^-/e^+ accelerator becomes strong enough to support so much $e^\pm$ production that the supply of pair plasma quenches further gap growth. For the Crab pulsar this is the result of a series of processes:

a. An $e^\pm$ pair produced in the gap is instantly separated by the large $\vec{E} \cdot \vec{B}$ there which accelerates the e^- and e^+ in opposite directions. Because of magnetic field line curvature each lepton radiates multi-GeV curvature γ-rays.

b. These are converted into $e^\pm$ pars in collisions with keV x-rays [from **d** below]. Pairs created in the gap repeat process **a**.

c. Pairs created beyond the gap boundary lose their energy to synchrotron radiation (optical to MeV) and to higher energy γ-rays from Compton scattering on the same x-ray flux responsible for the pair creation.

d. The x-ray flux from the synchrotron radiation of **c** is that which initially caused the curvature radiation γ-rays of **a** to materialize and the inverse Compton scattering of the pairs in **e**. Since the entire series of processes is powered by the extremely energetic e^- and e^+ of **a** moving in opposite directions within the gap, all of the resulting fluxes of photons and $e^{\pm}$ pairs are also oppositely directed. The pairs and γ-rays moving in one direction then interact mainly with the x-rays moving oppositely.

e. A third generation of $e^{\pm}$ pairs comes from partial materialization of the crossed γ-ray beams from **c**.

The above processes **a-d** "bootstrap" the creation of an $e^{\pm}$ plasma until enough is produced to form a gap boundary layer which quenches further growth. The charge depleted gaps where $\vec{E} \cdot \vec{B} \neq 0$ and e^-/e^+ are strongly accelerated along $\vec{B}$ are shown crosshatched in Fig. 9.4. Pair production from $\gamma + \gamma \rightarrow e^+ + e^-$ sufficiently fill all other regions of the outer magnetosphere to satisfy Eqs. 9.3 and 9.2 despite current flow. (Sustained current flow and gap formation are expected only on "open" $\vec{B}$ field lines–those that pierce the light cylinder. Most γ-rays from relativistically accelerated e^-/e^+ there are emitted almost tangentially to the local $\vec{B}$. Thus the γ-rays from the crosshatched accelerator region may supply $e^{\pm}$ to all other open field line regions of the outer magnetosphere but not vice-versa.) Most of the γ-ray emission which escapes the magnetosphere comes from synchrotron radiation and inverse Compton scattering along the border region between the accelerating gap and the rest of the open field lines of the outer magnetosphere. Because of the symmetry between the flow of e^- in one direction and e^+ in the other, together with that between the predominantly dipolar $\vec{B}$ from both poles near the light cylinder, the γ-ray emission is always in the form of four fan beams as shown in Fig. 9.4 and any observer will see two of them (latitutudinal width $\sim \frac{\pi}{2}$) with a phase separation (partly because of different travel times and partly from abberation) which depends on the tilt between the dipole ($\vec{M}$) and $\vec{\Omega}$ and on the direction to the observer. Crossed γ-ray beams fill the rest of the magnetosphere with pairs from $\gamma + \gamma \rightarrow e^+ + e^-$. A very small fraction of the e^- or e^+ are reversed by weak $\vec{E} \cdot \vec{B}$ to keep $\vec{E} \cdot \vec{B} \sim 0$ everywhere except in the accelerator gaps. The dotted line on which $\vec{\Omega} \cdot \vec{B} = 0$ vanishes separates the regions in which net current is carried by e^- moving in toward the star from those in which e^+ is moving out, the needed e^+/e^- source coming from local $e^{\pm}$ production. Photons of 10^{12} eV in beam 1 may reach an observer without absorbtion. Photons of this energy or higher in beam 3 which must pass through a stronger $\vec{B}$ before leaving the magnetosphere may be converted first into $e^{\pm}$ pairs so that the subpulse peak in Fig. 9.1b may be absent.

The $e^{\pm}$ production from mechanism **e** above has been estimated to produce about 10^{39} $e^{\pm}$ per second which suggests that these may be the source of those further accelerated and ultimately injected into the nebula

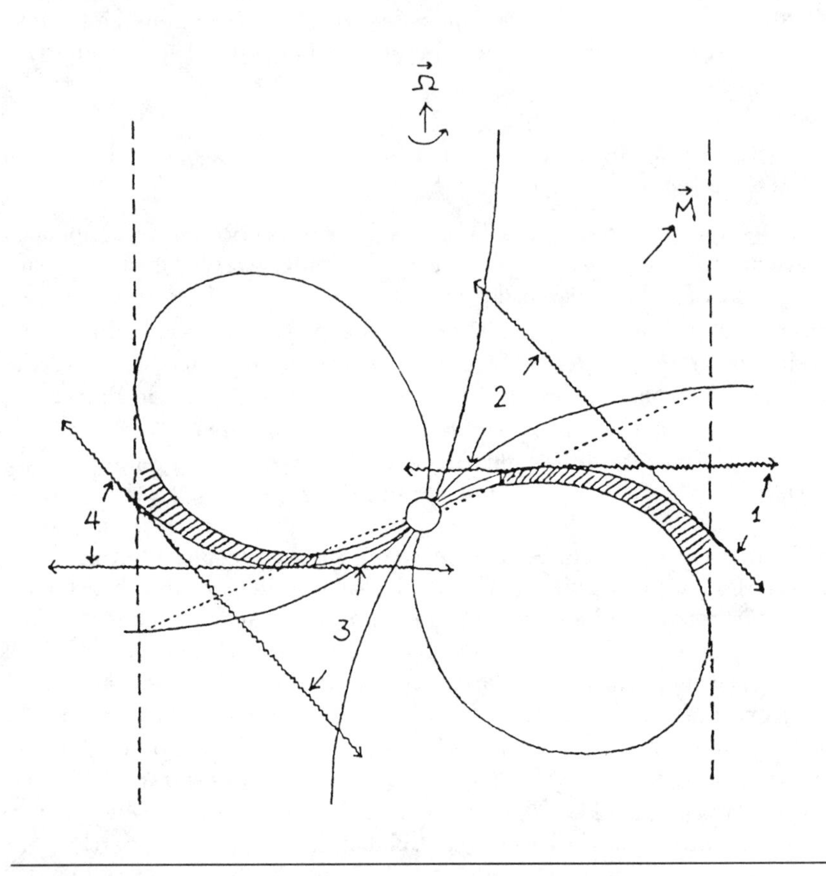

FIGURE 9.4
Outer magnetosphere "gap" accelerator regions from the current flow of Fig. 9.3 [10,14].

in which the pulsar is imbedded. The calculated spectral shape from $1-10^9$ eV which results from mechanisms **a-d** fits the crab pulsar observations in Fig. 9.5. This shape depends mainly only on the width of the gap in Fig. 9.4, but is not very sensitive to it or other parameters [14,15]. The local B is $\sim 10^6$ G. The regimes where synchrotron (SYN) and inverse compton scattering (ICS) occur are indicated. Some 10^{13} eV emission (not shown) is expected from inverse Compton scattering of gap e^-/e^+ on optical photons in the gap from process **c** and from the synchrotron radiation by the $e^\pm$ pairs these ultra high energy γ-rays make beyond the gap boundary.

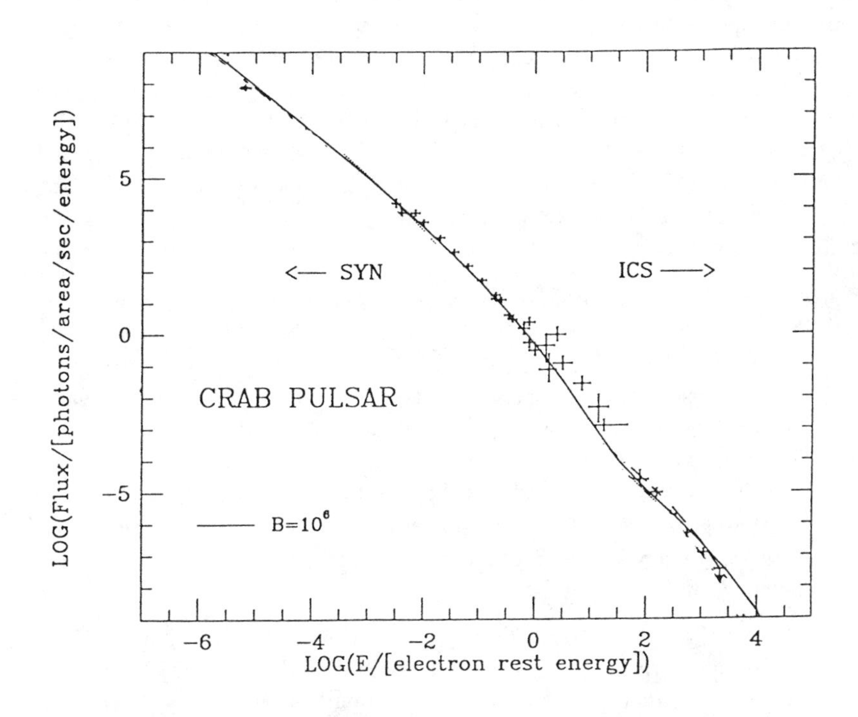

FIGURE 9.5
Calculated and observed spectra of the energetic pulsed photon emission from the Crab pulsar [14]. The absolute magnitude is arbitrarily adjusted.

(Radiation reaction limits gap e^-/e^+ to about 10^{13} eV. Higher energies would be given to protons which traverse the entire 10^{15} Volt gap potential drop. These 10^{15} eV protons can make π^0-mesons in collisions with the soft x-ray photons of the Crab pulsar, but 10^{15} eV γ-rays from π^0-decay could escape without conversion to $e^{\pm}$ by the magnetospheric B only if the π^0 is made well beyond the pulsar light cylinder. Nebular matter is another possible target for π^0 production).

9.5 From the Crab to the Vela Pulsar [16–18]

Observations of the Crab and Vela pulsars imply that these spinning neutron stars differ only in their spin periods: the Crab has a period $P \sim$

3×10^{-2} s while for Vela $P \sim 9 \times 10^{-2}$ s. Both pulsars appear to have the same dipole field strengths (surface dipole $B_s \sim 4 \times 10^{12} G$) and an almost identical double subpulse structure which suggests very similar relative orientations between their spins and magnetic dipoles and probably also with the line of sight to us. It seems very plausible, therefore, that in less than 10^4 years, when the Crab pulsar's spin will have slowed to that of Vela, its observed energetic radiation will have evolved into that seen today from Vela. This implies that certain large changes must occur as the neutron star spin slows by a factor of three:

1. The Crab and Vela energetic radiation spectra are similar at photon energies above 1 MeV as shown in the spectra of Figs. 9.5 and 9.6. From the Crab the emitted power remains strong down to optical frequencies. In Vela's spectrum, however, there is a sharp break at around one MeV and a very great relative intensity suppression at all lower photon energies. (In Fig. 9.6 the spectral break at three possible energies cover a range of possibilities for e^-/e^+ energies and angles with respect to the local $\vec{B}$ ($\sim 5 \cdot 10^3$ G) which give different $e^\pm$ exit times from the magnetosphere before strong synchrotron emission has reached the x-ray regime.)
2. The fraction of the total spindown power of the Crab which is radiated as energetic radiation ($\sim 10^{-3}$ if the radiated beams are much more greatly extended in latitude than in longitude as in Fig. 9.4) is only 10^{-1} of that from the more slowly spinning Vela.

Both of these features have a natural explanation in the outer gap accelerator model.

The series of mechanisms which limit the growth of an outer magnetosphere gap in the Crab pulsar are not nearly so efficient in the Vela pulsar. The synchrotron x-rays of Sec. 9.4 item **c**, necessary for the $e^\pm$ conversion of curvature γ-rays from the gap, are greatly suppressed in Vela. This is because the time it takes for a relativistic e^-/e^+ to synchrotron radiate down to a particular characteristic frequency is proportional to B^{-3}. Since Vela has about the same dipole moment as the Crab, but spins 3 times less rapidly, the local outer magnetosphere B is smaller by $3^3 = 27$. This is enough to suppress strong x-ray synchroton radiation since the radiating pairs will leave the magnetosphere, either directly or after reflection if spiralling inward along converging magnetic field lines, before such radiation is strong. Vela seems to limit its outer magnetosphere gap growth by mechanisms rather different from those of the Crab, and the gap grows to almost $\frac{1}{3}$ the available outer magnetosphere volume before this is accomplished. The calculated and observed spectra are compared in Fig. 9.6.

Because of the strong synchrotron x-ray emission from the Crab pul-

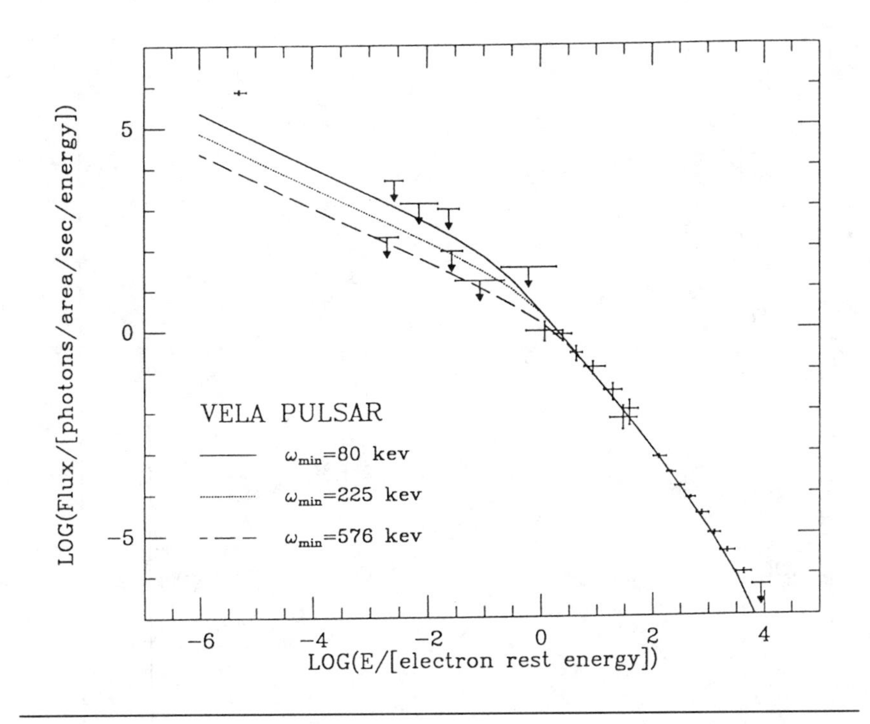

FIGURE 9.6
Calculated and observed spectra of energetic pulsed photon emission from the Vela radiopulsar [14].

sar's outer magnetosphere, $e^{\pm}$ production there can utilize more efficient mechanisms than those available for Vela. These Crab mechanisms become dominant for $P < 5 \times 10^{-2}$ s. In this faster spin regime

$$L_\gamma \sim \left(\frac{P}{3 \times 10^{-2}\, s}\right) 10^{36}\, \mathrm{erg\, s^{-1}}\,. \tag{9.4}$$

We note that for pulsars spinning more rapidly than the Crab but which are otherwise identical, Eq. (9.4) predicts that their L_γ should be less. This is because in the larger outer magnetosphere magnetic fields of faster spinning pulsars it becomes easier to make $e^{\pm}$ pairs. (If a residual pulsar of SN 1987A is in that family, it would therefore be expected to be much less bright in hard x-rays and γ-rays than the Crab pulsar.) The calculated evolution of γ-ray luminosities as a function of pulsar periods up to that of Vela is shown in Fig. 9.7. As a Vela pulsar slows to a period somewhat

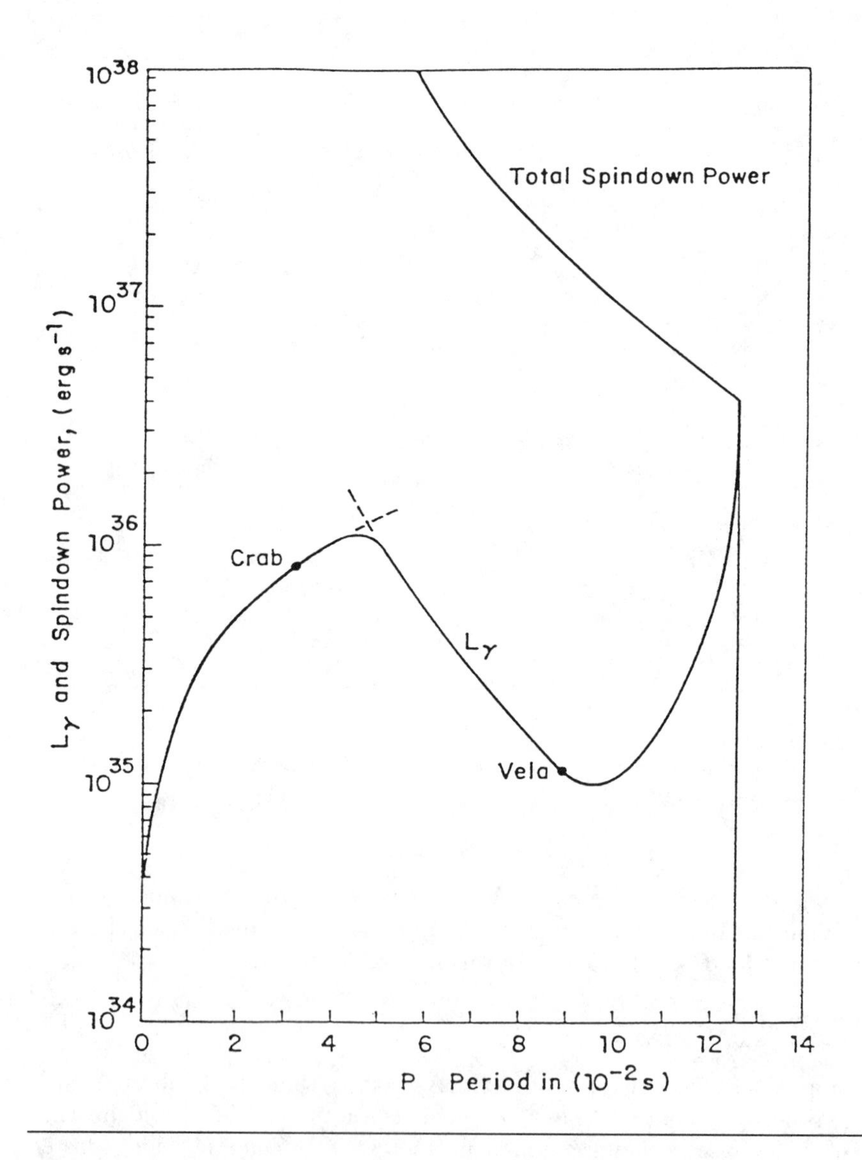

FIGURE 9.7
Calculated evolution of γ-ray luminosity (L_γ) and total spin-down power as a function of pulsar period for a pulsar with the magnetic dipole field of the Crab and Vela radiopulsars [18].

greater than 9×10^{-2} s, calculation of its expected L_γ indicates a change from one which diminishes with the decreasing spin-down power to one which increases sharply until almost all of the pulsar's spin-down power is being radiated away as energetic γ-rays. We turn next to that stage of the pulsar's evolution.

9.6 Future Evolution of Vela to a Stronger Cos B Source [16–18]

Key mechanisms in the symbiotic complex which sustains the $e^\pm$ pair production needed for Vela's outer magnetospheric current flow include $\gamma+\gamma \to e^- + e^+$ by crossing radiation beams which contain the observed γ-ray flux. Pair production from such crossed beam photon collisions is largest for photon center of mass energies between the threshold $2mc^2 \sim 10^6$ eV and several MeV. As a young pulsar's spin slows to around that of Vela, L_γ remains a fixed fraction ($\sim 10^{-2}$) of the pulsar's total spin-down power (*cf.* Fig. 9.7). However, as the spectral break exceeds several MeV, the number of γ-rays in the beams with energies in the most effective range to make $e^\pm$ pairs greatly decreases. Moreover the suppression of the lower end of the γ-ray spectrum leaves many fewer γ-ray photons in the beams. Because the spectral break rises so sharply with increasing pulsar period, a very much larger fraction of the pulsar's spin-down power must then be devoted to sustaining the needed outer magnetospheric pair production and thus to the γ-rays associated with it. In Vela the outer magnetosphere accelerator already occupies around $\frac{1}{3}$ the total open field line volume. As P increases even very modestly, that fraction must grow very considerably in order to maintain required outer magnetosphere $e^\pm$ production until, finally, all of the available volume is used for the accelerator. Then most of the pulsar's spin-down power will be dissipated in γ-rays produced near the pulsar light cylinder. For $P \sim 10^{-1}$ s the consequent upper bound to the pulsar's γ-ray luminosity is

$$L_\gamma \sim I\Omega\dot{\Omega} \sim 5 \times 10^{36}\,\mathrm{erg\,s^{-1}}\,. \qquad (9.5)$$

It has been estimated that this maximum L_γ will be achieved for Vela when its period reaches $P \sim 0.13$ s, as indicated in Fig. 9.7. In the interval between $P = P(\text{Vela}) \sim 9 \times 10^{-2}$ s and $P \sim 0.13$ s, the pulsar γ-ray luminosity could almost reach that of Eq. (9.5). The radiated hard γ-ray spectrum would be expected to remain that of Fig. 9.6 which can be approximated by a flux spectrum $N(\omega) \sim \omega^{-2}$ at the higher energy end of its range. The number of such luminous "post-Vela" γ-ray sources is the number of Vela-like pulsars whose periods lie between that of Vela and one a factor 1.4 greater. For a nominal birthrate of pulsars in this family of one per 10^2 years the total Vela-like Galactic population in this period interval would be 40, a large fraction of which could have L_γ approaching that of Eq. (9.5). The population, the γ-ray luminosity and spectrum, and the Vela-like suppression of x-ray emission are consistent with those of the 20 "unidentified" Cos B sources. It is tempting to suggest that the latter may consist of such post-Vela neutron stars with rising L_γ and $P \sim 10^{-1}$ s.

We turn next to the evolution of these γ-ray pulsars when their outer magnetospheres can no longer sustain the large $e^\pm$ production rates needed to sustain strong magnetospheric current flow.

9.7 Terminal Alignment of Post-Vela Gamma-ray Pulsars [16–18]

In the very luminous γ-ray emission of Sec. 9.6 the charge deficient open field line accelerator region of the outer magnetosphere had to grow with increasing P until it occupied such a large region that L_γ approached the entire spin-down power of the pulsar. That growth was necessary to sustain needed large $e^\pm$ creation rates there. When P lengthens still further the accelerator, already at its maximum relative volume, can no longer sustain the needed steady copious $e^\pm$ production. As a result we may expect that a very large part of the magnetosphere current flow between the polar cap and the light cylinder will also be quenched. A great suppression of current flow through the polar cap ($\vec{J}$) would give a corresponding reduction to spin-down torque from $\vec{J} \times \vec{B}$ inside the neutron star beneath the cap. Then any large further spin-down of the neutron star would have to be accomplished through the Maxwell torque from the radiation of the star's spinning non-aligned magnetic dipole moment. If the neutron star is approximated by a rigid sphere or, more realistically, as a spinning fluid whose shape is always axially symmetric about its spin axis, then that torque would not only spin-down a Vela-like star in about 10^4 years, it would also, on that same time scale, align its magnetic moment with $\vec{\Omega}$ [19,20] and thus quench the Maxwell spin-down torque. Long lived solid crust deformations and neutron superfluid vortex pinning may complicate estimates of alignment timescale [21], but crustal relaxation mechanisms would ultimately allow alignment. If such relaxation is within 10^4 years, then the range of final periods for the family of aligning post-Vela γ-ray pulsars is $P \sim (1-2) \times 10^{-1}$ s if the current flow through the stellar magnetosphere is largely quenched. (Proposed models for the magnetospheres of such aligned neutron stars have been given by Krause-Polstorff and Michel [22]. According to Michel a very small departure from exact alignment would result in enough magnetic dipole radiation for radiation pressure to prevent much ambient matter from being pulled to the star despite net electric charge on the star plus magnetosphere in these models.) Because they no longer spin-down, the expected number of solitary aligned rapidly spinning ($P \sim 1 - 2 \times 10^{-1}$ s) former γ-ray pulsars in the Galaxy could be of order the birthrate of Vela- like pulsars ($\sim 10^{-2}$ yr) times the age of the Galaxy. (This proposed evolutionary scenario is certainly very different from that which describes the evolution of conventional radiopulsars. These pulsars have periods in the range $3 \times 10^{-1} < P < 3$ s, and are much older than 10^4 years. There is almost certainly considerable open-field-line current flow through their polar caps which is the source of RF radiation. Most significantly, typical 10^6 year old radiopulsars are not yet aligned and are spinning down canonically. How and why does their evolution differ so strikingly from that proposed here for Vela-like γ-ray pulsars? One possibility is, of course, frozen crustal deformations which do not relax for 10^6 yrs in canonical pulsars. Another is variations in polar cap local magnetic field geometry. These can conduct $e^{\pm}$ pairs created above the polar caps in the very strong $\vec{B}$ there by $\gamma + \vec{B} \rightarrow e^- + e^+ + \vec{B}$ out to the outer magnetosphere regions. There they may replace the otherwise needed local creation of $e^{\pm}$ pairs so that strong current flow there continues at spin periods where the latter is no longer sustainable.)

A spinning neutron star with suppressed magnetosphere current flow and an aligned dipole would no longer spin down. Thus such a population of extinct Vela-like pulsars, despite an enormous rotational energy, is dead. But we are assured in the English Book of Common Prayer: "After death there is the sure and certain hope of ressurection" and we turn next to more detailed consideration of such a prediction.

9.8 Transient x-ray and γ-Ray Emission from Latent Aligned Post-Vela Pulsars: GRB's [16–18]

If the magnetic field of strongly magnetized neutron stars decays substantially after 10^7 years, then only 10^5 of the solitary aligned rapidly spinning neutron stars in the Galaxy are still Vela-like.. If strong dipole fields survive longer this population would be increased proportionately. It has been proposed that members of this large population of aligned latent pulsars can be re-ignited during certain brief transient events (*e.g.*, a modest flux of soft x-rays passing through their outer magnetospheres). Their temporary emissions would then resemble those of the Vela family except that the alignment of the pulsar dipole would greatly diminish strong modulation at the pulsar spin period. A possible 10^5 population is not inconsistent

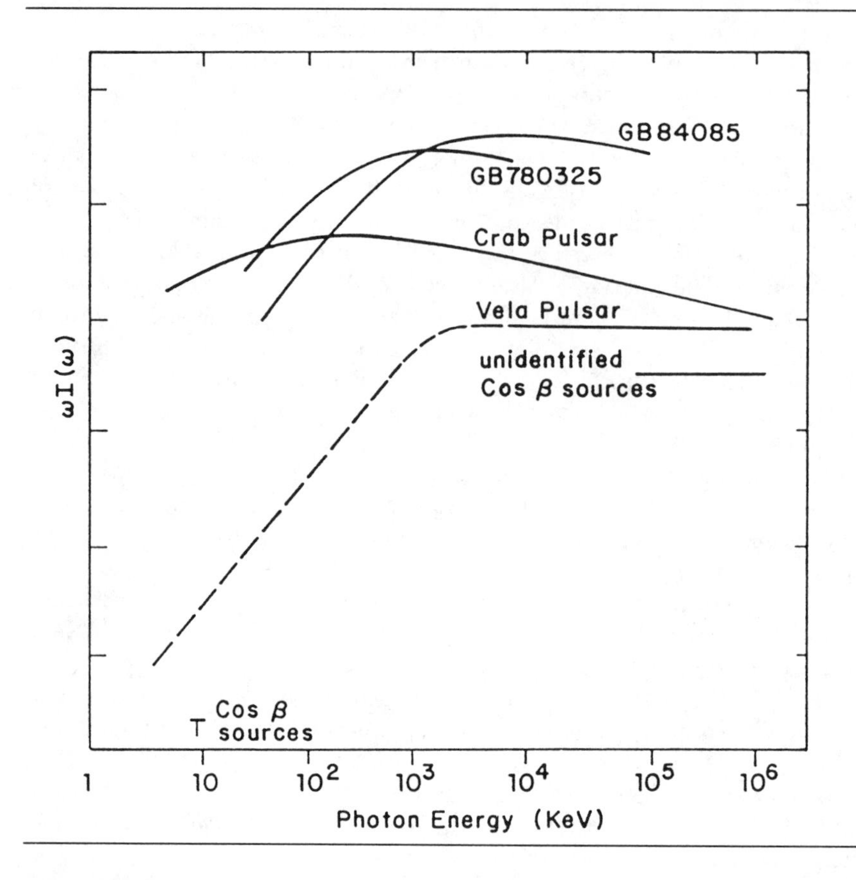

FIGURE 9.8
Spectra of two hard-spectrum GRB sources, GB 780325 and GB 84085 [26], the Crab pulsar, the Vela pulsar, unidentified Cos B sources (including Geminga) [18]. The intensity scale is arbitrary.

with the minimum number of distinct GRB sources of Sec. 9.2, item **6**. GRB spectra resembling that of Vela are suggested by Fig. 9.8. The dotted extrapolation of Vela's x-ray spectrum is the theoretical one of Fig. 9.6, based upon synchrotron radiation from e^-/e^+ which do not survive in Vela's outer magnetosphere long enough to lose all of their energy. A much more steeply decreasing x-ray spectrum is not achievable in such models, and a less steep one would exceed measured upper bounds. The designated x-ray Cos B upper bound assumes $L_x \lesssim 10^{-3} L_\gamma$ as the criterion for no x-ray detection. It is also consistent with the observation of a thermal-like x-ray emission from Geminga. A key feature of a Vela-like model is that

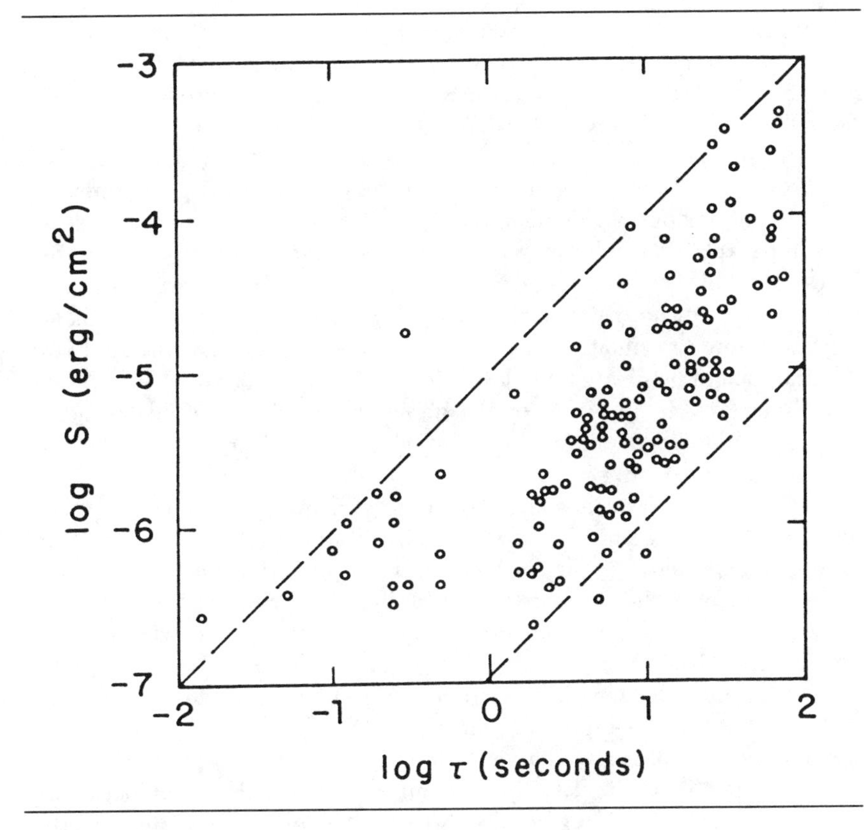

FIGURE 9.9
Gamma-ray burst fluence S (time-integrated intensity) as a function of burst duration time from the KONUS [23] experiment data which measured mainly x-rays of energy < 1 MeV [11].

the main source for L_γ is in the outer magnetosphere, far enough from the stellar surface that reprocessing of γ-rays into x-rays by stellar surface γ-ray absorption does not give $L_x/L_\gamma > 3 \times 10^{-2}$. A standard candle for energetic radiation up to 1 MeV of $L_{x,\gamma} \sim 10^{36}$ erg s^{-1} suggested by Eq. (9.5) is compared to observed GRB time integrated fluences in Fig. 9.9, where the two diagonals give the fluence of 10^{35} ergs s^{-1} sources at 10^2 and 10^3 lt-yr distances [18]. These data do not conflict with the possibility that most GRB's are temporarily reignited Velas.

Among many claimed but controversial GRB spectral features is an occasional (10^{-1} of GRB's) bump at photon energies near 450 KeV which contains over 10^{-2} of $L_{\gamma,x}$ [23]. If this feature is interpreted as a 510 KeV

γ-ray from $e^+ + e^- \to \gamma + \gamma$ gravitationally red-shifted on escaping from the neutron star surface, a 10^{36} erg s^{-1} GRB source must put about 10^{40} $e^\pm$ per second onto the stellar surface. The Crab pulsar produces $e^\pm$ at a rate $\sim 10^{39}$ s^{-1} mainly from crossed beam $\gamma + \gamma \to e^- + e^+$ in its outer magnetosphere [15]. A similar estimate for reignited latent Vela-like pulsars with $L_\gamma \sim 10^{36}$ erg s^{-1} gives $N_\pm \sim 10^{40}$ s^{-1}. Those $e^\pm$ pairs which are created moving out of the magnetosphere along open field lines will escape. A comparable flux will be created moving inward along converging field lines.. Most of these may be reflected and also escape. Some can synchrotron radiate rapidly enough to avoid such reflection and reach the neutron star surface where $e^\pm$ will annihilate with emission of a γ-ray pair. An unknown fraction would be created on closed field lines on which all e^+ would suffer the same fate. If, in either of the above ways, a substantial fraction of $N_\pm$ reach the stellar surface, red-shifted e^+ annihilation γ-rays with the claimed observed intensity would be emitted.

The first observations of GRB's were very quickly followed by a large number of suggested models in which neutron stars were the sources. Examples included neutron stars in binaries with occasional nuclear explosions of accreted material or accretion instabilities. Models with solitary neutron star sources invoked cometary or asteroidal impacts, internal structure readjustments, surface magnetic field reconnection and flares, and explosions from interstellar matter accretion [11]. A scaled down version of any of these could serve simply as a "match" which injects enough soft x-rays or plasma into the partially charge depleted outer magnetosphere of a latent γ-ray pulsar to reignite, temporarily, outer magnetosphere production and the γ-ray/x-ray emission associated with it. The match need supply only a modest power, just enough to support closing of the outer magnetospheric current circuit where it was opened by a charge deficient void, and allow the former pulsar to radiate again at nearly its full spin-down power as long as the match remains lit.

Quasiperiodic surface explosions of accreted interstellar matter may be the most plausible match candidate for igniting a solitary latent γ-ray pulsar. That accretion mass flow down onto the stellar polar cap should, however, not continually spin-down the neutron star prior to the transient explosion. It has been estimated that a polar cap x-ray burst from runaway nuclear burning of accreted interstellar matter may be repeated at intervals of order 10^2 years [24]. The relatively weak match with $L_x \sim 10^{34}$ erg s^{-1} could stay lit for up to 10^2 s and supply enough soft x-rays to keep a latent outer gap active with an $L_\gamma \sim 10^2 L_x$.

9.9 Evolution in a Millisecond Gamma-ray Pulsar Family

The magnetospheres of low magnetic field millisecond period pulsars are qualitatively similar to the outer magnetosphere of Vela. The maximum potential drop from charge depletion which can be realized along $\vec{B}$ and the maximum current flow are both proportional to $B_s P^{-2}$; the maximum luminosity $L \propto B_s^2 P^{-4}$. The position of the spectral break in the γ-ray/x-ray spectrum is proportional to $B_s^{-3} P^7$. These quantities are comparable for Vela and the millisecond pulsar PSR 1937. The γ-ray luminosity of the millisecond pulsar is expected to be much less than that from Vela despite the similarity in total spin-down power. The millisecond pulsars may be expected to contain the same two sub-families as those of the Vela-like pulsars of Sec. 9.6, one of which "soon" aligns. That millisecond pulsar family may also go through an evolution which passes through Cos B sources and ends as GRB sources. However, with the same spin-down power as a Vela-like pulsar the millisecond pulsar will stretch the duration of the Cos B evolutionary phase by the ratio of its spin energy to that of Vela, $(9 \times 10^{-2}/1.5 \times 10^{-3})^2 \sim 4 \times 10^3$. Therefore the formation rate of millisecond pulsars need be only 3×10^{-4} that of Vela-like pulsars for them to be a significant part of, and perhaps even most of, the Cos B sources. Such a birthrate does not conflict with the observed population of millisecond radiopulsars. If millisecond pulsars are almost 10^{-1} of all radiopulsars [25] and each survives as a radiopulsar for $10^8 - 10^{10}$ years, then they could constitute $10^{-2} - 10^{-4}$ of all neutron stars. If their aligned latent millisecond pulsar cousins are similarly abundant they would have a birthrate $10^{-2} - 10^{-4}$ that of Vela-like pulsars and evolve into $10^4 - 10^6$ potential GRB sources.

9.10 Prospects

Clearly the window for seeing γ-ray stars clearly is rather more opaque for theorists than for observers. However, this should change dramatically in the next several years. The Soviet-French "Gamma" observatory satellite, scheduled for a 1989 launching, will detect γ-rays of energy 50 MeV to 5 GeV. The Gamma Ray Observatory (GRO) will explore sources at γ-ray energies from 20 to 30 GeV. The entire spectrum of transient GRB's should soon be observed on Soviet platforms as the bursts evolve. Larger extensive air shower arrays are being built which should confirm and extend the remarkable, but not quite compelling, evidence for several 10^{15} eV sources in the Galaxy, or else show that statistical fluctuations in the present data have mislead us. Whatever the results, we theorists, with hindsight, will have explained them.

Acknowledgements

It is a pleasure to thank Professor S. Gasiorowicz and the Theoretical Physics Institute of the University of Minnesota for their kind hospitality. This research was supported in part by NSF grant AGT-86-02381.

Bibliography

[1] G. Bignami, Revista Nuovo Cimento **7** (1984) 1.

[2] T. Weekes, in *Proc. of 14th Conference on Relativistic Astrophysics*, ed. E. Feneyes (World Scientific, Singapore, 1989) and references therein.

[3] R. Protheroe, in *Proceedings of 20th International Cosmic Ray Conference* (Moscow, 1987) and references therein.

[4] P.Sokolsky, *Introduction to Ultrahigh Energy Cosmic Ray Physics* (Addison-Wesley, Reading, MA, 1989).

[5] T. Weekes, M. Cawley, D. Fegan, K. Gibbs, A. Hillas, P. Kwok, R. Lamb, D. Lewis, D. Macomb, N. Porter, P. Reynolds and G. Vacanti, Ap. J. (1989) in press.

[6] M. Levinthal, in *13th Texas Symposium on Relativistic Astrophysics*, ed. M. Ulmer (World Scientific, Singapore, 1987) p. 382 and references therein.

[7] J. McClintok and M. Levinthal, Ap. J. (1989) in press.

[8] J. Dowthwaite, A. Harrison, I. Kirkman, H. McCrae, T. McComb, K. Oxford, K. Turner and M. Walmsley, Ap. J. **286** (1984) L35.

[9] G. Bignami, in *High Energy Phenomena around Collapsed Stars*, ed. F. Pacini, NATO ASI; Cargese 1985 (Reidel, Dordrecht, 1987).

[10] M. Ruderman, in *High Energy Phenomena around Collapsed Stars*, ed. F. Pacini (Reidel, Dordrecht, 1987) p.145 and references therein.

[11] *Gamma Ray Bursts*, ed. E. Liang and V. Petrosian (AIP Conf. Proc. **141**) and references therein.

[12] J. Middleditch, C. Pennypacker, and M. Burns, Ap. J. **315** (1987) 142 and Nature **313** (1985) 659.

[13] K. Cheng, C. Ho, and M. Ruderman, Ap. J. **300** (1986) 495.

[14] K. Cheng, C. Ho, and M. Ruderman, Ap. J. **300** (1986) 522.

[15] C. Ho, Ap. J. (1989) in press.

[16] M. Ruderman, in *Timing Neutron Stars*-Proceedings of NATO Advanced Study Institute, Cesme, 1988, ed. H. Ogelman and E. van den Heuvel (North Holland, Amsterdam, 1989); Secs. 9.5–9.8 are largely taken from this paper.

[17] M. Ruderman, in *13th Texas Symposium on Relativistic Astrophysics, 1986*, ed. M. Ulmer (World Scientific, Singapore, 1987) p.448.

[18] M. Ruderman and K. Cheng, Ap. J. **355** (1988) 306.

[19] L. Davis and M. Goldstein, Ap. J. **5** (1970) 21.

[20] F. C. Michel and H. Goldwire, Ap. J. **5** (1970) L21.

[21] P. Goldreich, Ap. J. **160** (1970) L11.

[22] J. Krause-Polstorff and F. C. Michel, M.N.R.A.S. **144** (1985) 72 and **213** (1985) 43.

[23] E. Mazets and S. Golenetskii, Astrophys. and Space Phys. Rev. **1** (1981) 205.

[24] R. Taam, in *13th Texas Symposium on Relativistic Astrophysics, 1986*, ed. M. Ulmer (World Scientific, Singapore, 1987) p.546.

[25] J. Taylor, in *13th Texas Symposium on Relativistic Astrophysics, 1986*, ed. M. Ulmer (World Scientific, Singapore, 1987) p.546.

[26] G. Share, S. Matz, D. Messina, P. Nolan, E. Chupp, D. Forrest, and J. Cooper (1986) preprint.

10

Relaxed States in Driven, Dissipative Magnetohydrodynamics: Helical Distortions and Vortex Pairs

David Montgomery
Department of Physics & Astronomy
Dartmouth College
Hanover, NH 03755

For more than thirty years discussion of the confined states of a magnetized, current-carrying plasma column has been dominated by a paradigm of an ideal axisymmetric MHD equilibrium with no steady-state flow velocity. Anything else that has happened has been interpreted as a consequence of "instabilities," regarded as perturbations of such equilibria. We question this way of viewing the problem and suggest that it has only been possible to sustain it this long because of the absence of good internal diagnostics for the relevant MHD variables. Electrically-driven, magnetically-supported, cylindrical magnetofluids seem naturally to relax into states which contain a pair of low-mode-number, counterrotating, helical vortices. The current channel and magnetic topology are the result of an axisymmetric current distribution plus a smaller helical component. The evidence for the configuration is numerical and analytical. The analytical argument is based on the assumption that the relaxed states of driven, dissipative MHD are those of the minimum rate of energy dissipation.

10.1 Introduction

Magnetohydrodynamics (hereafter MHD) means [1–5] the hydrodynamics of an electrically conducting fluid with fractionally very small charge separation. It is a much narrower and more sharply defined subject than is plasma physics, of which it may be considered a branch. Plasma physics includes all the microscopic processes of charged-particle kinetics, atomic physics, and electromagnetic radiation, as well as the bulk hydrodynamic behavior. In real-life situations, the hardest part of plasma physics is knowing which processes to leave out, and most of them always do have to be left out.

MHD, on the other hand, is only an enlarged version of ordinary hydrodynamics, with which [6,7] we have a century and a half of solid, cumulative experience. Also, it is at present the only serious candidate for a global description by means of which we may follow the large-scale dynamics of a magnetized plasma: such problems are simply too difficult for kinetic theory to do much with. So, there is some justification for trying to bring to MHD theory some of the same logical precision that has long been the goal in Navier-Stokes hydrodynamics. Whether or not we immediately accomplish the practical tasks we have set ourselves in which MHD seems to be central (*e.g.*, build a fusion reactor or predict the relations between solar and magnetospheric activity), there would seem to be little doubt that in the long run it will do us some good to be able to make clear-cut statements about MHD behavior. It is a rich, classical, nonlinear field theory for which there is little or no reason to suspect pathologies or divergences: the union of classical low-frequency electrodynamics and hydrodynamics.

This article addresses an aspect of MHD behavior that might seem embarassingly simple if we knew what the answers were: namely, what are the elementary states of an electrically-driven, dissipative MHD fluid? The corresponding question for ordinary fluids [6,7] occurs early in any hydrodynamics course and leads to such familiar answers as the parabolic, axisymmetric pipe flow profile; the parabolic plane Poiseuille profile between infinite parallel planes; the linear profiles of plane Couette flow, and rotating Couette flow between concentric cylinders; and so on. There are not so many of these flow profiles. Most of them eventually become unstable as some dimensionless measure of the gradients, such as Reynolds numbers (involving length scales, gradients of flows, and dissipation coefficients), exceed certain threshold values; then they become turbulent flows [8]. These elementary flow profiles are the essential prototypes in terms of which we organize our understanding of more elaborate and involved hydrodynamic situations.

A slightly astonishing point is that we do not have in hand the analogous profiles for the driven, dissipative MHD problem. There are relatively few generic means for making a hydrodynamic fluid do something interesting: moving boundaries, imposed pressure drops, temperature drops, and a very few more. These will all work for MHD fluids too, and there is an additional important electrical means: voltage drops (which result in electric currents). As in ordinary hydrodynamics, there are boundary conditions to be met, which tend to tie down certain field variables at the boundaries. The ensuing compromises between the boundary conditions, the driving mechanisms, and the abilities of fluids and magnetofluids to dissipate large-scale coherent energy into heat determine what the systems will and will not do. It sounds simple, and it is at first surprising to outsiders when they discover there are big gaps in our understanding of a magnetofluid, at even this elementary level, about what these elementary, driven, dissipative states are.

The reason is that we have wasted a lot of time on *ideal* solutions of the equations of motion, with the dissipative terms such as resistivity and viscosity completely dropped. It is easy to find a lot of solutions in this way, but it is hard to argue that most of them could be set up or that they otherwise have physical significance. The mid-nineteenth century hydrodynamicists did the same thing. Repeatedly the viscous terms were ignored because order-of-magnitude estimates revealed that they should be "small" in interesting cases. What went unappreciated for a long while was that: (1) they could be locally large in small but important neighborhoods; and (2) they multiplied the highest-order spatial derivatives in the problem, and altered the required boundary conditions when they disappeared. Both for fluids and magnetofluids, ideal solutions are numerous and easy to come by, but most of them are in no sense close to solutions of the non-ideal equations with proper boundary conditions, no matter how small the dissipation coefficients become. This fact went unappreciated for a considerable time in 19th century hydrodynamics.

There are, to be sure, MHD examples where everything has been done right. Hartmann flow [3,4], for example, is a prototype solution that is relevant to MHD power generation where all the boundary conditions have apparently been correctly taken into account. It is controlled fusion research (and then, by osmosis, space physics) that has allowed wishful thinking about ideal equilibria to get out of hand. We are only beginning to appreciate how far these ideal equilibria may be from any (even slightly) non-ideal equilibria. The purpose here is to show how much difference even a little bit of dissipation can make to an equilibrium.

The outline of the article is as follows. Sec. 10.2 is devoted to describing what has been one central problem in toroidal confinement, probably the single problem in toroidal confinement that has inspired the greatest number of calculations and the largest quotient of jargon in plasma physics, terrestrial or extraterrestrial. Sec. 10.3 describes some numerical evidence that has been collected in 3D MHD computations for this problem. (The numerical picture is in some ways much more detailed than the laboratory picture.) Sec. 10.4 describes a perturbation-theoretic attempt to understand what the elementary solutions of this problem ought to be. Sec. 10.5 summarizes what has been done and identifies some things there still are to do. An Appendix outlines a calculational scheme that is under exploration for extending the subject non-perturbatively.

10.2 Voltage–driven Magnetized Plasmas

A theorist's cartoon of a toroidal plasma is a torus, or doughnut, consisting of an electrically-conducting shell full of plasma. The direction the long way around the doughnut is the "toroidal" direction, and the short way around is the "poloidal" direction. There is typically a dc magnetic field maintained in the toroidal direction by current-carrying windings in the poloidal direction. There are slits and slots cut in the conducting shell to permit electric and magnetic fields externally generated to penetrate into the plasma. The geometrical distortions from toroidal symmetry necessarily associated with the slits and slots are often idealized away. A sketch of the arrangement is shown in Fig. 10.1.

The plasma is produced inside the shell by filling the torus with gas and ionizing the gas by one means or another. A toroidal electric field is often introduced into the plasma by changing a magnetic flux through the hole in the doughnut as a function of time. The line integral of this inductive electric field the long way around the doughnut is the toroidal voltage or "loop voltage." (Sometimes there is a circuit that also provides a poloidal electric field, also inductively.) Since the plasma has an electrical conductivity that is neither zero nor infinite, the loop voltage produces a toroidal current, induced parallel to the dc magnetic field. This toroidal current produces a poloidal magnetic field which has to be added vectorially to the pre-existing toroidal dc magnetic field. The resulting magnetic field lines are, roughly speaking, nested helices. The electric current develops both toroidal and poloidal components and thus modifies both the toroidal and poloidal magnetic fields. The plasma physics question implicit in this situation, at its least biased level, is: What happens? How do the electromagnetic and mechanical field variables arrange themselves? The answer could be straightforwardly, if tediously, provided if, instead of a plasma inside the toroidal conducting shell, there were a rigid metal. Then a straight application of Maxwell's equations with time-dependent boundary conditions would determine the evolution and spatial distribution of the

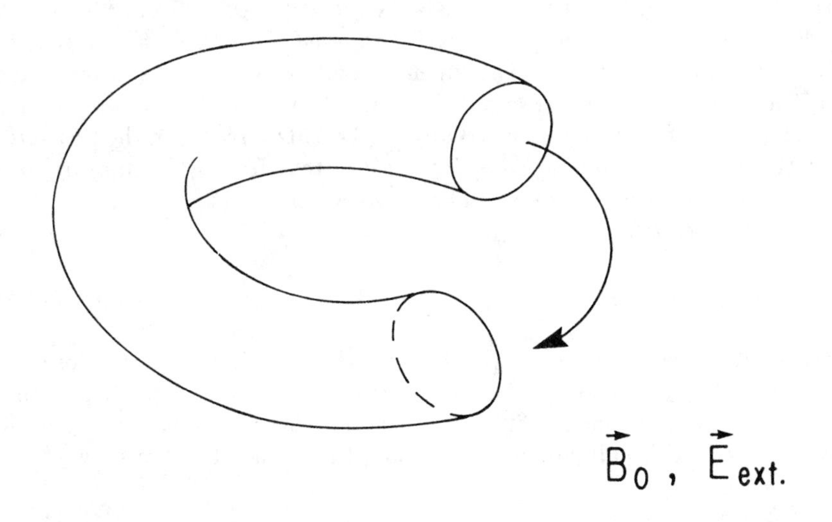

FIGURE 10.1
A sketch of a voltage-driven, plasma-filled torus. Current windings around the doughnut the short way (not shown) produce a toroidal dc magnetic field $\vec{B}_0$. Magnetic flux through the hole in the doughnut (upward in the figure, not shown) increases with time and induces a toroidal driving electric field $\vec{E}_{ext}$. For theoretical simplicity, the torus is "unwrapped" into a straight cylinder with periodically identified ends.

current density $\vec{j}$ and the magnetic field $\vec{B}$. However, because the plasma behaves like a fluid, it can move in response to electromagnetic stress and begins to thrash around. This produces a velocity field $\vec{v}$ and a vorticity field $\vec{\omega} = \nabla \times \vec{v}$ in an interactive, nonlinear way.

At this point, it is probably well to write down the simplest set of equations of motion that will contain only those features of the plasma behavior that we want to discuss here (*e.g.*, [2]). First, there is an equation of motion

$$\rho \frac{d\vec{v}}{dt} = \rho \left(\frac{\partial \vec{v}}{\partial t} + \vec{v} \cdot \nabla \vec{v} \right) = \frac{\vec{j} \times \vec{B}}{c} - \nabla p + \rho \nu \nabla^2 \vec{v} \, . \tag{10.1}$$

Eq. (10.1) is just Newton's second law of motion in differential form and the following symbols have been introduced. ρ is the mass density; a severe approximation that we will make, that ultimately should be lifted, is that ρ remains constant and uniform inside the plasma, or that we deal with incompressible flow. p is the mechanical pressure, determined by taking the divergence of Eq. (10.1) and using $\nabla \cdot \partial\vec{v}/\partial t = \partial/\partial t(\nabla \cdot \vec{v}) = 0$, which follows from the incompressibility. What is left is a Poisson equation to be solved for $p = p(\vec{v}, \vec{B})$. The kinematic viscosity ν will be taken to be uniform and constant, for simplicity, though it is really a tensor with a strong temperature dependence. Only the Lorentz force term, the first term on the right-hand side of Eq. (10.1), is different from what appears in a garden-variety Navier-Stokes fluid such as water.

How does $\vec{B}$ evolve? From the Maxwell equation

$$\frac{\partial \vec{B}}{\partial t} = -c\nabla \times \vec{E} \,, \tag{10.2}$$

we see that we need some expression for the electric field $\vec{E}$ to close the system. The easy way to get it is to *assume* that in the local rest-frame of the plasma, or magnetofluid as it has now become, there is a relation between electric field and current density, the so-called "Ohm's law":

$$\vec{E} + \frac{\vec{v}}{c} \times \vec{B} = \sigma^{-1}\vec{j} = \eta_R \vec{j} \,, \tag{10.3}$$

where the electrical conductivity σ will, like ν, be taken to be constant and uniform for simplicity, even though there is again a strong local temperature dependence.

One last approximation is made on the Maxwell equation

$$\frac{1}{c}\frac{\partial \vec{E}}{\partial t} = -\frac{4\pi}{c}\vec{j} + \nabla \times \vec{B} \,, \tag{10.4}$$

where the displacement current $\partial\vec{E}/\partial t$ is neglected compared to the conduction current $\vec{j}$. This neglect is justified in the same way as for a piece of copper wire: there are so many conduction electrons around, and the time scales of the phenomena are so slow, that we make the "pre-Maxwell" approximation, $\nabla \times \vec{B} \cong 4\pi\vec{j}/c$. Improving upon these arguments is not so easy. In any case, Eqs. (10.2) and (10.3) then become

$$\frac{\partial \vec{B}}{\partial t} = \nabla \times (\vec{v} \times \vec{B}) + \eta\nabla^2\vec{B} \,, \tag{10.5}$$

where $\eta \equiv c^2/(4\pi\sigma)$ is the "magnetic diffusivity," and has the same dimensions as ν, $(\text{length})^2/(\text{time})$. Equations (10.1) and (10.5), supplemented by the conditions $\nabla \cdot \vec{v} = 0 = \nabla \cdot \vec{B}$, are the simplest version of MHD. They apply to a liquid metal as well as to a plasma. If one goes through and sets $\vec{v} = 0$, they apply to a rigid metal (if p is replaced by the appropriate tensor); and if one goes through and sets $\vec{B} = 0$, they apply to a Navier-Stokes fluid. It is very easy to make a list of important physical effects that are left out of Eqs. (10.1) and (10.5), but they contain the answers to many, many questions which we don't now understand. Hereafter, we will be discussing only such a restricted MHD description, despite its limitations.

Understanding the implications of Eqs. (10.1) and (10.5) for the problem we have posed leads us to "unwrap" the torus, for simplicity, into a straight cylinder with periodically identified ends. This is not an obvious thing to do, but it is certain that if we cannot understand the problem in the periodic cylinder, we cannot understand it in the torus. Hereafter, we go into dimensionless units, which can be formally achieved by setting $\rho = 1$ and $c = 1$ and dropping the 4π. In these dimensionless units, ν and η have the interpretation of being reciprocal Reynolds-like numbers, and for cases of interest ("nearly ideal" cases), ν and η are both $\ll 1$. The ideal MHD equations are obtained by setting $\nu \equiv 0$ and $\eta \equiv 0$. As in the case of a Navier-Stokes fluid, there is an immense difference between "ideal" and "slightly non-ideal" or "nearly ideal;" the similarities are only semantic and the terminology is seriously misleading.

The problem is beginning to be almost well-posed. We put the rigid cylindrical wall at $r = a$, say, in cylindrical coordinates (r, φ, z) and call the periodicity length (in z) L_z. A delicate point is a mathematically satisfactory idealization of the applied electric field, the general direction of which is along the z-axis. As previously remarked, the electric field must be admitted through gaps in the conducting shell, which are at this point prohibitively complicated to include accurately in the mathematics. The electric field is typically turned on at a definite instant in time, ramped up to a finite value, and held there at a fixed value until the available "volt-seconds" are used up, an engineering limitation. Even the problem of calculating the electric field inside a slotted conductor in such a situation without any magnetofluid inside is not trivial. The vacuum field will be neither axisymmetric nor uni-directional. Eq. (10.2) makes clear that any $\vec{E}$ field, which depends upon r and points in the z-direction, is *not* time-independent. Finally, there is the question of boundary conditions at the conducting shell at $r = a$. If the shell is idealized as a perfect conductor, $\hat{n} \times \vec{E} = 0$ needs to be obeyed at its surface where $\hat{n} = \hat{e}_r$ is the unit normal to the surface. Alternatively, one may imagine the conductor as coated with a very thin layer of insulating dielectric, which then permits a finite tangential component of $\vec{E}$ but demands then that $\vec{j} \cdot \hat{n} = 0$. The upshot of all these complications is that the introduction of the driving electric field requires some bold assumptions and simplifications that may be quite hard to justify completely.

Integrating the poloidal (azimuthal) component of Eq. (10.2) over a tiny rectangular contour which crosses $r = a$ and using Stokes' theorem, we can see that the effect of a tangential z-directed electric field near the wall is to provide a non-zero rate of supply of poloidal magnetic flux. Interior to the magnetofluid, $\vec{E}$ is given by Eq. (10.3), while $\vec{v}$ and $\vec{B}$ are advanced by

Eqs. (10.1) and (10.5). Technically, the problem can be reduced to finding a satisfactory idealized way of supplying poloidal magnetic flux in the vicinity of $r = a$ at a prescribed rate and letting Eqs. (10.1) and (10.5) deal with everything away from $r = a$. In mathematical terms, it will suffice to have a delta-function in radius, directed in the $\hat{e}_\varphi$ direction, added to the right-hand side of Eq. (10.5), so that integration over a small rectangle, normal to $\hat{e}_\varphi$ and crossing $r = a$, will give a time rate of change of azimuthal magnetic flux that is proportional to the length of the rectangle.

From a computational point of view, it is equivalent to write $\vec{E}$ and $\vec{B}$ in terms of scalar and vector potentials, $\vec{B} = \nabla \times \vec{A}$, $\vec{E} = -\partial\vec{A}/\partial t + \nabla\Phi$, use Eq. (10.3), and replace Eq. (10.5) by (dimensionless units, $c = 1$):

$$\frac{\partial \vec{A}}{\partial t} = \vec{v} \times \vec{B} - \eta\vec{j} + \nabla\Phi \,, \tag{10.6}$$

then advancing $\vec{A}$ instead of $\vec{B}$. Introducing the external electric field then comes down to how we choose to treat $\vec{A}$ or Φ at the wall. One way that seems to work is the following. First, work in the Coulomb gauge so that $\nabla \cdot \vec{A} = 0$. Taking the divergence of Eq. (10.6) leads to a Poisson equation to solve for Φ:

$$\nabla^2\Phi = -\nabla \cdot (\vec{v} \times \vec{B} - \eta\vec{j}) = -\nabla \cdot (\vec{v} \times \vec{B}), \qquad \text{if } \eta = \text{const.} \,, \tag{10.7}$$

so that $\Phi = \Phi(\vec{v}, \vec{B})$ is to be substituted into the right-hand side of Eq. (10.6). The external electric field may be introduced now by replacing

$$\nabla\Phi = \nabla\tilde{\Phi} + E_{ext}\hat{e}_z \,, \tag{10.8}$$

where the spatial average of $\nabla\tilde{\Phi}$ is zero and E_{ext} is a constant (and therefore curl-free) except in the immediate neighborhood of $r = a$, where it plummets sharply to zero. Taking the curl of Eq. (10.6) then provides a delta function at $r = a$, of the type previously described, in Eq. (10.5) and leaves Eq. (10.5) alone away from $r = a$.

It remains to specify boundary conditions on the various fields. If the vacuum dc magnetic field is $\vec{B}_0 = B_0\hat{e}_z$ (flux $\pi a^2 B_0$), it is useful to decompose $\vec{A} = \vec{A}_0 + \tilde{\vec{A}}$ where $\nabla \times \vec{A}_0 = \vec{B}_0$, and then require $\tilde{\vec{A}} \times \hat{n} = 0$ at $r = a$. If we choose to have an uncoated perfect conductor at $r = a$, $\vec{j} \times \hat{n} = 0$, which makes the boundary condition on $\tilde{\Phi}$, $\hat{n} \times \nabla\tilde{\Phi} = 0$. $\vec{B} \cdot \hat{n}$ then equals zero and $\vec{j} \cdot \hat{n}$ is anything it wants to be. In the presence of the insulating dielectric coating, $\vec{j} \cdot \hat{n} = 0$, and a radial boundary condition on $\nabla\tilde{\Phi} \cdot \hat{n}$ is provided by Eq. (10.6).

These boundary conditions are a highly imperfect representation of the laboratory situation. Among the most important effects left out is the strong radial temperature dependence present in most experiments. Ohmic heat is being deposited in the column at the volume-averaged rate $\langle \eta j^2 \rangle$, where $\langle\ \rangle$ means a spatial average, and big radial temperature gradients are maintained between the hot center and the cool perimeter. Kinetic theory models of a plasma reveal a strong temperature dependence of ν and η (as $T^{-\frac{1}{2}}$ and $T^{-\frac{3}{2}}$, in many cases) which greatly alters the relative strength of the dissipative terms and the nonlinear terms at different radii. We are still feeling our way toward manageable idealizations that will reflect this strong temperature dependence. Including it properly in the dynamics, of course, means adding a temperature equation with a variable thermal conductivity to Eqs. (10.1) and (10.5).

The mechanical boundary conditions are more difficult since they incorporate all the difficulties of Navier-Stokes boundary conditions and do not represent the physical processes which go on (such as chemical interaction with the walls) as well. There seem to be two simple choices: "no-slip" ($\vec{v} = 0$ at $r = a$, all components) and "stress-free" or "free-slip" ($\vec{v} \cdot \hat{n} = 0$ at $r = a$ and $\vec{\vec{\sigma}} \cdot \hat{n} = 0$ at $r = a$, where $\sigma_{ij} = \frac{1}{2}\nu(\partial v_i/\partial x_j + \partial v_j/\partial x_i)$ is the viscous stress tensor).

The problem as stated is formidable, but barely manageable. The complications come from the nonlinearities and the boundary conditions. Historically, they have been swept under the rug in the following way.

It is asserted, without very convincing evidence, that the magnetofluid gets itself into an *equilibrium* when the E_{ext} is switched on. However, this is not a true equilibrium, but rather an ideal, static equilibrium. (This means $\nu = 0$ and $\eta = 0$, as well as $\vec{v} = 0$.)

If we go through Eqs. (10.1) and (10.5) and set $\partial/\partial t = 0$, $\vec{v} = 0$, $\nu = 0$, and $\eta = 0$, all that is left is, from Eq. (10.1) (*e.g.* [9]):

$$\vec{j} \times \vec{B} = \nabla p\,, \tag{10.9}$$

as the condition necessary for an ideal equilibrium. (Note from Eq. (10.3) that $\vec{E}$ must $\rightarrow 0$). Any $\vec{B} = (0, B_\varphi(r), B_z(r))$, $\vec{j} = \nabla \times \vec{B} = (0, j_\varphi(r), j_z(r))$ satisfy Eq. (10.9) in the sense that a $p = p(r)$ can be found which will make Eq. (10.9) true. This is the generic, axisymmetric, zero-flow equilibrium around which confinement research and a substantial fraction of space and astrophysical plasma theory has revolved for decades. There are in laboratory situations virtually never convincing measurements of the field quantities that do not already presuppose the adequacy of the axisymmetric model in their interpretation, and there is virtually never even a qualitative dynamical scenario to explain how the magnetofluid is supposed to get itself into the configuration of Eq. (10.9). Any hydrodynamicist to whom one explains this program is usually appalled because he knows that an analogous hydrodynamic procedure would lead nowhere.

In fact, many time-dependent, non-axisymmetric processes are observed in toroidally-confined plasmas, and the way they are usually discussed is in terms of "instabilities" on ideal, axisymmetric zero-flow equilibria of the type given by Eq. (10.9). That is, one writes [9]:

$$\begin{aligned} \vec{B} &= \vec{B}^{(0)} + \vec{B}^{(1)} + \dots \\ \vec{v} &= \qquad\ \vec{v}^{(1)} + \dots \\ \vec{j} &= \vec{j}^{(0)} + \vec{j}^{(1)} + \dots, etc., \end{aligned} \tag{10.10}$$

where the zeroth-order quantities are the ideal, axisymmetric, zero-flow equilibrium and the $\vec{B}^{(1)}$, $\vec{v}^{(1)}$, $\vec{j}^{(1)}$, *etc.* are perturbations about it, assumed small. The ansatz of Eq. (10.10) is then substituted in the dynamical equations and products of perturbation quantities are discarded. What remains is a linear system of dynamical equations for the perturbation quantities, with spatially-dependent but time-independent coefficients.

Solutions to the time-dependent linearized equations are sought which have an overall multiplicative time dependence of the form $\exp(-i\Omega t)$. Imposing a set of boundary conditions then defines an eigenvalue problem for Ω. Any solution for which Ω has a positive imaginary part is interpreted as an instability which, if it grows from a small but non-zero initial amplitude, will eventually become dynamically important.

Instabilities are not hard to find, but it is hard to be sure one has found them all. Books have been written and literally many thousands of papers have been published about their fate [10]. The possibilities are essentially limitless because the ideal equilibria which are solutions to Eq. (10.9) are themselves limitless. In calculations which try to show how the instabilities "saturate," the effects of the other instabilities (which each require specialization and approximation methods that are not applicable to the others) are usually ignored.

The instabilities can be either ideal ($\nu = 0 = \eta$) or non-ideal ($\eta \neq 0$, $\nu \neq 0$). It is the latter class that have been the focus of most of the theoretical activity of the last two decades. This is curious because they are *non-ideal* instabilities of *ideal* equilibria. The zeroth order dissipative terms, such as $\eta\nabla^2\vec{B}^{(0)}$, are ignored at all stages of the perturbation theory, while the first order terms, such as $\eta\nabla^2\vec{B}^{(1)}$, are retained and are dynamically important even though they are surely smaller, at least initially, than $\eta\nabla^2\vec{B}^{(0)}$. The debris associated with these stability calculations and their putative nonlinear effects surround the theory of plasma confinement like a barricade, one that is more than adequate to turn back the curious from even adjacent specialties of physics.

It is hard not to believe that if this program (of following the evolution of small dissipative perturbations on non-dissipative, static equilibria) were going to lead anywhere conclusive, it would have done so by now. This is a minority view. Since I cannot defend the majority view, I will now proceed in a tangential direction and inquire into: (1) what numerical computations reveal that are fully nonlinear and proceed from initial conditions other than ideal, axisymmetric, zero-flow equilibria; and (2) what the possibilities seem to be for interpreting the results of those computations in an analytical framework.

10.3 Numerical Solutions of the Equations

The experiments on toroidal confinement [10] which anyone has been willing to fund for the last fifteen or twenty years have been hot enough to burn up internal probes which are capable of measuring the MHD variables $\vec{B}$, $\vec{v}$, and $\vec{j}$. This has led to ambiguities and data shortages in interpreting the experiments, and long strings of inferences have been necessary to interpret such data as have been available (x-rays, light scattering, external magnetic fields). The best information we have about the solutions of Eqs. (10.1), (10.5) and (10.6) (if not about what goes on in a real plasma) comes from numerical solutions. I would now like to summarize some of these [11–16].

The codes which generated the numerical results which I shall summarize were mostly written and operated by Dr. J. P. Dahlburg. Timely early suggestions were provided by Drs. T. A. Zang and W. H. Matthaeus. Dr. M. L. Theobald added some important, more recent, innovations. Dr. G. D. Doolen provided some of the more advanced computational hardware available and knowledge of how to operate it. The methods are all of the "pseudo-spectral" variety [18,19], in which one goes back and forth from physical space to transform space at each time step: the former to advance the fields in time, carry out multiplications, and to look at the fields; the latter to carry out differentiations, enforce boundary conditions, and solve Poisson equations for p and $\tilde{\Phi}$. The boundary conditions enforced were of the perfectly-conducting, stress-free variety as described in the last section. The equations solved were the full 3D MHD [13–15] system previously described and, in addition the 3D Strauss equations of "reduced" MHD [11,12,16] which are also three-dimensional but which we will not digress to describe further here (they are essentially a limit in which the mean toroidal current $\langle j_z \rangle$ is $\ll a^{-1} B_0$). A square, rather than a circular, poloidal boundary was used in order to take advantage of the speed and convenience of rectangular fast Fourier transforms. The expansion functions used were Fourier transforms for the Strauss equations and Turner-Christensen [17] functions for the full MHD equations. (Turner-Christensen functions are derived from solutions of the vector wave equation $(\nabla^2 + \lambda^2)\vec{A} = 0$ and are related to wave-guide modes.)

The codes suffer from all the difficulties associated with numerical solution of the Navier-Stokes equations [18,19]. The most acute of these concerns the necessary spatial resolution as the coefficients ν and η become

$\ll 1$. It is in the nature of both the Navier-Stokes and the MHD equations to develop more and more spatial fine structure as the Reynolds-like numbers (ν^{-1} and η^{-1}) become larger and larger. Inability to resolve the small scales leads to great inaccuracies in the code's output. The current laboratory values are nominally numbers like ν, $\eta \lesssim 10^{-5}$, and even numbers like 10^{-3} exceed the capacity of present codes to resolve all the small scales that will develop in any but a highly artificially symmetrized situation. We have usually ended up running with values of ν and η between $\frac{1}{50}$ and $\frac{1}{500}$, with spatial resolution in any direction between 16 and 64 cells. This limits us to values of ν and η that are, from the point of view of experimental comparisons, unrealistically low, and we have no sure way of knowing that the phenomena computed do not change drastically if ν and η are decreased by two or three orders of magnitude or more. The situation is no different in hydrodynamic turbulence computations and may be expected to improve only slowly. The runs now require several tens of hours of CRAY time each, and each factor of 2 in a 3D computation's spatial resolution may be expected to multiply the running time by a factor of 4 to 8. It is therefore foolish to imagine that numerical solutions are capable of telling us everything that good internal experimental diagnostics would.

A variety of situations have been computed. Undriven, initial-value, decay [13,14] computations have been performed as well as driven ones with a finite E_{ext} [15]. Strauss equation runs have been done [11,12,16] for both classes of problems, as have the full-scale 3D MHD runs. Runs have been performed with and without viscosity. Resistivities have been allowed to be constant in time and space, or constant in time but variable in space, and also variable in both time and space (for this last case it has been necessary to introduce a temperature equation and a thermal conductivity, and this has been done [16] so far only in the Strauss approximation). It is necessary to inspect literally hundreds of graphs, contour plots, 3D perspective plots, computer-drawn streamlines, and spectra to acquire any kind of perspective on what the computer is saying— far more pictures than any journal will publish. The original papers [11–16] and Ph.D. theses [20,21] should be consulted for more details, but some are still omitted there.

Certain features emerged in virtually all 3D MHD cases, which can be summarized. These include the following:

1. The first few tens of time units were occupied with first the development of, and then the considerable decay of, a substantial amount of MHD turbulence of a rather disordered and not normal-mode-specific character. Times are conveniently measured in units of poloidal Alfvén transit times, or the time an Alfvén wave takes to travel a distance equal to the radius a. An Alfvén wave propagates with speed $C_A = B_0/\sqrt{4\pi\rho}$ (cgs units) or just B_0 (dimensionless units). There are typ-

ically 250 time steps per Alfvén transit time. This turbulent period occupies several tens of Alfvén transit times.

2. At the end of the time-dependent turbulent period, a quasi-time-independent "relaxed" state occurred which, though not completely quiescent, was much more so than the turbulent period which preceded it.

3. The magnetic field in the relaxed state consisted of two nearly time-dependent parts plus some (considerably smaller) variable parts. The larger part was as axisymmetric as the square poloidal boundary would allow. The smaller part, which was occasionally as much as 15 to 20 percent of the axisymmetric part in amplitude (2 to 4 percent in energy), had a helical spatial dependence dominated by a single pair of m, n numbers. By this it is meant that if a cylindrical Fourier decomposition is performed,

$$\vec{B} = \sum_{m,n} \vec{B}_{m,n}(r) \exp(im\varphi - i(2\pi nz/L_z)) , \tag{10.11}$$

a single pair of values of the two integers m, n (and $-m, -n$) has the largest $\langle |\vec{B}_{m,n}(r)|^2 \rangle$ associated with it.

4. The wall magnetic field $B_z(r = a)$ would be reduced below B_0 by an amount that increased as E_{ext} increases and would go negative if E_{ext} were strong enough ("reversed field pinch," or RFP). However, $j_z(r = a)$ did not reverse. (By "$r = a$" in these computational statements we mean at the location of the square poloidal boundary.)

5. The interior of the magnetofluid would develop a "force-free" region in the sense that $|\vec{j} \times \vec{B}| \ll |\vec{j}| \cdot |\vec{B}|$. However, as the conducting wall was approached, a boundary layer would be reached in which $\vec{j}$ and $\vec{B}$ were no longer aligned.

6. The ratio $|\vec{j}|/|\vec{B}|$ depended strongly on space. (The result **5** seems to confirm Taylor's "relaxation theory" [21–25] but the result **6** does not.)

7. The velocity field $\vec{v}$ would never go away but would contain roughly the same amount of energy as the helical contribution to the magnetic and current channels. The velocity field would be associated with a pair of counterrotating helical vortices with the same dominant m and n numbers that dominated the magnetic field (see **3**).

8. The total toroidal current would be reduced below the $\vec{v} = 0$ value that would be inferred from $j_z = E_{ext}/\eta$ alone, indicating that the $\vec{v} \times \vec{B}$ in Eq. (10.3) was directed so as to provide a mean emf that opposed the applied one, causing the plasma activity to act as an "anomalous

resistance" in the circuit. (This statement applied to the 3D MHD computations but not to the Strauss equation ones.)

9. Once the partially helical relaxed state was achieved, the magnetofluid was grossly stable from that point on. Whatever MHD activity that might have resulted from subsequent instabilities there was, it remained limited to a very few percent of the total energy and produced no more qualitative changes in the field distributions.

These features occurred often enough in enough different contexts that we ceased to attribute them to numerical malfunctions and began to believe them to be real. What finally convinced us was when we saw sawtooth oscillations for the variable temperature case [16]. The oscillations were not oscillations on an axisymmetric background as the accepted theory said they should be, but rather on a helically deformed one of the type just described! At that point we began to search hard for an alternative vantage point from which to view the whole body of results, other than perturbations of ideal, axisymmetric, zero-flow equilibria. The results of that search so far are described in Sec. 10.4. Figure 10.2 is a schematic drawing which indicates what the relaxed-state magnetic field lines and velocity field lines look like when projected into a typical poloidal plane (plane of constant z).

10.4 The Driven, Dissipative States of MHD; Minimum Energy Dissipation

Textbooks probably make the basic steady-state hydrodynamic flows, such as pipe flow and Couette flow, seem more obvious and deductive than they seemed at the time, when the equations and boundary conditions were still arguable and the flows were not well measured. We have perhaps neglected our homework in not having established the corresponding basic states of driven, dissipative MHD before proceeding to fancier applications (an exception is the well-documented case of Hartmann flow [3,4]). In the absence of fundamental, directly-diagnosable MHD experiments (which no one seems in a hurry to do), the question arises as to what the possibilities are for rendering analytically manageable the kinds of MHD states, so manifestly non-ideal and non-axisymmetric, that have appeared in the computations. As in hydrodynamics, the effort involved in constructing explicit solutions increases steeply as symmetries are discarded and boundary conditions (required by dissipative terms) multiply.

We have only begun the task of identifying the driven, dissipative states of MHD. It is slightly ironical that we have been driven to the ones

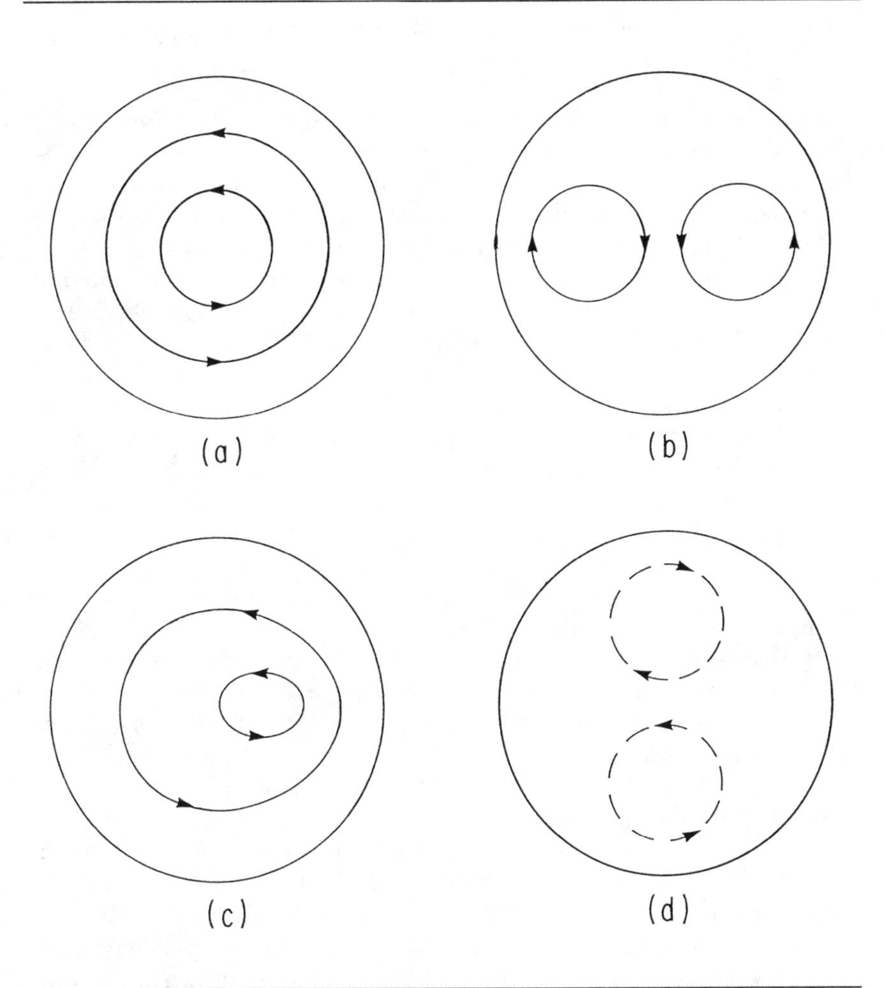

FIGURE 10.2
Four sketches showing projections of field lines into a plane of constant z, for the computationally-observed relaxed state. (a) The axisymmetric part of $\vec{B}$ only; $\vec{E}_{ext}$ and $\vec{B}_0$ are out of the plane of the paper. (b) The (smaller) helical part of $\vec{B}$ only; the pattern would be rotated in a plane of different z. (c) Field lines for the sum of the magnetic fields in (a) and (b). The null of the poloidal magnetic field follows a helical trajectory as z increases. (d) Projections of the velocity field $\vec{v}$ (associated with the helical vortex pair) into a plane of constant z. The pattern is basically the same as in (b), but rotated, so that the emf $(\vec{v} \times \vec{B}) \cdot \hat{e}_z$ opposes $\vec{E}_{ext}$.

we have been driven to by trying to do turbulent relaxation computations. We have found useful, as an organizing principle, a search for states of *minimum rate of energy dissipation* [26].

It has been known since the time of Kirchoff that the electric current distribution in a non-uniform rigid conductor will distribute itself in such a way as to minimize [27] the overall Ohmic dissipation, $\int \eta j^2 d^3x$, if the electrostatic potential is given over the boundary of the conductor. Later in the 19th century, Helmholtz, Korteweg and Lord Rayleigh [26] showed how to calculate the elementary hydrodynamic profiles by seeking states of minimum viscous dissipation, $\rho\nu \int \omega^2 d^3x$. The limitations of the method for hydrodynamics, for which the method has never been fully proved, are uncertain. It is sometimes asserted that the method only works for situations in which $\vec{v} \cdot \nabla \vec{v} = 0$ ("Stokes flow"), but it seems to give the correct answer for rotating Couette flow. Certainly it does not have any clear interpretation when Reynolds numbers exceed instability thresholds. But it has strong intuitive appeal for weakly driven flows in which one may regard such gradients as there are as the result of a thwarted attempt to reach the gradient-free thermal equilibrium state, where only the boundary conditions demand the presence of some non-zero gradients. In this mean-square sense, the minimum energy dissipation state is as close to a "minimum gradient" state as the boundary conditions will allow.

It is possible to begin to explore those MHD states which are minimum-dissipation states in the sense that, in the dimensionless units, $\int(\nu\omega^2 + \eta j^2)d^3x \equiv R$ is as small as it can be, compatible with boundary conditions and constraints [28].

Deciding what the boundary conditions and constraints ought to be is a task that requires patience and ingenuity. For the problem described in Secs. 10.2 and 10.3 there are two kinds of constraints. The first is exemplified by the constant rate of supply and dissipation of magnetic helicity proportional to $\eta \int \vec{j} \cdot \vec{B} d^3x = \int \vec{E}_{ext} \cdot \vec{B} d^3x$, which the boundary conditions alone determine. Minimizing R subject to such global integral constraints for $\vec{v} = 0$ states and the boundary conditions stated, is an exercise in variational calculus of a familiar type. It tends to lead to quiescent ($\vec{v} = 0$) magnetic profiles which for weak enough driving are axisymmetric [28]. However, they are not time independent steady states because, lacking a $\vec{v}$, they do not satisfy Eq. (10.3), the Ohm's law, or more specifically, they are *not* time-independent solutions of Eqs. (10.6) and (10.8).

The problem is that the Ohm's law is itself a constraint, pointwise rather than global, and in effect determines a $\vec{v}$ locally in terms of $\vec{B}$, $\vec{E}_{ext}$ and $\nabla \times \vec{B}$. This $\vec{v}$ has, in general, an associated $\vec{\omega}$ which contributes to R. Once this pointwise constraint is introduced, it becomes a much more serious matter to minimize R; it is no longer a standard problem.

We have not yet succeeded fully in solving this problem. What we *have* done [29] is to construct an explicit solution to the equations of motion that has a finite $\vec{v}$ which lowers the dissipation rate R *below* the minimum value that it can acquire if $\vec{v} = 0$ is assumed at the outset. This proves that whatever the minimum dissipation state is, it must have some associated flow. Moreover, the explicitly-constructed state bears more than a little similarity to the computationally-identified states described in Sec. 10.3.

The construction is perturbation-theoretic; the original paper should be consulted for details [29]. A series expansion

$$\begin{aligned} \vec{B} &= \vec{B}^{(0)} + \epsilon \vec{B}^{(1)} + \epsilon^2 \vec{B}^{(2)} + \ldots \\ \vec{v} &= \epsilon \vec{v}^{(1)} + \epsilon^2 \vec{v}^{(2)} + \ldots \end{aligned} \tag{10.12}$$

is carried out through terms of $0(\epsilon^2)$. ϵ is a formal expansion parameter which indicates the smallness of the helical components $\vec{B}^{(1)}$, $\vec{v}^{(1)}$, relative to the axisymmetric component $\vec{B}^{(0)}$.

$\vec{B}^{(0)}$ is the *only* zero-flow, static, axisymmetric solution of Eqs. (10.3) and (10.5): $\vec{B}^{(0)} = B_0 \hat{e}_z + \frac{1}{2} j_0 r \hat{e}_\varphi$, where B_0 and j_0 are uniform constants. j_0 is E_{ext}/η. The insulator-coated conducting boundary conditions are necessary, since j_0=constant is not a solution for a bare, perfectly-conducting wall.

$\vec{B}^{(1)}$ and $\vec{v}^{(1)}$ are formally a linear normal-mode solution about $\vec{B}^{(0)}$ at the threshold where the first normal mode becomes linearly unstable. Explicitly, $\vec{B}^{(1)}$ is an (arbitrary multiplier times) the function

$$\vec{B}^{(1)} = -A \sin\psi \hat{e}_r + B \cos\psi \hat{e}_\phi + C \cos\psi \hat{e}_z \ , \tag{10.13}$$

while $\vec{v}^{(1)}$ is (the same arbitrary multiplier times)

$$\vec{v}^{(1)} = \left(\frac{\eta\lambda^2}{\frac{1}{2} m j_0 - k_n B_0} \right) [A \cos\psi \hat{e}_r + B \sin\psi \hat{e}_\varphi + C \sin\psi \hat{e}_z] \ . \tag{10.14}$$

The notation must be explained. First $\psi \equiv m\varphi - k_n z$, where m and $n = k_n L_z / 2\pi$ are from this point on the integers which characterize the first normal mode that crosses the zero-frequency threshold into instability. The quantity $A \equiv ((\lambda m/r) J_m - k_n dJ_m/dr)$, $B \equiv (-\lambda dJ_m/dr + (mk_n/r) J_m)$, and $C \equiv \gamma^2 J_m$, where $J_m(\gamma r)$ is the Bessel function of the first kind and $\lambda^2 = \gamma^2 + k_n^2$. Both λ and γ are determined by the requirement that $A = 0$ at $r = a$. This has the implication that $\vec{B} \cdot \hat{n} = 0$, $\vec{j} \cdot \hat{n} = 0$, $\vec{v} \cdot \hat{n} = 0$, and $\vec{\omega} \cdot \hat{n} = 0$ at $r = a$ at this order, but we are left with finite values of $\vec{j} \times \hat{n}$ and $\vec{v} \times \hat{n}$ there. (The finite value of $\vec{v} \times \hat{n}$ does not permit no-slip boundary conditions and since $\vec{\vec{\sigma}} \cdot \hat{n} \neq 0$ at $r = a$, stress-free boundary conditions are not obeyed either. This is a matter about which more will have to be said later.)

The second order part of the solution has

$$\vec{v}^{(2)} = 0 \tag{10.15}$$

and

$$\vec{j}^{(2)} = \frac{1}{\eta}\vec{v}^{(1)} \times \vec{B}^{(1)} = \frac{1}{\eta}\left(\frac{\eta\lambda^2}{\frac{1}{2}mj_0 - k_n B_0}\right)\left[-AC\hat{e}_\varphi + AB\hat{e}_z\right] , \tag{10.16}$$

with $\nabla \times \vec{B}^{(2)} = \vec{j}^{(2)}$ determining $\vec{B}^{(2)}$. Of course, if the first-order solution is multiplied by an arbitrary number, the second-order solution is multiplied by the square of that number and so on in the usual manner of perturbation series. However, $\vec{B}^{(0)}$, and $\vec{v}^{(0)} = 0$, are still an exact solution for the same value of E_{ext} without any $\vec{v}^{(1)}$, $\vec{B}^{(1)}$. The point is that if the total dissipation rate R is computed accurately through terms of second order, it is *lowered* by the presence of a finite $\vec{v}^{(1)}$ and $\vec{B}^{(1)}$ *below* the value of R for $\vec{v}^{(1)} = 0 = \vec{B}^{(1)}$ [29]. When various perspective plots and contour plots of the second-order solution are drawn, they look in some detail like the computed plots arrived at by numerical simulation as described in Sec. 10.3 [29]. The similarity is closest to the Strauss equation plots since the perturbation-theoretic nature of the solution makes it impossible to believe in large-scale modifications of the zeroth-order such as field reversal (a sign change of B_z at $r = a$).

The solution has two less-than-fully-satisfactory features. First, there is at this point no determination of the limiting amplitude or the multiplier by which $\vec{v}^{(1)}$, $\vec{B}^{(1)}$ are absolutely determined. All we know is that the presence of the helical components lowers the total dissipation. Probably the perturbation theory would have to be taken through fourth order to provide a positive-definite expression for the expanded R, thus determining a genuine minimum for it. So far, this has proved prohibitively complicated.

Second, the boundary conditions on $\vec{v}$ are less than satisfactory. Neither no-slip nor stress-free boundary conditions have been easy to implement. The best we have been able to do is to characterize the neighborhood of the wall as a thin, viscous sublayer in which ν rises steeply to infinity at $r = a$ and in which $\vec{v} \to 0$. We do not have an explicit boundary layer solution in the sublayer. It is the case that, because of the rapidly falling temperature, both ν and η may be expected to rise rapidly near the wall in real-life situations. The matter of realistic velocity boundary conditions clearly requires more attention.

Perhaps the most pointed lesson to be drawn from these studies is the need for robust enough computer codes, unrestricted by computationally-convenient symmetries, to permit the unexpected and unplanned to occur without causing a breakdown in the numerical method. The theoretical framework in which previous analytical treatments of voltage-driven cylinders of magnetofluid had operated simply did not, and could not, anticipate the steady-state presence of these helical vortex pairs. There is no more indefensible use of large blocks of computer time than confirming previously-existing "party-line" scenarios for what is *supposed* to happen.

10.5 Summary

The elementary states of driven, dissipative MHD with realistic boundary conditions have not been explored at a level of detail that could be considered an adequate theoretical foundation for the operation of confinement devices which have dominated plasma physics for a generation. The stability studies of ideal, axisymmetric, zero-flow equilibria are an inadequate dynamical framework to encompass the time-dependent MHD processes that go on in a real plasma. Obvious lessons from hydrodynamics have been inadequately learned, in that the resistive, driven steady states of a magnetofluid have not been satisfactorily identified.

Three-dimensional MHD computation, unrestricted by overly specialized or unnaturally symmetric initial conditions, has provided some evidence concerning nearly-stationary driven, dissipative states of bounded magnetofluids. These are not axisymmetric, but partially helical, and involve helical vortex pairs that account for a few percent of the energy of the magnetofluid.

An alternative framework (*i.e.*, alternative to the ideal, axisymmetric, zero-flow equilibria with linearly unstable normal modes) is provided by the century-old principle of minimum energy dissipation rate. It is a somewhat surprising result that states of lower dissipation rate than those of the axisymmetric, zero-flow states can be achieved by allowing a helical departure from axisymmetry and permitting helical vortex pairs. The resulting $\vec{v} \times \vec{B}$ emf that follows from the helical components is axisymmetric and opposes the applied emf, resulting in an overall reduction of the toroidal current below its expected classical value. A second (pleasant) surprise is the extent to which such states can be constructed analytically and then turn out to resemble the computer solutions.

The present major obstacle to a non-perturbation-theoretic determination of the minimum-dissipation states is the need to incorporate the MHD Ohm's law in the variational calculation which minimizes the dissipation rate. This is a *pointwise* constraint relating $\vec{v}$, $\vec{B}$ and $\nabla \times \vec{B}$ at all interior points of the magnetofluid and thereby fails to fit into the standard mathematical machinery for including global, integral constraints in variational problems. A possible scheme for dealing with this difficulty is outlined in the Appendix.

In summary, we are only beginning to approach the level of theoretical understanding of the elementary, driven, dissipative states of a magnetofluid that it might have been reasonable to require in the 1960s, before the present elephantine program of "reactor construction" was embarked upon. And much of such theoretical understanding as we do have has been acquired on the margins of the subject, somewhat in defiance of the mainstream program.

Acknowledgements

Thanks for their valuable contributions are expressed to Drs. J. P. Dahlburg, G. D. Doolen, W. H. Matthaeus, L. Phillips, M. L. Theobald, and L. Turner. This work was supported in part by U.S. Department of Energy, Grant DE-FG02-85ER53194, and National Aeronautics and Space Administration, Grant NAG-W-710.

Bibliography

[1] H. Alfvén, *Cosmical Electrodynamics* (Oxford, Clarendon Press, 1950).

[2] T.G. Cowling, *Magnetohydrodynamics* (New York, Interscience, 1957).

[3] V.C.A. Ferraro and C. Plumpton, *Magneto-fluid Mechanics* (Oxford, Clarendon Press, 1961).

[4] J.A. Shercliff, *A Textbook of Magnetohydrodynamics* (Oxford, Pergamon Press, 1965).

[5] S.I. Braginskii, in *Reviews of Plasma Physics*, **1**, ed. M.A. Leontovich (New York, Consultant's Bureau, 1965) p. 205ff.

[6] L.D. Landau and E.M. Lifshitz, *Fluid Mechanics* (London, Pergamon Press, 1969).

[7] G.K. Batchelor, *An Introduction to Fluid Dynamics* (Cambridge, U. K., Cambridge University Press, 1967).

[8] G.K. Batchelor, *The Theory of Homogeneous Turbulence* (Cambridge, U. K., Cambridge University Press, 1970).

[9] G. Bateman, *MHD Instabilities* (Cambridge, MA, MIT Press, 1978).

[10] *Proceedings of the International Conferences on Plasma Physics and Controlled Thermonuclear Fusion Research* (These conferences occur biennially; for example, the eleventh such conference was held in Kyoto in 1986. They are published by the International Atomic Energy Agency, Vienna, frequently as a supplement to *Nuclear Fusion*. They are perhaps the most specialized concentration of reported controlled fusion research from the USA, the USSR, Western Europe and Japan.)

[11] J.P. Dahlburg, D. Montgomery and W.H. Matthaeus, J. Plasma Phys. **34** (1985) 1.

[12] J.P. Dahlburg, D. Montgomery, G.D. Doolen and W.H. Matthaeus, J. Plasma Phys. **35** (1986) 1.

[13] J.P. Dahlburg, D. Montgomery, G.D. Doolen and L. Turner, Phys. Rev. Lett. **57** (1986) 428.

[14] J.P. Dahlburg, D. Montgomery, G.D. Doolen and L. Turner, J. Plasma Phys. **37** (1987) 299.

[15] J.P. Dahlburg, D. Montgomery, G.D. Doolen and L. Turner, J. Plasma Phys. **40** (1988) 39.

[16] M.L. Theobald, D. Montgomery, G.D. Doolen and J.P. Dahlburg, Phys. Fluids **B1** (1989) 766.

[17] L. Turner and J.P. Christiansen, Phys. Fluids **24** (1981) 893.

[18] D. Gottlieb and S.A. Orszag, *Numerical Analysis of Spectral Methods: Theory and Applications* (Philadelphia, Society for Industrial and Applied Mathematics, 1977).

[19] C. Canuto, M.Y. Hussaini, A. Quarteroni and T.A. Zang, *Spectral Methods in Fluid Dynamics* (New York, Springer-Verlag, 1988).

[20] J.P. Dahlburg, *Turbulent Disruptions from the Strauss Equations*, Ph. D. Thesis, College of William and Mary, Williamsburg, VA, (1985).

[21] M.L. Theobald, *A Numerical Simulation of Tokamak Sawteeth*, Ph. D. Thesis, Dartmouth College, Hanover, NH (1988).

[22] J.B. Taylor, Phys. Rev. Lett. **33** (1974) 1139.

[23] J.B. Taylor, in *Proc. 5th Int. Conf. on Plasma Phys. and Contr. Fusion Research*, Tokyo, 1974 (Vienna, IAEA, 1975) **1**, pp. 161 ff.

[24] J.B. Taylor, Revs. Mod. Phys. **58** (1986) 741.

[25] A.C. Ting, W.H. Matthaeus and D. Montgomery, Phys. Fluids **29** (1986) 3261.

[26] H. Lamb, *Hydrodynamics*, 6th ed. (New York, Dover, 1945) pp. 617–619.

[27] E.T. Jaynes, Ann. Rev. Phys. Chem. **31** (1980) 579.

[28] D. Montgomery and L. Phillips, Phys. Rev. A **38** (1988) 2953.

[29] D. Montgomery, L. Phillips and M.L. Theobald, "Helical, Dissipative, Magnetohydrodynamic States with Flow," Phys. Rev. A **40** (1989) to appear in August issue.

[30] S. Chandrasekhar and P.C. Kendall, Astrophys. J. **126** (1957) 457.

[31] L. Turner, Ann. Phys. (N.Y.) **149** (1983) 58.

CHAPTER APPENDIX
Non–perturbative, Minimum–dissipation States

In the light of Secs. 10.3 and 10.4, it is natural to ask how to construct the partially-helical states with their associated vortex pairs when the perturbation-theoretic assumptions employed in Sec. 10.4 fail, as they do for the RFP computations [11–13]. The perturbation theory is intrinsically limited to modifications of the uniform-current profile which are in some sense "small." The approximation may predict correctly some features of tokamak operation, but clearly cannot deal with phenomena like those in the reversed-field pinch [11–13] where the poloidal currents are large enough to reverse the sign of the dc magnetic field at the wall.

A non-perturbative approach to calculating minimum dissipation states which satisfy the equations of motion is described, but not developed in detail, in this appendix. The method incorporates the well-documented tendency [13–15] for a strongly-driven magnetofluid to relax to a state which is close to "force free" in the sense that $\vec{j} \times \vec{B} \cong 0$. This fact, combined with the observation that the kinetic energy remains small relative to the magnetic energy, suggests a successive-approximations procedure based upon expansions in *Chandrasekhar-Kendall functions.*

Chandrasekhar-Kendall functions [30,31] are solenoidal vector eigenfunctions of the curl: *e.g.*, $\nabla \times \vec{A} = \lambda \vec{A}$, where λ is a constant. Since $\vec{j} = \nabla \times \vec{B}$, a magnetic field which is a Chandrasekhar-Kendall function will be "force-free." It is to be noted that the helical functions $\vec{v}^{(1)}$, $\vec{B}^{(1)}$ of Sec. 10.4 are Chandrasekhar-Kendall functions. They may be constructed, for cylindrical geometry, out of solutions to the scalar Helmholtz equation $(\nabla^2 + \lambda^2)\chi = 0$. In terms of χ,

$$\vec{A} = \nabla \times \hat{e}_z \chi + \lambda \nabla \times (\nabla \times \hat{e}_z \chi) \,. \tag{10.17}$$

Explicitly, $\chi = J_m(\gamma_{nmq} r) \exp(im\varphi - ik_n z)$ where $k_n = 2\pi n / L_z$ and n, m, q are integers. J_m is the Bessel function of the first kind, and $\lambda^2_{nmq} = k_n^2 + \gamma^2_{nmq}$, with the $\lambda = \lambda_{nmq}$ determined by boundary conditions. It thus requires three integer indices, n, m, q, to identify a particular $\vec{A} = \vec{A}(n, m, q)$. Both signs of λ are in general allowed and Eq. (10.17) defines two real eigenfunctions. The set may be assumed complete.

We may expand

$$\vec{B} = \sum \xi_b \vec{A}$$
$$\vec{v} = \sum \xi_v \vec{A} \qquad (10.18)$$

in a highly abbreviated notation. The sums in Eq. (10.18) are over n, m, q and the $\vec{A}$'s are the orthonormal set of real eigenfunctions defined by Eq. (10.17). The ξ_b and ξ_v are amplitudes for the magnetic field and velocity field, real numbers whose specification will completely determine $\vec{B}$ and $\vec{v}$. ξ_b and ξ_v are likewise bearers of the indices m, n, q.

In terms of these amplitudes ξ_b and ξ_v, the dissipation rate R to be minimized becomes the quadratic form

$$R = \sum (\eta \lambda^2 \xi_b^2 + \nu \lambda^2 \xi_v^2) , \qquad (10.19)$$

with the variables being the ξ_b, ξ_v, and the summations running over n, m, q.

The constraints imposed keep the ξ_b and ξ_v from being zero. These are of two types: (1) kinematic constraints such as the constancy of toroidal magnetic flux and the boundary conditions and these may be built into the expansion functions themselves; (2) those implied by the Ohm's law, Eq. (10.3).

For the static case $\vec{E}$ in Eq. (10.3) can be written as $\vec{E}_{ext} + \nabla\tilde{\Phi}$, and $\tilde{\Phi}$ can be expressed in terms of $\vec{v}$ and $\vec{B}$ by taking the divergence of Eq. (10.3) and solving the resulting Poisson equation. As previously remarked, it has been the pointwise (rather than integral) nature of the constraint imposed by the Ohm's law which has been the principal obstacle to the application of standard minimization procedures based upon Lagrange undetermined multipliers.

The expansion of Eq. (10.18) may now be used to replace the Ohm's law with a sequence of standard-variety integral constraints. If we insert Eq. (10.18) for $\vec{v}$ and $\vec{B}$ into

$$\vec{E}_{ext} + \nabla\tilde{\Phi} + \vec{v} \times \vec{B} = \eta \nabla \times \vec{B} , \qquad (10.20)$$

we may then take the inner product of Eq. (10.20) with the Chandrasekhar-Kendall functions in turn (and with a constant vector, which is also a Chandrasekhar-Kendall function) and obtain a sequence of algebraic constraints which relate the ξ_b and ξ_v to each other. The coefficients of the ξ_b and ξ_v are performable, if tedious, integrals involving the Chandrasekhar-Kendall functions.

Of course the sequence of constraints so obtained is in general infinite, as is the number of terms in Eq. (10.18). The utility of the method will depend upon rapid convergence because the difficulty of evaluating the ξ_b

and ξ_v will increase rapidly as the number of terms and retained constraints increases.

The rapidity of the convergence has not been explored at the time of this writing (June, 1989). But a set of fields which already considerably resembles those obtained in the computations [11–15] results from taking a very restrictive truncation: two non-zero ξ_b's, one non-zero ξ_v, and one integral of Eq. (10.20). The two non-zero ξ_b's correspond to an axisymmetric ($m = 0 = n$) Chandrasekhar-Kendall function or "Taylor state" [22–24] and a smaller ($m \neq 0$, $n \neq 0$) helical contribution, while the non-vanishing ξ_v corresponds to a helical contribution with the same m and n. The constraint is simply the integral of Eq. (10.20) dotted into $\hat{e}_z$. The resultant fields are the same as those obtained in perturbation theory in Sec. 10.4, except for the important difference that the zeroth-order uniform-current fields have been replaced by the force-free field, proportional to

$$\vec{B}^{(0)} = \hat{e}_\varphi J_1(\lambda_0 r) + \hat{e}_z J_0(\lambda_0 r) \,. \tag{10.21}$$

The resulting magnetic field at the lowest nontrivial approximation is thus the vector sum of two Chandrasekhar-Kendall fields: one axisymmetric and one helical.

11

Color Transparency and the Structure of the Proton in Quantum Chromodynamics

Stanley J. Brodsky
Stanford Linear Accelerator Center
Stanford University
Stanford, CA 94305

11.1 Introduction

One of the most remarkable claims of theoretical physics, is that the Lagrangian density of Quantum Chromodynamics (QCD),

$$\begin{aligned} \mathcal{L}_{QCD} &= -\frac{1}{2}\mathrm{Tr}\,[F^{\mu\nu}F_{\mu\nu}] + \bar{\psi}(i\not{D} - m)\psi \\ F^{\mu\nu} &= \partial^\mu A^\nu - \partial^\nu A^\mu + ig[A^\mu, A^\nu] \\ D^\mu &= \partial^\mu + ig\,A^\mu \end{aligned} \tag{11.1}$$

describes all aspects of the hadron and nuclear physics. This elegant expression compactly describes a renormalizable theory of color-triplet spin-$\frac{1}{2}$ quark fields ψ and color-octet spin-1 gluon fields A^μ with an exact symmetry under SU(3)-color local gauge transformations. According to QCD, the elementary degrees of freedom of hadrons and nuclei and their strong interactions are the quark and gluon quanta of these fields. The theory is, in fact, consistent with a vast array of experiments, particularly high momentum transfer phenomena, where because of the smallness of the effective coupling constant and factorization theorems for both inclusive and exclusive processes, the theory has high predictability [1]. (The term "exclusive" refers to reactions in which all particles are measured in the final state.)

The general structure of QCD indeed meshes remarkably well with the facts of the hadronic world, especially quark-based spectroscopy, current algebra, the approximate point-like structure of large momentum transfer inclusive reactions, and the logarithmic violation of scale invariance in deep inelastic lepton-hadron reactions. QCD has been successful in predicting the features of electron-positron and photon-photon annihilation into hadrons, including the magnitude and scaling of the cross sections, the complete form of the photon structure function, the production of hadronic jets with patterns conforming to elementary quark and gluon subprocesses. Recent Monte Carlo studies incorporating coherence (angle-ordering) have been successful in reproducing the detailed features of the two-jet ($q\bar{q}$) and three-jet ($q\bar{q}g$) reactions. The experimental measurements appear to be consistent with the basic postulates of QCD, that the charge and weak currents within hadrons are carried by fractionally charged quarks, and that the strength of the interactions between the quarks and gluons becomes weak at short distances, consistent with asymptotic freedom.

Nevertheless in some very striking cases, the predictions of QCD appear to be in dramatic conflict with experiment:

1. The spin dependence of large angle pp elastic scattering has an extraordinarily rich structure— particularly at center of mass energies $E_{\rm cm} \simeq 5$ GeV. The observed behavior is quite different than the structureless predictions of the perturbative QCD theory of exclusive processes.

2. QCD predicts a rather novel feature: instead of the traditional Glauber theory of initial and final state interactions, QCD predicts negligible absorptive corrections, *i.e.*, the "color transparency" of high momentum transfer quasi-elastic processes in nuclei. A recent experiment at Brookhaven National laboratory seems to confirm this prediction, at least at low energies, but the data show, that at the same energy where the anomalous spin correlations are observed in pp elastic scattering, the color transparency prediction unexpectedly fails.

3. Recent measurements by the European Muon Collaboration of the deep inelastic structure functions on a polarized proton show a number of unexpected features; a strong positive correlation of the up quark spin with the proton, a strong negative polarization of the down quark, and a significant strange quark content of the proton. The EMC data indicate that the net spin of the proton is carried by gluons and orbital angular momentum, rather than the quarks themselves.

4. The J/ψ and ψ' are supposed to be simple S-wave n=1 and n=2 QCD bound states of the charm and anti-charm quarks. Yet these two states

have anomalously different two-body decays into vector and pseudoscalar hadrons.

5. The hadroproduction of charm states and charmonium is supposed to be predictable from the simple fusion subprocess $gg \rightarrow c\bar{c}$. However, recent measurements indicate that charm particles are produced at higher momentum fractions than allowed by the fusion mechanism, and they show a much more complex nuclear dependence than simple additivity in nucleon number predicted by the model.

All of these anomalies suggest that the proton itself is a much more complex object than suggested by simple non-relativistic quark models. Recent analyses of the proton distribution amplitude using QCD sum rules points to highly nontrivial proton structure. Solutions to QCD in one space and one time dimension suggest that the momentum distributions of non-valence quarks in the hadrons have a non-trivial oscillatory structure. The data seems also to be suggesting that the "intrinsic" bound state structure of the proton has a non-negligible strange and charm quark content, in addition to the "extrinsic" sources of heavy quarks created in the collision itself. As we shall see in this lecture, the apparent discrepancies with experiment are not so much a failure of QCD, but rather symptoms of the complexity and richness of the theory. An important tool for analyzing this complexity is the light-cone Fock state representation of hadron wavefunctions, which provides a consistent but convenient framework for encoding the features of relativistic many-body systems in quantum field theory.

11.1.1 General Features of QCD

The quark fields of QCD carry flavor quantum numbers as well as the electromagnetic and weak currents. The charge and other quantum numbers of the hadrons thus reflect the quantum numbers of their quark constituents. However, the proton in QCD is only to first approximation a bound state of two u and one d quark. Because of quantum fluctuations, one expect that a highly relativistic bound state contains admixtures of $|uudg\rangle$, $|uuds\bar{s}\rangle$ and other higher particle number Fock components which match the proton's global quantum numbers. There is some evidence that the proton wavefunction even contains $|uudc\bar{c}\rangle$ Fock states at the half-percent level.

The exchange of the spin-one color-octet gluons between the quarks and other gluons leads to strong confining forces at large distances, but progressively weaker forces at short distances. This is the "asymptotic freedom" property of QCD which allows perturbative calculations of large momentum transfer processes. The gluons are neutral with respect to the flavor and electroweak charges. In principle, QCD should give just as accurate a description of hadronic phenomena as quantum electrodynamics provides for the interactions of leptons. However, because of its non-Abelian structure, calculations in QCD are much more complex. The central feature of the theory is, in fact, its non-perturbative nature which evidently leads to the confinement of quarks and gluons in color-singlet bound states. Rigorous proofs of confinement, however, have not been given. Because of the postulated confinement of the colored quanta, observables always involve the dynamics of bound systems; hadron-hadron interactions are thus as complicated as the Van der Waals and covalent exchange forces of neutral atoms.

Unlike atomic physics, the constituents of light hadrons in QCD are highly relativistic; because the forces are non-static, a hadron cannot be represented as a state of fixed number of quanta at a fixed time. The vacuum structure of the QCD Hamiltonian quantized at fixed time relative to the perturbative basis is also complex; it is believed that virtually every local color-singlet operator constructed from the product of quark and gluon fields may have a non-zero vacuum condensate expectation value. In the light-cone framework, the vacuum itself is trivial since it is an eigenstate of the bare Hamiltonian; the complexity of the vacuum at equal time gets shifted to the complexity of the Fock representation when one quantizes the theory on the light cone.

Despite the complexity of the theory, QCD has several key properties which make calculations tractable and systematic, at least in the short-distance, high momentum-transfer domain. The critical feature is asymptotic freedom: the effective coupling constant $\alpha_s(Q^2)$ which controls the interactions of quarks and gluons at momentum transfer Q^2 vanishes logarithmically at high Q^2:

$$\alpha_s(Q^2) = \frac{4\pi}{\beta \log(Q^2/\Lambda^2_{\mathrm{QCD}})} , \qquad (Q^2 \gg \Lambda^2_{\mathrm{QCD}}) . \tag{11.2}$$

(Here $\beta = 11 - \frac{2}{3} n_f$ is derived from the gluonic and quark loop corrections to the effective coupling constant; n_f is the number of quark contributions to the vacuum polarization with $m_f^2 \lesssim Q^2$.) The parameter Λ_{QCD} normalizes the value of $\alpha_s(Q_0^2)$ at a given momentum transfer Q_0^2, given a specific renormalization or cutoff scheme. The value of α_s can be determined fairly unambiguously using the measured branching ratio for upsilon radiative decay $\Upsilon(b\bar{b}) \to \gamma X$ [2]:

$$\alpha_s(0.157\ M_\Upsilon) = \alpha_s(1.5\ \mathrm{GeV}) = 0.23 \pm 0.03 . \tag{11.3}$$

Taking the standard $\overline{MS}$ dimensional regularization scheme, this gives $\Lambda_{\overline{MS}} = 119^{+52}_{-34}$ MeV. A recent analysis of logarithmic scale-breaking of the isoscalar nucleon structure functions $F_2(x, Q^2)$ and $xF_3(x, Q^2)$ from deep inelastic neutrino and anti-neutrino interactions in neon by the BEBC

WA59 collaboration [3] gives values for $\Lambda_{\overline{MS}}$ in the neighborhood of 100 MeV. The observed multijet distributions [6] in e^+e^- annihilation also suggest that $\Lambda_{\overline{MS}}$ is below 200 MeV and is perhaps as small as 100 MeV. In order to determine the absolute value of $\Lambda_{\overline{MS}}$ one must know the correct argument Q^* of the running coupling constant appropriate to the measurement. The above determinations of $\Lambda_{\overline{MS}}$ use the method of Ref. [4] in which this scale is determined "automatically" by requiring that light fermion pairs contributions are summed by the running coupling constant, just as is done in Abelian QED.

In more physical terms, the effective potential between infinitely heavy quarks has the form ($C_F = \frac{4}{3}$ for $n_c = 3$) [4],

$$\begin{aligned} V(Q^2) &= -C_F \, \frac{4\pi\alpha_V(Q^2)}{Q^2} \\ \alpha_V(Q^2) &= \frac{4\pi}{\beta \log(Q^2/\Lambda_V^2)} \, , \qquad (Q^2 \gg \Lambda_V^2) \end{aligned} \tag{11.4}$$

where $\Lambda_V = \Lambda_{\overline{MS}} e^{\frac{5}{6}} \simeq 270 \pm 100$ MeV. Thus the effective physical scale of QCD is ~ 1 fm^{-1}. At momentum transfers beyond this scale, α_s becomes small, QCD perturbation theory should begin to become applicable, and a microscopic description of short-distance hadronic and nuclear phenomena in terms of quark and gluon subprocesses is expected to become viable.

The above argument is the main basis for the reliability of perturbative calculations for processes in which all of the interacting particles are forced to exchange large momentum transfer (a few GeV). Complimentary to asymptotic freedom is the existence of factorization theorems for both exclusive and inclusive processes at large momentum transfer which are valid for all gauge theories. In the case of exclusive processes (in which the kinematics of all the final state hadrons are fixed), any hadronic scattering amplitude can be represented as the product of a hard-scattering amplitude for the constituent quarks, convoluted with a distribution amplitude for each in-going or out-going hadron. The distribution amplitude contains all of the bound-state dynamics and specifies the momentum distribution of the quarks in each hadron independent of the process. The hard scattering amplitude can be calculated perturbatively in powers of $\alpha_s(Q^2)$. The predictions can be applied to form factors, exclusive photon-photon reactions, photoproduction, fixed-angle scattering, *etc.*

In the case of high momentum transfer inclusive reactions (in which final state hadrons are summed over), the hadronic cross section can be computed from the product of a perturbatively-calculable hard-scattering subprocess cross section involving quarks and gluons convoluted with the appropriate quark and gluon structure functions which incorporate all of bound-state dynamics. Since the distribution amplitudes and structure functions only depend on the composition of the respective hadron, but not the nature of the high momentum transfer reaction, the complicated non-perturbative QCD dynamics is factorized out as universal quantities. Recently there has been encouraging progress in actually calculating these fundamental quantities, which I shall briefly review here. Eventually these calculations can be compared with the phenomenological parameterization extracted from inclusive and exclusive experiments.

11.1.2 Hadronic Structure in QCD

The central unknown in the QCD predictions is the composition of the hadrons in terms of their quark and gluon quanta. Recently several important tools have been developed which allow specific predictions for the hadronic wave functions directly from the theory. A primary tool is the use of light-cone quantization to construct a consistent relativistic Fock state basis for the hadrons in terms of quark and gluon quanta. The distribution amplitude and the structure functions are defined directly in terms of these light-cone wave functions. The form factor of a hadron can be computed exactly in terms of a convolution of initial and final light-cone Fock state wave functions.

A second important tool is the use of QCD sum rules to constrain moments of the hadron distribution amplitudes [5]. This method, developed by Chernyak and Zhitnitskii (CZ), has yielded important information on the possible momentum space structure of hadrons. A particularly important advance is the construction of nucleon distribution amplitudes, which together with the QCD factorization formulae, predicts the correct sign and magnitude as well as scaling behavior of the proton and neutron form factors. A recent analysis by King and Sachrajda [7] has confirmed these results.

Another recent advance has been the development of a formalism to calculate the moments of the meson distribution amplitude using lattice gauge theory. The most recent analysis, by Martinelli and Sachrajda [8], gives moments for the pion distribution amplitude in good agreement with the QCD sum rule calculation. The results from both the lattice calculations and QCD sum rules also demonstrate that the light quarks are highly relativistic in the bound state wave functions. This gives further indication that while potential models are useful for enumerating the spectrum of hadrons (because they express the relevant degrees of freedom), they are not reliable predicting wave function structure. However, in the case of the proton, the lattice calculation [8] of the lowest moments suggests equal partition of momentum among the three valence quarks.

11.1.3 Fock State Expansion on the Light Cone

A key problem in the application of QCD to hadron and nuclear physics is how to determine the wave function of a relativistic multi-particle composite system. It is not possible to represent a relativistic field-theoretic bound system limited to a fixed number of constituents at a given time since the interactions create new quanta from the vacuum. Although relativistic wave functions can be represented formally in terms of the covariant Bethe-Salpeter formalism, calculations beyond ladder approximation appear intractable. Unfortunately, the Bethe-Salpeter ladder approximation is often inadequate. For example, in order to derive the Dirac equation for the electron in a static Coulomb field from the Bethe-Salpeter equation for muonium with $m_\mu/m_e \to \infty$ one requires an infinite number of irreducible kernel contributions to the QED potential. Matrix elements of currents and the wave function normalization also require, at least formally, the consideration of an infinite sum of irreducible kernels. The relative-time dependence of the Bethe Salpeter amplitudes for states with three or more constituent fields adds severe complexities.

A different and more intuitive procedure would be to extend the Schrödinger wave function description of bound states to the relativistic domain by developing a relativistic many-body Fock expansion for the hadronic state. Formally this can be done by quantizing QCD at equal time, and calculating matrix elements from the time-ordered expansion of the S-matrix. However, the calculation of each covariant Feynman diagram with n-vertices requires the calculation of $n!$ frame-dependent time-ordered amplitudes. Even worse, the calculation of the normalization of a bound state wave function (or the matrix element of a charge or current operator) requires the computation of contributions from all amplitudes involving particle production from the vacuum. (Note that even after normal-ordering, the interaction Hamiltonian density for QED, $H_I = e : \bar{\psi}\gamma_\mu\psi A^\mu :$, contains contributions $b^\dagger d^\dagger a^\dagger$ which create particles from the perturbative vacuum.)

Fortunately, there is a natural and consistent covariant framework, originally due to Dirac [10], (quantization on the "light front ") for describing bound states in gauge theory analogous to the Fock state in non-relativistic physics. This framework is the light-cone quantization formalism in which

$$
\begin{aligned}
|\pi\rangle &= |q\bar{q}\rangle\psi^\pi_{q\bar{q}} + |q\bar{q}g\rangle\psi^\pi_{q\bar{q}g} + \cdots \\
|p\rangle &= |qqq\rangle\psi^p_{qqq} + |qqqg\rangle\psi^p_{qqqg} + \cdots .
\end{aligned}
\tag{11.5}
$$

Each wave function component ψ_n, *etc.* describes a state of fixed number of quark and gluon quanta evaluated in the interaction picture at equal light-cone "time" $\tau = t + z/c$. Given the $\{\psi_n\}$, virtually any hadronic property can be computed, including anomalous moments, form factors, structure functions for inclusive processes, distribution amplitudes for exclusive processes, *etc.*

The use of light-cone quantization and equal τ wave functions, rather than equal t wave functions, is necessary for a sensible Fock state expansion. It is also convenient to use τ-ordered light-cone perturbation theory (LCPTh) in place of covariant perturbation theory for much of the analysis of light-cone dominated processes such as deep inelastic scattering, or large-$p_\perp$ exclusive reactions.

The use of quark and gluon degrees of freedom to represent hadron dynamics seems paradoxical since free quark and gluon quanta have not been observed. Nevertheless, we can use a complete orthonormal Fock basis of free quarks and gluons, color-singlet eigenstates of the free part H_0^{QCD} of the QCD Hamiltonian to expand any hadronic state at a given time t. It is particularly advantageous to quantize the theory at a fixed light-cone time $\tau = t + z/c$ and choose the light-cone $A^+ = A^0 + A^z = 0$ gauge since the formulation has simple properties under Lorentz transformations, there are no ghost (negative metric) gluonic degrees of freedom, and complications due to vacuum fluctuations are minimized. Thus in e^+e^- annihilation into hadrons at high energies it is vastly simpler to use the quark and gluon Fock basis rather than the set of $J = 1,\ J_z = 1,\ Q = 0$ multi-particle hadronic basis to represent the final state. Notice that the complete hadronic basis must include gluonium and other hadronic states with exotic quantum numbers. Empirically, the perturbative QCD calculations of the final state based on jets or clusters of quarks and gluons, have been shown to give a very successful representation of the observed energy and momentum distributions.

Since both the hadronic and quark-gluon bases are complete, either can be used to represent the evolution of a QCD system. For example, the proton QCD eigenstate can be defined in terms of its projections on the free quark and gluon momentum space basis to define Fock wavefunctions; the sum of squares of these quantities then defines the structure functions measured in deep inelastic scattering.

In the case of large momentum transfer exclusive reactions such as the elastic proton form factor, the state formed immediately after the hard collision is most simply described as a valence Fock state with the quarks at small relative impact parameter $b_\perp \sim 1/Q$, where $Q = p_T$ is the momentum transfer scale. Such a state has a small color-dipole moment and thus can penetrate a nuclear medium with minimal interaction. The small impact parameter state eventually evolves to the final recoil hadron, but at high energies this occurs outside the nuclear volume. Thus quasi-elastic hard exclusive reactions are predicted to have cross sections which are additive in the number of nucleons in the nucleus. This is the phenomenon of "color transparency" which is in striking contrast to Glauber and other calculations based on strong initial and final state absorption corrections. Alternatively, the small impact parameter state can be represented as a coherent sum of all hadrons with the same conserved quantum numbers. At high energies, the phase coherence of the state can be maintained through the nucleus, and the coherent state can penetrate the nucleus without interaction. This is the dual representation of color transparency.

In the following sections I will discuss recent developments in hadron and nuclear physics which make use of the quark/gluon light-cone Fock representation of hadronic systems. The method of discretized light-cone quantization (DLCQ) provides a numerical method for solving gauge theories in the light-cone Fock basis. Recent results for QCD in one space and one time are presented in Sec. 11.12. The most important tool for examining the structure of hadrons is deep inelastic and elastic lepton scattering. I give a survey of tests of QCD in exclusive and inclusive electroproduction in Sec. 11.3, especially experiments which use a nuclear target to filter or modify the hadronic state. I also give a brief review of what is known about proton structure in QCD. A new approach to shadowing and anti-shadowing of nuclear structure functions is also presented. The distinction between intrinsic and extrinsic contributions to the nucleon structure function is emphasized,

One of the most important challenges to the validity of the QCD description of proton interactions is the extraordinary sensitivity of high energy large angle proton-proton scattering to the spin correlations of the incident protons. A solution to this problem based on heavy quark thresholds is described in Sec. 11.11. A prediction for a new form of quasi-stable nuclear matter is also briefly discussed.

11.1.4 Probes of Hadron Structure

A scanning transmission electron microscope [11] provides an image of a specimen by combining information from both the elastically and inelastically scattered electrons that emerge after passing through the target. A high energy electroproduction experiment which measures both exclusive and inclusive reactions is a close analog of an electron microscope, providing images of the nucleon and nucleus at a resolution scale $\lambda \sim 1/Q$ where $Q^2 = -(p_e - p'_e)^2$ is the momentum transfer squared. At the most basic level, Bjorken scaling of deep inelastic structure functions implies the production of a single quark jet, recoiling against the scattered lepton. The spectator system— the remnant of the target remaining after the scattered quark is removed— is a colored $\overline{3}$ system. (See Fig. 11.1.) According to QCD factorization, the recoiling quark jet, together with the gluonic radiation produced in the scattering process, produces hadrons in a universal way, independent of the target or particular hard scattering reaction. This

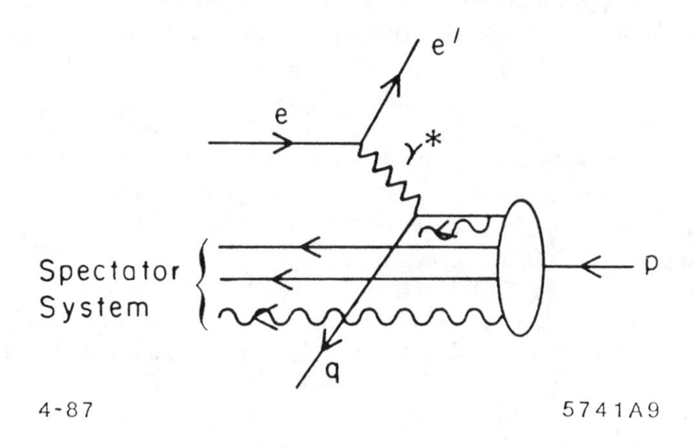

FIGURE 11.1
Struck quark and spectator systems in electroproduction.

jet should be identical to the light quark jets produced in e^+e^- annihilation. A very close analogy can be made between soft radiation from colored quark and gluon quanta, and soft photon radiation from charged particles in QED. In contrast, the hadronization of the spectator system depends in detail on the target properties. Unlike the quark jet, the leading particles of the target spectator system do not evolve and thus should not depend on the momentum transfer Q^2 (at fixed $W^2 = (q+p)^2$). Measurements of the final state radiation pattern in ep collisions at HERA should be able to discriminate between these different types of QCD radiators.

Measurements of the nucleon and nuclear structure functions have not only tested the short-distance properties of the theory (such as the scaling properties of structure functions and their logarithmic evolution with momentum transfer), but they have also illuminated the nonperturbative bound state structure of the nucleon and nuclei in terms of their quark and gluon degrees of freedom. For the most part, this information has been obtained from single-arm inclusive experiments where only the recoil lepton was detected.

In the future we can expect to see much more extensive measurements of the structure of the nucleon and nucleus by utilizing an internal target facility in an electron storage ring, such as PEP, Tristan, HERA, or LEP. The entire final state of electroproduction can be measured in coincidence with the scattered electron with close to 4π acceptance. In the case of the planned gas jet PEGASYS experiment at PEP ($E(e^{\pm}) \sim 15$ GeV), measurements can be performed well above the onset of Bjorken scaling. Both polarized and unpolarized hydrogen and nuclear targets may be feasible, and eventually even polarized electron beams may be available. High precision comparisons between electron and positron scattering would allow the study of higher order QED and electroweak interference effects. The asymmetry in the cross sections for $e^{\pm}p \to e^{\pm}\gamma X$ can be sizeable [12], providing a sum rule for the cube of the charges of the quarks in the target. The PEGASYS kinematic range interpolates between the lower energy CEBAF domain where quark degrees of freedom begin to become manifest, and the much higher energies of HERA, which is far into the perturbative QCD regime of logarithmic evolution and multi-jet structure.

Since the intrinsic mass scales of QCD $\Lambda_{\overline{MS}}$, $\langle k_{\perp}^2 \rangle^{\frac{1}{2}}$, and $m_q(q = u, d, s)$ are less than a few hundred MeV, quark and gluon degrees of freedom should become evident at momentum transfers as low as a few GeV. The observation of Bjorken scaling at Q^2 as low as 1 GeV^2 supports this argument. At larger momentum transfer, one studies logarithmic structure function evolution, the onset of new quark flavors, and multi-jet production. However, the dynamics of hadrons and nuclei in terms of their light quark and gluon degrees of freedom can be studied at moderate energies. At a more detailed level, the features of the standard leading twist description are modified by coherent or non-perturbative effects. For example, higher twist–power-law suppressed contributions arise when two or more quarks recoil against the scattered lepton.

The study of QCD phenomena in the intermediate energy range can also be carried out at $\bar{p}p$ facilities such as LEAR and the proposed AMPLE facility at Fermilab, designed to measure $p\bar{p}$ annihilation at anti-proton laboratory energies up to 10 GeV. The $\gamma\gamma$ reactions (for real and virtual photons from tagged $e^{\pm}$) provide some of the cleanest tests of QCD. Presently these reactions can be studied at the PEP, Cornell, and Tristan e^+e^- storage rings. One can test the scaling laws of QCD in exclusive reactions involving two large momentum scales, the virtual photon mass and the p_T of the reaction. An interesting feature of QCD is that at large p_T^2, the Q^2 dependence of each exclusive virtual photoproduction amplitude becomes minimal for $Q^2 \ll p_T^2$. This can be contrasted with the vector meson dominance model which predicts a universal fall-off in Q^2 at any p_T. This feature is due to the photon's point-like direct local coupling to the quark current in QCD. I will discuss QCD tests in photon-photon reactions in Sec. 11.8.

11.2 Nuclear Effects in QCD

The study of electroproduction in nuclear targets gives the experimentalist the extraordinary ability to modify the environment in which hadronization occurs. The essential question is how the nucleus changes or influences the mechanism in which the struck quark and the spectator system of the target nucleon form final state hadrons. The nucleus acts as a background field modifying the dynamics in interesting, though possibly subtle, ways. In particular, the observation of non-additivity of the nuclear structure functions as measured by the EMC and SLAC/American University collaborations have opened up a whole range of new physics questions:

1. What is the effect of simple potential-model nuclear binding, as predicted, for example, by the shell model? What is the associated modification of meson distributions required by momentum sum rules?

2. Is there a physical change in the nucleon size, and hence the shape of quark momentum distributions?

3. Are there nuclear modifications of the nucleonic and mesonic degrees of freedom, such as induced mesonic currents, isobars, six-quark states, or even "hidden color" degrees of freedom?

4. Does the nuclear environment modify the starting momentum scale evolution scale for gluonic radiative corrections?

5. What are the effects of diffractive contributions to deep inelastic structure functions which leave the nucleon or nuclear target intact?

6. Are there shadowing and possibly anti-shadowing coherence effects influencing the propagation of virtual photons or redistributing the nuclear constituents? Do these appear at leading twist?

7. How important are interference effects between quark currents in different nucleons? [13]

It seems likely that all of these non-additive effects occur at some level in the nuclear environment. In particular it will be important to examine the A-dependence of each reaction channel by channel.

The use of nuclear targets in electroproduction allows one to probe effects specific to the physics of the nucleus itself such as the short-distance structure of the deuteron, high momentum nucleon-nucleon components, and coherent effects such as shadowing, anti-shadowing, and $x > 1$ behavior. However, perhaps the most interesting aspect for high energy physics is the use of the nucleus to modify the environment in which quark hadronization and particle formation occurs.

11.2.1 The Target Length Condition

There are several general properties of the effect of the nuclear environment which follow from quantum mechanics and the structure of gauge theory. The first effect is the "formation zone" which reflects the principle that a quark or hadron can change state only after a finite intrinsic time in its rest system. This implies that the scattered quark in electroproduction cannot suffer an inelastic reaction with mass squared change ΔM^2 while propagating a distance L if its laboratory energy is greater than $\Delta M^2 L$. Thus at high energies, the quark jet does not change its state or hadronize over a distance scale proportional to its energy; inelastic or absorptive processes cannot occur inside a nucleus–at least for the very fast hadronic fragments. The energy condition is called the target length condition [16, 17]. However the outgoing quark can still scatter elastically as it traverses the nuclear volume, thus spreading its transverse momentum due to multiple scattering. Recently Bodwin and Lepage and I have explained the quantum mechanical origin of formation zone physics in terms of the destructive interference of inelastic amplitudes that occur on two different scattering centers in the nuclear target [19]. The discussion in that paper for the suppression of inelastic interactions of the incoming anti-quark in Drell-Yan massive lepton pair reactions can be carried over directly to the suppression of final state interactions of the struck quark in electroproduction.

11.2.2 Color Transparency

One can also use a nuclear target to test an important principle of gauge theory controlling quark hadronization into exclusive channels inside nuclei: "color transparency" [14]. Suppose that a hadronic state has a small transverse size $b_\perp$. Because of the cancellation of gluonic interactions with wavelength smaller than $b_\perp$, such a small color-singlet hadronic state will propagate through the nucleus with a small cross section for interacting either elastically or inelastically. In particular, the recoil proton in large momentum transfer electron-proton scattering is produced initially as a small color singlet three-quark state of transverse size $b_\perp \sim 1/Q$. If the electron-proton scattering occurs inside a nuclear target (quasi-elastic scattering) then the recoil nucleon can propagate through the nuclear volume without significant final-state interactions. This perturbative QCD prediction is in striking contrast to standard treatments of quasi-elastic scattering which predict significant final state scattering and absorption in the nucleus due to large elastic and inelastic nucleon-nucleon cross sections. The theoretical calculations of the color transparency effect must also take into account the expansion of the state as it evolves to a normal proton of normal transverse size while it traverses the nucleus. I will discuss color transparency further in Sec. 11.10.

11.2.3 Shadowing and Anti-Shadowing

One of the most striking nuclear effects seen in the deep inelastic structure functions is the depletion of the effective number of nucleons F_2^A/F_2^N in the region of low $x = x_{bj}$. The results from the EMC collaboration indicate that the effect is roughly Q^2-independent; *i.e.*, shadowing is a leading twist in the operator product analysis. In contrast, the shadowing of the real photo-absorption cross section due to ρ-dominance falls away as an inverse power of Q^2.

Shadowing is a destructive interference effect which causes a diminished flux and interactions in the interior and back face of the nucleus. The Glauber analysis of hadron-nucleus scattering corresponds to the following: the incident hadron scatters elastically on a nucleon N_1 on the front face of the nucleus. At high energies the phase of the amplitude is imaginary. The hadron then propagates through the nucleus to nucleon N_2 where it interacts inelastically. The accumulated phase of the propagator is also i so that this multi-scattering amplitude is coherent and opposite in phase to the amplitude where the beam hadron interacts directly on N_2 without initial-state interactions. Thus the target nucleon N_2 sees less incoming flux; it is shadowed by elastic interactions on the front face of the nucleus. If the hadron-nucleon cross section is large, then the effective number of nucleons participating in the inelastic interactions is reduced to $\sim A^{2/3}$, the number of surface nucleons.

In the case of virtual photo-absorption, the photon converts to a $q\bar{q}$ pair at a distance proportional to $\omega = x^{-1} = 2p \cdot q/Q^2$ laboratory frame. The nuclear structure function F_2^A can then be written as an integral over the inelastic cross section $\sigma_{\bar{q}A}(s')$ where s' grows as $1/x$ for fixed space-like $\bar{q}$ mass. Thus the $A-$dependence of the cross section is equivalent to the shadowing of the $\bar{q}$ interactions in the nucleus. Recently Hung Lu and I have applied the standard Glauber multi-scattering theory, assuming that formalism can be taken over to off-shell $\bar{q}$ interactions [20]. Our results show that for reasonable values of the $\bar{q}$-nucleon cross section, one can easily understand the magnitude of the shadowing effect at small x. Moreover, if one introduces an $\alpha_R \simeq \frac{1}{2}$ Reggeon contribution to the $\bar{q}N$ amplitude, the real part of the phase introduced by such a contribution automatically leads to "anti-shadowing" at $x \sim 0.1$ (effective number of nucleons $F_2^A(x,Q)/F_2^N(x,Q) > A$) of the few percent magnitude seen by the SLAC and EMC experiments. Since the Reggeon term is non-singlet, anti-shadowing is associated with a redistribution of the valence quarks in the nucleus.

Our analysis provides the input or starting point for the $\log Q^2$ evolution of the deep inelastic structure functions. The parameters for the effective quark-nucleon cross section required to understand shadowing phenomena provide important information on the interactions of quarks and gluons in nuclear matter.

The above analysis also has implications for the nature of particle production for virtual photo-absorption in nuclei. At high Q^2 and $x > 0.3$, hadron production should be uniform throughout the nucleus. At low x or at low Q^2, where shadowing occurs the inelastic reaction occurs mainly at the front surface. These features can be examined in detail by studying non-additive multiparticle correlations in both the target and current fragmentation regions.

11.3 Proton Structure and Electroproduction

11.3.1 Spin Effects in Deep Inelastic Scattering

The EMC and SLAC data on polarized structure functions imply significant correlations between the spin of the target proton with the spin of the gluons and strange quarks. Thus there should be significant correlations between the target spin and spin observables in the electroproduction final state, both in the current and target fragmentation region. It thus would be interesting to measure the spin of specific hadrons which are helicity self-analyzing through their decay products such as the ρ and the Λ.

It is useful to keep in mind the following simple model for the helicity parallel and helicity anti-parallel gluon distributions in the nucleon: $G^+_{g/N}(x) = \frac{3}{2}(1-x)^4/x$ and $G^-_{g/N}(x) = \frac{3}{2}(1-x)^6/x$, respectively. This model is consistent with the momentum fraction carried by gluons in the proton, correct crossing behavior, dimensional counting rules at $x \to 1$, and Regge behavior at small x. Integrating over x, one finds that the gluon carries, on the average, $\frac{11}{24}$ of the total nucleon J_z. It is thus consistent with experiment and the Skryme model prediction that more of the nucleon spin is carried by gluons rather than quarks [21].

The analyses of the EMC and SLAC spin-dependent structure functions as well as elastic neutrino-proton scattering imply substantial strange and anti-strange quarks in the proton, highly spin-correlated with the proton spin. The usual description of the strange sea assumes that $s\bar{s}$ is strictly due to the simple gluon splitting process. The spin correlation of the strange quarks then requires a very large gluon spin correlation, much stronger than the simple model given above. Alternatively the strange sea may be "intrinsic" to the bound state equation of motion of the nucleon and thus the strong strange spin correlation may be a non-perturbative phenomenon. One expects contributions at order $1/m_s^2$ to the strange sea from cuts of strange loops quark loops in the wavefunction with 2, 3, and 4 gluons connecting to the other quark and gluon constituents of the nucleon. Alternatively, one can regard the strange sea as a manifestation of intermediate $K - \Lambda$ and other virtual meson-baryon pair states in the fluctuations of the proton ground state.

Experiments which examine the entire final state in electroproduction can discriminate between these extrinsic and intrinsic components to the strange sea. For example, consider events in which a strange hadron is observed at large z in the fragmentation region of the recoil jet, signifying the production and tagging of a strange quark. In the case of intrinsic strangeness, the associated $\bar{s}$ will be in the target fragmentation region. In the case that the strange quark is created extrinsically via $\gamma^* g \to s\bar{s}$, both the tagged s quark and the $\bar{s}$ hadrons will be found predominantly in the current fragmentation region.

11.3.2 "Extrinsic" versus "Intrinsic" Contributions to the Proton Structure Functions

The central focus of inelastic electroproduction is the electron-quark interaction, which at large momentum transfer can be calculated as an incoherent sum of individual quark contributions. The deep inelastic electron-proton cross section is thus given by the convolution of the electron-quark cross section times the structure functions, or equivalently the probability distributions $G_{q/p}(x, Q^2)$. In the "infinite momentum frame" where the proton has large momentum P^μ and the virtual photon momentum is in the transverse direction, $G_{q/p}(x, Q^2)$ is the probability of finding a quark q with momentum fraction $x = Q^2/2p \cdot q$ in the proton. However in the rest frame of the target, many different physical processes occur: the photon can scatter out a quark as in the atomic physics photoelectric effect, it can hit a quark which created from a vacuum fluctuation near the proton, or the photon can first make a $q\bar{q}$ pair, either of which can interact in the target. Thus the electron interacts with quarks which are both *intrinsic* to the proton's structure itself, or quarks which are *extrinsic*, *i.e.*, created in the electron-proton collision itself. Much of the phenomena at small values of x such as Regge behavior, sea distributions associated with photon-gluon fusion processes, and shadowing in nuclear structure functions can be identified with the extrinsic interactions, rather than processes directly connected with the proton's intrinsic structure.

There is an amusing, though *gedanken* way to (in principle) separate the extrinsic and intrinsic contributions to the proton's structure functions. For example, suppose that one wishes to isolate the intrinsic contribution $G^I_{d/p}(x, Q)$ to the d-quark distribution in the proton. Let us imagine that there exists another set of quarks $\{q_o\} = u_o, d_o, s_o, c_o, \ldots$ identical in all respects to the usual set of quarks but carrying zero electromagnetic and weak charges. The experimentalist could then measure the difference in scattering of electrons on protons versus electrons scattering on a new baryon with valence quarks $|uud_o\rangle$. This is analogous to an "empty target" subtraction. Contributions from $q\bar{q}$ pair production in the gluonic field of the target (photon-gluon fusion) effectively cancel, so that one can then identify the difference in scattering with the intrinsic d-quark distribution of the nucleon. Because of the Pauli exclusion principle, $d\bar{d}$ production on the proton where the d is produced in the same quantum state as the d in the nucleon is absent, but the corresponding contribution is allowed in the case of the $|uud_o\rangle$ target. Because of this extra subtraction, the contributions associated with Reggeon exchange also cancel in the difference, and thus the intrinsic structure function $G^I(x,Q)$ vanishes at $x \to 0$. The intrinsic contribution gives finite expectation values for the light-cone kinetic energy operator, "sigma" terms, and the $J = 0$ fixed poles associated with $\langle \frac{1}{x} \rangle$ [22].

11.3.3 Higher Twist and other QCD Contributions to Electroproduction

Although there have been extensive measurements of the deep inelastic structure functions, some aspects remain to be verified, and will require data over a large range of Q^2. For example, how much of the scale violation is due to power-law (higher twist) contributions [23] versus logarithmic perturbative quantum chromodynamics (PQCD) evolution? Does the Bjorken-scaling non-isosinglet structure function $F_2(x,Q)$ behave as $Cx^{1-\alpha_\rho}$ as $x \to 0$ as dictated by Regge exchange and duality or is this a manifestation of higher twist contributions to the virtual photo-absorption cross section which falls as $1/Q^2$? Are the non-additive shadowing and anti-shadowing nuclear effects really leading twist or are they Q^2 dependent?

Electron-proton scattering also involves additional processes such as photoproduction, Compton processes, QED radiative corrections, *etc.* Electroproduction reactions in which large transverse momentum photons appear are particularly interesting. In the exclusive process $e^\pm p \to e^\pm \gamma p$ one can isolate the virtual Compton cross section as well as the real part of the Compton amplitude. In the inclusive reaction $e^\pm p \to e^\pm \gamma X$ one can determine reactions and sum rules proportional to the quark charge cubed.

It is thus interesting to consider inclusive electron-proton collisions from a general point of view. As long as there is at least one particle detected at large transverse momentum, whether it is a scattered electron, or a produced hard photon, or a hadron at large P_T, one can use the factorization formula [24]

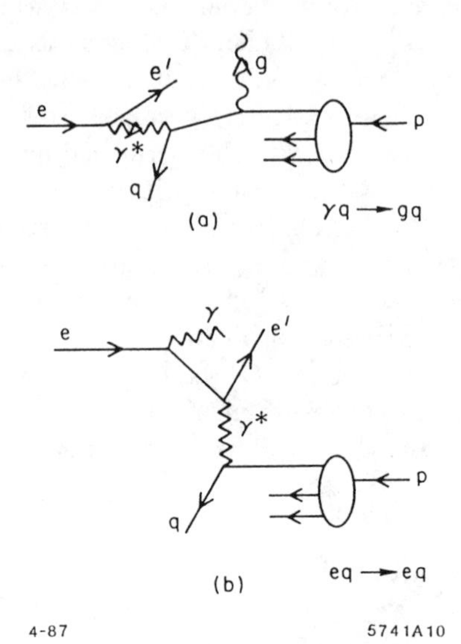

FIGURE 11.2
Application of gauge theory factorization to electroproduction. (a) The $\gamma q \to gq$ subprocess produces hadron jets at high p_T. (b) The $eq \to eq$ produces one quark jet and one recoil electron jet at high p_T. The QED radiative corrections are incorporated into the electron and photon QED structure functions.

$$\frac{d\sigma(AB \to CX)}{d^3p_c/E_c} \cong \sum_{ab,cd} \int_0^1 dx_a \int_0^1 dx_b \int_0^1 \frac{dx_c}{x_c^2} \times G_{a/A}(x_a, Q) G_{b/B}(x_b, Q) \widetilde{G}_{C/c}(x_c, Q) \times \delta(s' + t' + u') \, \frac{s'}{\pi} \frac{d\sigma}{dt'} \, (ab \to cd) \tag{11.6}$$

which has general validity in gauge theory. The systems A, B, C can be leptons, photons, hadrons, or nuclei. The primary subprocess in electroproduction is $eq \to eq$. The electron structure function $G_{e/e}(x, Q)$ automatically provides the (leading logarithmic) QED radiative corrections. The spectrum of the electron beam plays the role of the non-perturbative or initial structure function. (See Fig. 11.2 (b).) The subprocess $\gamma^* q \to gq$ corresponds to photon-induced two-jet production. (See Fig. 11.2 (a).) This subprocess dominates reactions in which the large transverse momentum trigger is a hadron rather than the scattered lepton. Thus one sees that conventional deep inelastic $eq \to eq$ scattering subprocess is just one of the several modes of electroproduction.

The dominant contribution to the meson semi-inclusive cross section is predicted by QCD factorization to be due to jet fragmentation from the recoil quark and spectator diquark jets.

When the momentum transfer is in the intermediate range $1 \lesssim Q^2 \lesssim 10\ \text{GeV}^2$, several other contributions for meson production are expected to become important in $eN \to e'MX$. These include:

1. Higher twist contributions to jet fragmentation:

$$\frac{dN_\pi}{dz} = D_{\pi/q}(z, Q^2) \cong A(1-z)^2 + \frac{C}{Q^2} \quad \text{for} \quad z \to 1 . \tag{11.7}$$

The scaling term reflects the behavior of the pion fragmentation function at large fractional momentum ($z \to 1$) as predicted by perturbative QCD (one-gluon exchange). (See Fig. 11.3 (a).) The C/Q^2 term [25] is computed from the same perturbative diagrams. For large z where this term dominates, we predict that the deep inelastic cross section will be dominantly longitudinal rather than transverse $R = \sigma_L/\sigma_T > 1$.

2. "Direct" meson production. Isolated pions may also be created by elastic scattering off of an effective pion current (See Fig. 11.3 (b).):

$$\begin{aligned} \frac{d\sigma}{dQ^2 dx_\pi} &= G_{\pi/p}(x_\pi) \left.\frac{d\sigma}{dQ^2}\right|_{e\pi\to e\pi} \\ \left.\frac{d\sigma}{dy dQ^2}\right|_{e\pi\to e\pi} &= \frac{4\pi\alpha^2}{(Q^2)^2}\, |F_\pi(Q^2)|^2 (1-y) . \end{aligned} \tag{11.8}$$

Here $y = q\cdot p/p_e\cdot p$. In the case of a nuclear target, one can test for non-additivity of virtual pions due to nuclear effects, as predicted in models [26] for the EMC effect [27] at small x_{Bj}. Jaffe and Hoodbhoy [13] have shown that the existence of quark exchange diagrams involving quarks of different nucleons in the nucleus invalidates general applicability of the simplest convolution formulae conventionally used in such analyses. The $G_{\pi/p}(x, Q)$ structure function at large x is predicted to behave roughly as $(1-x)^5$, as predicted from spectator quark counting rules [24, 29]. Applications of these rules to other off-shell nucleon processes are discussed in Refs. [28] and [30].

3. Exclusive channels. (See Fig. 11.3 (c).) The mesons can of course be produced in exclusive channels; *e.g.* $\gamma^* p \to \pi^+ n$, $\gamma^* p \to \rho^0 p$. Pion electroproduction extrapolated to $t = m_\pi^2$ provides the basic knowledge of the pion form factor at spacelike Q^2. With the advent of

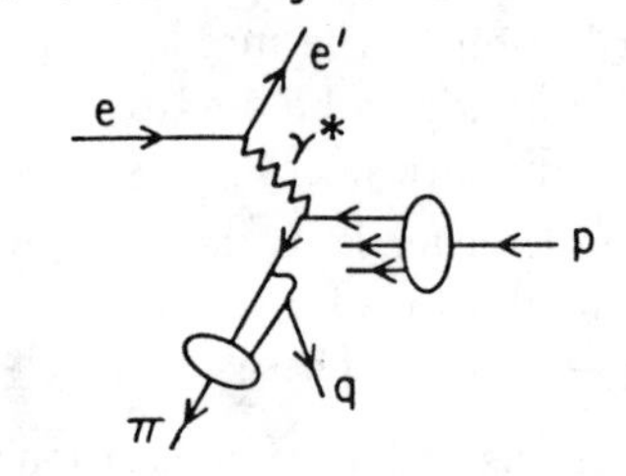

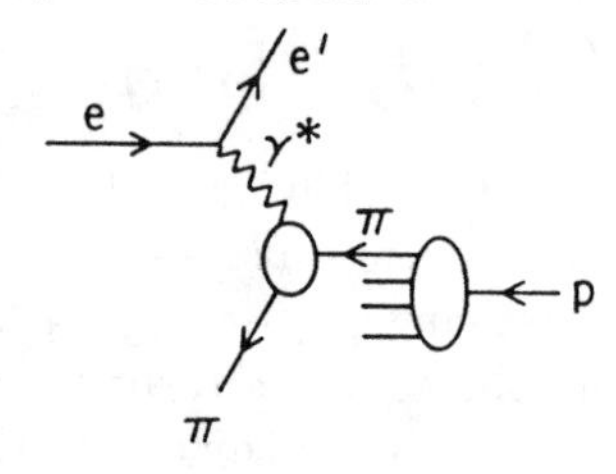

FIGURE 11.3
QCD contributions to pion electroproduction. (a) Jet fragmentation, including leading and $1/Q^2$ higher twist contributions. (b) Isolated pion contributions at order $1/Q^4$. (c) Exclusive production. (d) Primakoff contribution.

the perturbative QCD analyses of large momentum transfer exclusive reactions, predictions can be given over the whole range of large t and Q^2. Exclusive processes are discussed in more detail in Secs. 11.5–11.8.

4. Another meson production channel is the Primakoff reaction $\gamma^*\gamma \to \pi^0$, *etc.*, which dominates over other events at very low target recoil momentum. (See Fig. 11.3 (d).) Such measurements would allow the determination of the $\gamma \to \pi^0$ transition form factor. This quantity, combined with the QCD analysis of the pion form factor leads to a method to determine the QCD running coupling constant $\alpha_s(Q^2)$ solely from exclusive measurements [31].

11.3.4 Hadronization of the Quark and Spectator Systems

At the most basic level, Bjorken scaling of deep inelastic structure functions implies the production of a single quark jet, recoiling against the scattered lepton. The spectator system— the remnant of the target remaining after the scattered quark is removed— is a color-$\overline{3}$ system. The struck quark is sensitive to the magnitude of the momentum transfer Q and logarithmically evolves by radiated gluons with relative transverse momentum controlled by Q^2 and the available phase-space. According to QCD factorization, the recoiling quark jet, together with the gluonic radiation produced in the scattering process, produces hadrons in a universal way, independent of the target or particular hard scattering reaction. This jet should be identical to the light quark jets produced in e^+e^- annihilation. In contrast, the hadronization of the spectator system depends in detail on the target properties. Unlike the quark jet, the leading particles of the target spectator system do not evolve and thus should not depend on the momentum transfer Q^2 (at fixed $W^2 = (q+p)^2$). At present we do not have a basic understanding of the physics of hadronization, although phenomenological approaches, such as the Lund string model, have been successful in parameterizing many features of the data.

11.3.5 Analogs between QCD and QED

Many of the novel features expected in QCD are also apparent in Quantum Electrodynamics (QED). It is thus often useful to keep a QED analog in mind, replacing the target by a neutral atom such as positronium. Even in QED where there is no confinement, one expects in certain kinematic regions significant corrections to the Bjorken scaling associated with positron or electron knockout, in addition to the logarithmic evolution of the QED structure functions associated with induced photon radiation. For example, at low Q^2, the interference between amplitudes where different constituents are struck become important. Near threshold, where charged particles emerge at low relative velocities, there are strong Coulomb distortions, as summarized by the Sommerfeld [18] factor. In QCD these have their analog in a phenomenon called "jet coalescence" [32]. The Coulomb distortion factor must be included if one wants to maintain duality between the inelastic continuum and a summation over exclusive channels in electroproduction [33].

11.4 Applications of QCD to the Phenomenology of Exclusive Reactions

In this section I will discuss the application of QCD to exclusive reactions at large momentum transfer. The primary processes of interest are those in which one learns new information on the structure of the proton and other hadrons. The wavefunctions involved in such reactions are also relevant to the understanding of jet hadronization and the computation of hadron matrix elements for weak decays, *etc.* This includes form factors at large momentum transfer Q and large angle scattering reactions. Specific examples are reactions such as $e^-p \to e^-p$, $e^+e^- \to p\bar{p}$, elastic scattering reactions at large angles and energies such as $\pi^+p \to \pi^+p$ and $pp \to pp$, two-photon annihilation processes such as $\gamma\gamma \to K^+K^-$ or $\bar{p}p \to \gamma\gamma$, exclusive nuclear processes such as deuteron photo-disintegration $\gamma d \to np$, and exclusive decays such as $\pi^+ \to \mu^+\nu$ or $J/\psi \to \pi^+\pi^-\pi^0$.

As discussed in the introduction, QCD has two essential properties which make calculations of processes at short distance or high-momentum transfer tractable and systematic. The critical feature is asymptotic freedom: the effective coupling constant $\alpha_s(Q^2)$ which controls the interactions of quarks and gluons at momentum transfer Q^2 vanishes logarithmically at large Q^2 since it allows perturbative expansions in $\alpha_s(Q^2)$. Complementary to asymptotic freedom is the existence of *factorization theorems* for both exclusive and inclusive processes at large momentum transfer. In the case of "hard" exclusive processes (in which the kinematics of all the final state hadrons are fixed at large invariant mass), the hadronic amplitude can be represented as the product of a process-dependent hard-scattering amplitude $T_H(x_i, Q)$ for the scattering of the constituent quarks convoluted with a process-independent *distribution amplitude* $\phi(x, Q)$ for each incoming or outgoing hadron [34]. When Q^2 is large, T_H is computable in perturbation theory as is the Q-dependence of $\phi(x, Q)$.

In Table 11.1 we give a summary of the main scaling laws and properties of large momentum transfer exclusive and inclusive cross sections which are derivable starting from the light-cone Fock space basis and the perturbative expansion for QCD.

TABLE 11.1
Comparison of Exclusive and Inclusive Cross Sections

Exclusive Amplitudes	Inclusive Cross Sections
$\mathcal{M} \sim \Pi\, \phi(x_i, Q) \otimes T_H(x_i, Q)$	$d\sigma \sim \Pi\, G(x_a, Q) \otimes d\hat{\sigma}(x_a, Q)$
$\phi(x, Q) = \int^{Q} [d^2 k_\perp]\, \psi^{Q}_{\text{val}}(x, k_\perp)$	$G(x, Q) = \sum_n \int^{Q} [d^2 k_\perp]\, [dx]'\, \lvert\psi^{Q}_{n}(x, k_\perp)\rvert^2$
Measure ϕ in $\gamma \to M\overline{M}$	Measure G in $\ell p \to \ell X$
$\sum_{i \in H} \lambda_i = \lambda_H$	$\sum_{i \in H} \lambda_i \neq \lambda_H$
Evolution	
$\frac{\partial \phi(x, Q)}{\partial \log Q^2} = \alpha_s \int [dy]\, V(x, y)\, \phi(y)$	$\frac{\partial G(x, Q)}{\partial \log Q^2} = \alpha_s \int dy\, P(x/y)\, G(y)$
$\lim_{Q \to \infty} \phi(x, Q) = \prod_i x_i \cdot C_{\text{flavor}}$	$\lim_{Q \to \infty} G(x, Q) = \delta(x)\, C$
Power Law Behavior	
$\frac{d\sigma}{dx}(A + B \to C + D) \cong \frac{1}{s^{n-2}} f(\theta_{\text{cm}})$	$\frac{d\sigma}{d^2p/E}(AB \to CX) \cong \sum \frac{(1 - x_T)^{2n_s - 1}}{(Q^2)^{n_{act} - 2}} f(\theta_{\text{cm}})$
$n = n_A + n_B + n_C + n_D$	$n_{act} = n_a + n_b + n_c + n_d$
T_H : expansion in $\alpha_s(Q^2)$	$d\hat{\sigma}$: expansion in $\alpha_s(Q^2)$
Complications	
End point singularities	Multiple scales
Pinch singularities	Phase-space limits on evolution
High Fock states	Heavy quark thresholds
	Higher twist multiparticle processes
	Initial and final state interactions

As emphasized in Sec. 11.1, a convenient relativistic description of hadron wavefunctions is given by the set of n-body momentum space amplitudes, $\psi_n(x_i, k_{\perp_i}, \lambda_i)$, $i = 1, 2, \ldots n$, defined on the free quark and gluon Fock basis at equal "light-cone time" $\tau = t + z/c$ in the physical "light-cone" gauge $A^+ \equiv A^0 + A^3 = 0$. (Here $x_i = k_i^+/p^+$, $\sum_i x_i = 1$, is the light-cone momentum fraction of quark or gluon i in the n-particle Fock state; $k_{\perp_i}$, with $\sum_i k_{\perp_i} = 0$, is its transverse momentum relative to the total momentum p^μ; and λ_i is its helicity.) The quark and gluon structure functions $G_{q/H}(x, Q)$ and $G_{g/H}(x, Q)$ which control hard inclusive reactions and the hadron distribution amplitudes $\phi_H(x, Q)$ which control hard exclusive reactions are simply related to these wavefunctions:

$$G_{q/H}(x, Q) \propto \sum_n \int^Q \Pi d^2 k_{\perp_i} \int \Pi dx_i \, |\psi_n(x_i, k_{\perp_i})|^2 \delta(x_q - x) \, , \text{ and}$$

$$\phi_H(x_i, Q) \propto \int^Q \Pi d^2 k_{\perp_i} \, \psi_{\mathrm{valence}}(x_i, k_{\perp_i}) \quad . \tag{11.9}$$

In the case of inclusive reactions, such as deep inelastic lepton scattering, two basic aspects of QCD are relevant: (1) the scale invariance of the underlying lepton-quark subprocess cross section, and (2) the form and evolution of the structure functions. A structure function is a sum of squares of the light-cone wavefunctions. The logarithmic evolution of $G_q(x, Q^2)$ is controlled by the wavefunctions which fall off as $|\psi(x, \vec{k}_\perp)|^2 \sim \alpha_s(k_\perp^2)/k_\perp^2$ at large $k_\perp^2$. This form is a consequence of the pointlike $q \to gq$, $g \to gg$, and $g \to q\bar{q}$ splitting. By taking the logarithmic derivative of G with respect to Q one derives the evolution equations of the structure function. All of the hadron's Fock states generally participate; the necessity for taking into account the (non-valence) higher-particle Fock states in the proton is apparent from two facts: (1) the proton's large gluon momentum fraction and (2) the recent results from the EMC collaboration [35] suggesting that, on the average, little of the proton's helicity is carried by the light quarks [21].

In the case of exclusive electroproduction reactions such as the baryon form factor, again two basic aspects of QCD are relevant: (1) the scaling of the underlying hard scattering amplitude (such as $l + qqq \to l + qqq$), and (2) the form and evolution of the hadron distribution amplitudes. The distribution amplitude is defined as an integral over the lowest (valence) light-cone Fock state. The logarithmic variation of $\phi(x, Q^2)$ is derived from the integration at large $k_\perp$, *i.e.*, wavefunctions which behave as $\psi(x, \vec{k}_\perp) \sim \alpha_s(k_\perp^2)/k_\perp^2$ at large $k_\perp^2$. This behavior follows from the simple one-gluon exchange contribution to the tail of the valence wavefunction. By taking the logarithmic derivative, one then obtains the evolution equation for the hadron distribution amplitude.

The form factor of a hadron at any momentum transfer can be computed exactly in terms of a convolution of initial and final light-cone Fock state wavefunctions [36]. In general, all of the Fock states contribute. In contrast, exclusive reactions with high momentum transfer Q, perturbative QCD predicts that only the lowest particle number (valence) Fock state is required to compute the contribution to the amplitude to leading order in $1/Q$. For example, in the light-cone Fock expansion the proton is represented as a column vector of states ψ_{qqq}, ψ_{qqqg}, $\psi_{qqq\bar{q}q}$, ... In the light-cone gauge, $A^+ = A^0 + A^3 = 0$, only the minimal "valence" three-quark Fock state needs to be considered at large momentum transfer since any additional quark or gluon forced to absorb large momentum transfer yields a power-law suppressed contribution to the hadronic amplitude. Thus at large Q^2, the baryon form factor can be systematically computed by iterating the equation of motion for its valence Fock state wherever large relative momentum occurs. To leading order the kernel is effectively one-gluon exchange. The sum of the hard gluon exchange contributions can be arranged as the gauge invariant amplitude T_H, the final form factor having the form

$$F_B(Q^2) = \int_0^1 [dy] \int_0^1 [dx]\, \phi_B^\dagger(y_j, Q) T_H(x_i, y_j, Q) \phi_B(x_i, Q) \,. \qquad (11.10)$$

The essential gauge-invariant input for hard exclusive processes is the distribution amplitude $\phi_H(x, Q)$. For example $\phi_\pi(x, Q)$ is the amplitude for finding a quark and antiquark in the pion carrying momentum fractions x and $1 - x$ at impact (transverse space) separations less than $b_\perp < 1/Q$. The distribution amplitude thus plays the role of the "wavefunction at the origin" in analogous non-relativistic calculations of form factors. In the relativistic theory, its dependence on $\log Q$ is controlled by evolution equations derivable from perturbation theory or the operator product expansion. A complete discussion may be found in the papers by Lepage and myself [38], and our recent review in Ref. [1]. A discussion of the light-cone Fock state wavefunctions and their relation to observables is given in Ref. [37].

The distribution amplitude contains all of the bound-state dynamics and specifies the momentum distribution of the quarks in the hadron. The hard-scattering amplitude for a given exclusive process can be calculated perturbatively as a function of $\alpha_s(Q^2)$. Similar analyses can be applied to form factors, exclusive photon-photon reactions, and with increasing degrees of complication, to photoproduction, fixed-angle scattering, *etc.* In the case of the simplest processes, $\gamma\gamma \to M\overline{M}$ and the meson form factors, the leading order analysis can be readily extended to all-orders in perturbation theory.

In the case of exclusive processes such as photo-production, Compton scattering, meson-baryon scattering, *etc.*, the leading hard scattering QCD contribution at large momentum transfer $Q^2 = tu/s$ has the form (helicity labels are suppressed) (see Fig. 11.4)

$$\mathcal{M}_{A+B\to C+D}(Q^2, \theta_{\rm cm}) = \int [dx] \phi_C(x_c, \widetilde{Q}) \phi_D(x_d, \widetilde{Q})\; T_H(x_i; Q^2, \theta_{\rm cm}) \times \phi_A(x_a, \widetilde{Q}) \phi_B(x_b, \widetilde{Q}) \,. \qquad (11.11)$$

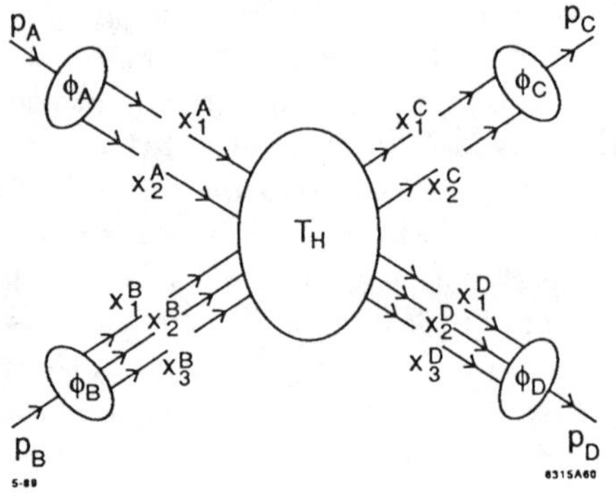

FIGURE 11.4
QCD factorization for two-body amplitudes at large momentum transfer.

In general the distribution amplitude is evaluated at the characteristic scale $\widetilde{Q}$ set by the effective virtuality of the quark propagators.

By definition, the hard scattering amplitude T_H for a given exclusive process is constructed by replacing each external hadron with its massless, collinear valence partons, each carrying a finite fraction x_i of the hadron's momentum. Thus T_H is the scattering amplitude for the constituents. The essential behavior of the amplitude is determined by T_H, computed where each hadron is replaced by its (collinear) quark constituents. We note that T_H is "collinear irreducible," *i.e.*, the transverse momentum integrations of all reducible loop integration are restricted to $k_\perp^2 > \mathcal{O}(Q^2)$ since the small $k_\perp$ region is already contained in ϕ. If the internal propagators in T_H are all far-off-shell $\mathcal{O}(Q^2)$, then a perturbative expansion in $\alpha_s(Q^2)$ can be carried out.

Higher twist corrections to the quark and gluon propagator due to mass terms and intrinsic transverse momenta of a few hundred MeV give nominal corrections of higher order in $1/Q^2$. These finite mass corrections combine with the leading twist results to give a smooth approach to small Q^2. It is thus reasonable that PQCD scaling laws become valid at relatively low momentum transfer of order of a few GeV.

11.4.1 General Features of Exclusive Processes in QCD

The factorization theorem for large-momentum-transfer exclusive reactions separates the dynamics of hard-scattering quark and gluon amplitudes T_H from process-independent distribution amplitudes $\phi_H(x, Q)$ which isolates all of the bound state dynamics. However, as seen from Table 11.1, even without complete information on the hadronic wave functions, it is still possible to make predictions at large momentum transfer directly from QCD.

Although detailed calculations of the hard-scattering amplitude have not been carried out in all of the hadron-hadron scattering cases, one can abstract some general features of QCD common to all exclusive processes at large momentum transfer:

1. Since the distribution amplitude ϕ_H is the $L_z = 0$ orbital-angular-momentum projection of the hadron wave function, the sum of the interacting constituents' spin along the hadron's momentum equals the hadron spin [39]:

$$\sum_{i \in H} s_i^z = s_H^z . \tag{11.12}$$

In contrast, there are any number of non-interacting spectator constituents in inclusive reactions, and the spin of the active quarks or gluons is only statistically related to the hadron spin (except at the edge of phase space $x \longrightarrow 1$).

2. Since all loop integrations in T_H are of order $\widetilde{Q}$, the quark and hadron masses can be neglected at large Q^2 up to corrections of order $\sim m/\widetilde{Q}$. The vector-gluon coupling conserves quark helicity when all masses are neglected— *i.e.*, $\bar{u}_\downarrow \gamma^\mu u_\uparrow = 0$. Thus total quark helicity is conserved in T_H. In addition, because of item **1** above, each hadron's helicity is the sum of the helicities of its valence quarks in T_H. We thus have the selection rule

$$\sum_{\text{initial}} \lambda_H - \sum_{\text{final}} \lambda_H = 0, \tag{11.13}$$

i.e., total hadronic helicity is conserved up to corrections of order m/Q or higher. Only (flavor-singlet) mesons in the 0^{-+} nonet can have a two-gluon valence component and thus even for these states the quark helicity equals the hadronic helicity. Consequently hadronic-helicity conservation applies for all amplitudes involving light meson and baryons [40]. Exclusive reactions which involve hadrons with quarks or gluons in higher orbital angular states are suppressed by powers.

3. The nominal power-law behavior of an exclusive amplitude at fixed θ_{cm} is $(1/Q)^{n-4}$, where n is the number of external elementary particles (quarks, gluons, leptons, photons, ...) in T_H. This dimensional-counting rule [41] is modified by the Q^2 dependence of the factors of $\alpha_s(Q^2)$ in T_H, by the Q^2 evolution of the distribution amplitudes, and possibly by a small power correction associated with the Sudakov suppression of pinch singularities in hadron-hadron scattering.

The dimensional-counting rules for the power-law falloff appear to be experimentally well established for a wide variety of processes [24, 42]. The helicity-conservation rule is also one of the most characteristic features of QCD, being a direct consequence of the gluon's spin. A scalar-or tensor-gluon-quark coupling flips the quark's helicity. Thus, for such theories, helicity may or may not be conserved in any given diagram contribution to T_H depending upon the number of interactions involved. Only for a vector theory, such as QCD, can one have a helicity selection rule valid to all orders in perturbation theory.

11.5 Electromagnetic Form Factors

Any helicity conserving baryon form factor at large Q^2 has the form (see Fig. 11.5 (a)):

$$F_B(Q^2) = \int_0^1 [dy] \int_0^1 [dx]\ \phi_B^\dagger(y_j, Q) T_H(x_i, y_j, Q) \phi_B(x_i, Q)\ , \quad (11.14)$$

where to leading order in $\alpha_s(Q^2)$, T_H is computed from $3q + \gamma^* \to 3q$ tree graph amplitudes (Fig. 11.5 (b)):

$$T_H = \left[\frac{\alpha_s(Q^2)}{Q^2}\right]^2 f(x_i, y_j)$$

and

$$\phi_B(x_i, Q) = \int [d^2k_\perp]\ \psi_V(x_i, \vec{k}_{\perp i}) \theta(k_{\perp i}^2 < Q^2) \quad (11.15)$$

is the valence three-quark wavefunction (Fig. 11.5 (c)) evaluated at quark impact separation $b_\perp \sim \mathcal{O}(Q^{-1})$. Since ϕ_B only depends logarithmically on Q^2 in QCD, the main dynamical dependence of $F_B(Q^2)$ is the power behavior $(Q^2)^{-2}$ derived from scaling of the elementary propagators in T_H. More explicitly, the proton's magnetic form factor has the form [38]:

$$G_M(Q^2) = \left[\frac{\alpha_s(Q^2)}{Q^2}\right]^2 \sum_{n,m} a_{nm} \left(\log \frac{Q^2}{\Lambda^2}\right)^{-\gamma_n - \gamma_m} \quad (11.16)$$
$$\times \left[1 + \mathcal{O}(\alpha_s(Q)) + \mathcal{O}\left(\frac{1}{Q}\right)\right] .$$

The first factor, in agreement with the quark counting rule, is due to the

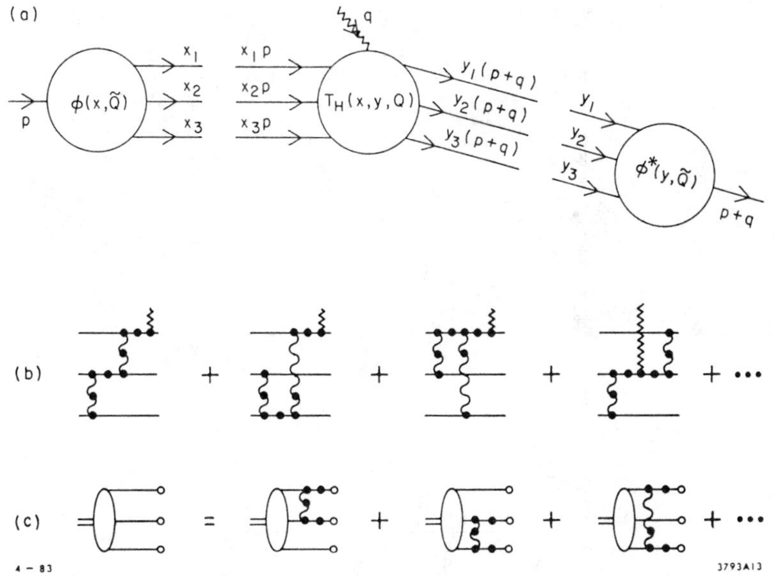

FIGURE 11.5
(a) Factorization of the nucleon form factor at large Q^2 in QCD. (b) The leading order diagrams for the hard scattering amplitude T_H. The dots indicate insertions which enter the renormalization of the coupling constant. (c) The leading order diagrams which determine the Q^2 dependence of the distribution amplitude $\phi(x, Q)$.

hard scattering of the three valence quarks from the initial to final nucleon direction. Higher Fock states lead to form factor contributions of successively higher order in $1/Q^2$. The logarithmic corrections derive from an evolution equation for the nucleon distribution amplitude. The γ_n are the computed anomalous dimensions, reflecting the short distance scaling of three-quark composite operators [44]. The results hold for any baryon to baryon vector or axial vector transition amplitude that conserves the baryon helicity. Helicity non-conserving form factors should fall as an additional power of $1/Q^2$ [39]. Measurements [43] of the transition form factor to the $J = \frac{3}{2}$, $N(1520)$ nucleon resonance are consistent with $J_z = \pm\frac{1}{2}$ dominance, as predicted by the helicity conservation rule [39]. A review of the data on spin effects in electron nucleon scattering in the resonance region is given in Ref. [43]. It is important to explicitly verify that $F_2(Q^2)/F_1(Q^2)$ decreases at large Q^2. The angular distribution decay of the $J/\psi \rightarrow p\bar{p}$ is consistent with the QCD prediction $\lambda_p + \lambda_{\bar{p}} = 0$.

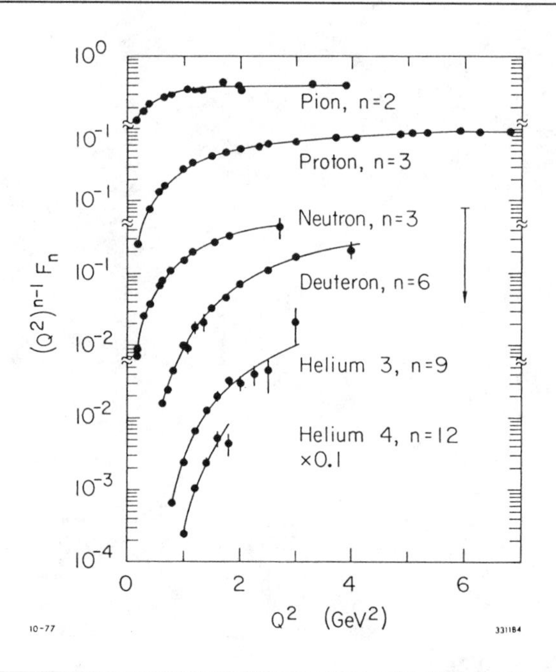

FIGURE 11.6
Comparison of experiment [45] with the QCD dimensional counting rule $(Q^2)^{n-1}F(Q^2) \sim$const for form factors. The proton data extends beyond 30 GeV^2.

Thus, modulo logarithmic factors, one obtains a dimensional counting rule for any hadronic or nuclear form factor at large Q^2 ($\lambda = \lambda' = 0$ or $\frac{1}{2}$)

$$F(Q^2) \sim \left(\frac{1}{Q^2}\right)^{n-1} ,$$

$$F_1^N \sim \frac{1}{Q^4} , \quad F_\pi \sim \frac{1}{Q^2} , \quad F_d \sim \frac{1}{Q^{10}} , \tag{11.17}$$

where n is the minimum number of fields in the hadron. Since quark helicity is conserved in T_H and $\phi(x_i, Q)$ is the $L_z = 0$ projection of the wavefunction, total hadronic helicity is conserved at large momentum transfer for any QCD exclusive reaction. The dominant nucleon form factor thus corresponds to $F_1(Q^2)$ or $G_M(Q^2)$; the Pauli form factor $F_2(Q^2)$ is suppressed by an extra power of Q^2. Similarly, in the case of the deuteron, the dominant form factor has helicity $\lambda = \lambda' = 0$, corresponding to $\sqrt{A(Q^2)}$.

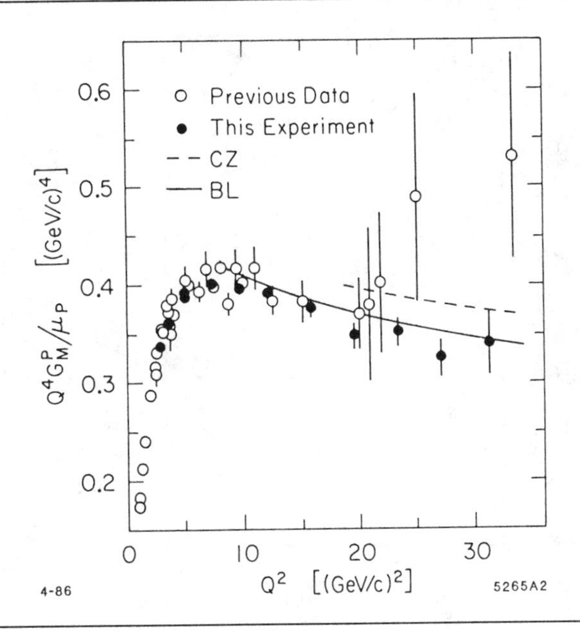

FIGURE 11.7
Comparison of the scaling behavior of the proton magnetic form factor with the theoretical predictions of Refs. [5] and [38]. The CZ predictions [5] are normalized in sign and magnitude. The data are from Ref. [47].

The comparison of experimental form factors with the predicted nominal power-law behavior is shown in Fig. 11.6. The general form of the logarithmic corrections to the leading power contributions form factors can be derived from the operator product expansion at short distance [44, 46] or by solving an evolution equation [38] for the distribution amplitude computed from gluon exchange (Fig. 11.5 (c)), the only QCD contribution which falls sufficiently slowly at large transverse momentum to affect the large Q^2 dependence.

The comparison of the proton form factor data with the QCD prediction arbitrarily normalized is shown in Fig. 11.7. The fall-off of $Q^4\, G_M(Q^2)$ with Q^2 is consistent with the logarithmic fall-off of the square of QCD running coupling constant. As we shall discuss below, the QCD sum rule [5] model form for the nucleon distribution amplitude together with the QCD factorization formulae, predicts the correct sign and magnitude as well as scaling behavior of the proton and neutron form factors [47].

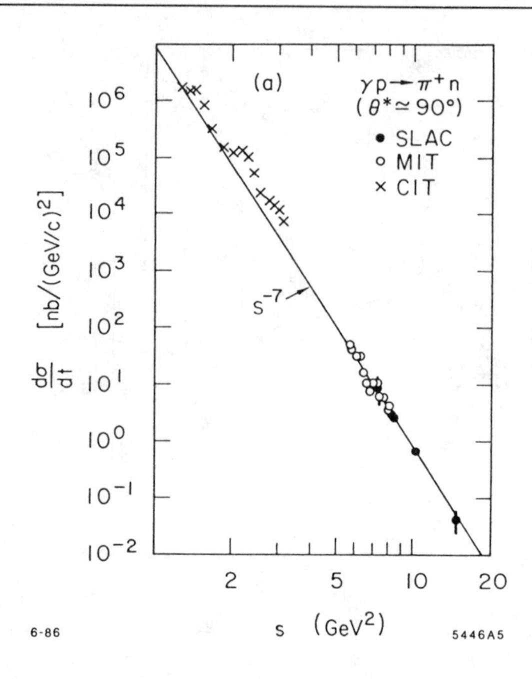

FIGURE 11.8
Comparison of photoproduction data with the dimensional counting power-law prediction. The data are summarized in Ref. [48].

11.5.1 Comparison of QCD Scaling for Exclusive Processes with Experiment

Phenomenologically the dimensional counting power laws appear consistent with measurements of form factors, photon-induced amplitudes, and elastic hadron-hadron scattering at large angles and momentum transfer [42]. The successes of the quark counting rules can be taken as strong evidence for QCD since the derivation of the counting rules require scale invariant tree graphs, soft corrections from higher loop corrections to the hard scattering amplitude, and strong suppression of pinch singularities. QCD is the only field theory of spin $\frac{1}{2}$ fields that has all of these properties.

As shown in Fig. 11.8, the data for $\gamma p \to \pi^+ n$ cross section at $\theta_{\rm cm} = \pi/2$ are consistent with the normalization and scaling $d\sigma/dt\ (\gamma p \to \pi^+ n) \simeq [1\ {\rm nb}/(s/10\ {\rm GeV})^7]\ f(t/s)$.

The check of fixed angle scaling in proton-proton elastic scattering is shown in Fig. 11.9. Extensive measurements of the $pp \to pp$ cross sec-

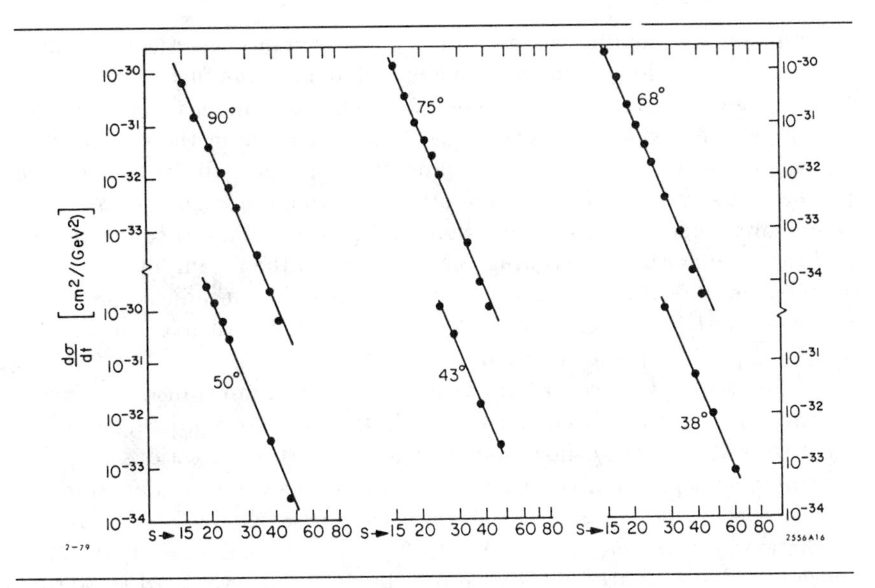

FIGURE 11.9
Test of fixed $\theta_{\rm cm}$ scaling for elastic pp scattering. The data compilation is from Landshoff and Polkinghorne.

tion have been made at ANL, BNL and other laboratories. The scaling law $s^{10} d\sigma/dt(pp \to pp) \simeq$ constant, predicted by QCD, seems to work quite well over a large range of energy and angle. The best fit gives the power $N = 9.7 \pm 0.5$ compared to the dimensional counting prediction $N = 10$. There are, however, measurable deviations from fixed power dependence which are not readily apparent on the log-log plot. As emphasized by Hendry [49] the $s^{10} d\sigma/dt$ cross section exhibits oscillatory behavior with p_T (see Sec. 11.11). Even more serious is the fact that polarization measurements [51] show significant spin-spin correlations (A_{NN}), and the single spin asymmetry (A_N) is not consistent with predictions based on hadron helicity conservation which is expected to be valid for the leading power behavior [39]. Recent discussions of these effects have been given by Farrar [52] and Lipkin [53]. We discuss a new explanation of all of these effects in Sec. 11.11.

As emphasized by Landshoff, the ISR data for high energy elastic pp scattering at small $|t|/s$ can be parameterized in the form $d\sigma/dt \sim \text{const}/t^8$ for 2 GeV$^2 < |t| < 10$ GeV^2. This suggests a role for triple gluon exchange pinch contributions at large energies where multiple vector exchange diagrams could dominate. However, from Mueller's analysis [54] one expects stronger fall-off in t due to the Sudakov form factor suppression. This paradox implies that the role of the pinch singularity in large momentum transfer exclusive reactions is not well understood and deserve further attention. Pinch singularities are also expected to modify the dimensional counting scaling laws for wide-angle scattering, but the change in the exponent of s is small and hard to detect experimentally. However, Ralston and Pire [15] have suggested that the oscillatory behavior in the wide-angle pp scattering amplitude results from interference between the pinch contributions and the ordinary hard-scattering contributions to the pp amplitude. Pinch singularities do not arise in form factors, or such photon-induced processes as $\gamma\gamma \to M\overline{M}$ [5], $\gamma^* + \gamma \to M$ [38], $\gamma^* \to M_1 \ldots M_N$ at fixed angle [55], $\gamma\gamma \to B\overline{B}$, $\gamma B \to \gamma B$, *etc.* [56, 57].

The role of pinch contributions in large momentum transfer exclusive reactions has recently been clarified by Botts and Sterman [58]. In agreement with Mueller they show that the Sudakov vertex corrections suppress large impact separation contributions from multi-scattering diagrams, reducing the net power to a value very close to the dimensional counting prediction, *e.g.* $d\sigma/dt(pp \to pp) = f(\theta_{\rm cm})/s^{9.66}$ rather than $1/s^{10}$. The pinch contributions are thus asymptotically dominant over hard scattering diagrams which carry five powers of the running coupling constant. Furthermore the effective quark separation is of order $b_\perp \sim 1/Q^{1-\epsilon}$, where ϵ is small, so that the predictions for color transparency in quasi-elastic scattering in nuclei will hold for pinch contributions as well as the usual hard-scattering diagrams. The contributions to the pinch amplitude coming from regions of integration where one or more of the exchanged gluons carries soft momentum, is suppressed because of the presence of four powers of the hadron distribution amplitude, so the region of validity of the QCD scaling is extended to quite low momentum transfer. Botts and Sterman do not find an energy-dependent structure in the pinch analysis of the type required to account for the observed "oscillations" about $1/s^{10}$ behavior seen in the pp scattering data. It is also apparent from the structure of the pinch contributions that they do not have large spin-spin correlations of the type observed in the data of Krisch et al. (See Sec. 11.11)

11.6 Hadronic Wavefunction Phenomenology

Let us now return to the question of the normalization of exclusive amplitudes in QCD. It should be emphasized that because of the uncertain magnitude of corrections of higher order in $\alpha_s(Q^2)$, comparisons with the normalization of experiment with model predictions could be misleading. Nevertheless, in this section it shall be assumed that the leading order normalization is at least approximately accurate. If the higher order corrections are indeed small, then the normalization of the proton form factor at large

Q^2 is a non-trivial test of the distribution amplitude shape; for example, if the proton wave function has a non-relativistic shape peaked at $x_i \sim \frac{1}{3}$ then one obtains the wrong sign for the nucleon form factor. Furthermore symmetrical distribution amplitudes predict a very small magnitude for $Q^4\, G^p_M(Q^2)$ at large Q^2.

The phenomenology of hadron wavefunctions in QCD is now just beginning. Constraints on the baryon and meson distribution amplitudes have been recently obtained using QCD sum rules and lattice gauge theory. The results are expressed in terms of gauge-invariant moments $\langle x_j^m \rangle = \int \Pi dx_i\, x_j^m\, \phi(x_i, \mu)$ of the hadron's distribution amplitude. A particularly important challenge is the construction of the baryon distribution amplitude. In the case of the proton form factor, the constants a_{nm} in the QCD prediction for G_M must be computed from moments of the nucleon's distribution amplitude $\phi(x_i, Q)$. There are now extensive theoretical efforts to compute this nonperturbative input directly from QCD. The QCD sum rule analysis of Chernyak *et al.* [5, 59] provides constraints on the first 12 moments of $\phi(x, Q)$. Using as a basis the polynomials which are eigenstates of the nucleon evolution equation, one gets a model representation of the nucleon distribution amplitude, as well as its evolution with the momentum transfer scale. The moments of the proton distribution amplitude computed by Chernyak *et al.*, have now been confirmed in an independent analysis by Sachrajda and King [60].

A three-dimensional "snapshot" of the proton's *uud* wavefunction at equal light-cone time as deduced from QCD sum rules at $\mu \sim 1$ GeV by Chernyak *et al.* [59] and King and Sachrajda [60] is shown in Fig. 11.10. The QCD sum rule analysis predicts a surprising feature: strong flavor asymmetry in the nucleon's momentum distribution. The computed moments of the distribution amplitude imply that 65% of the proton's momentum in its 3-quark valence state is carried by the u-quark which has the same helicity as the parent hadron.

Dziembowski and Mankiewicz [64] have recently shown that the asymmetric form of the CZ distribution amplitude can result from a rotationally-invariant cm wave function transformed to the light cone using free quark dynamics. They find that one can simultaneously fit low energy phenomena (charge radii, magnetic moments, *etc.*), the measured high momentum transfer hadron form factors, and the CZ distribution amplitudes with a self-consistent ansatz for the quark wave functions. Thus for the first time one has a somewhat complete model for the relativistic three-quark structure of the hadrons. In the model the transverse size of the valence wave function is not found to be significantly smaller than the mean radius of the proton–averaged over all Fock states as argued in Ref. [61]. Dziembowski *et al.* also find that the perturbative QCD contribution to the form factors in

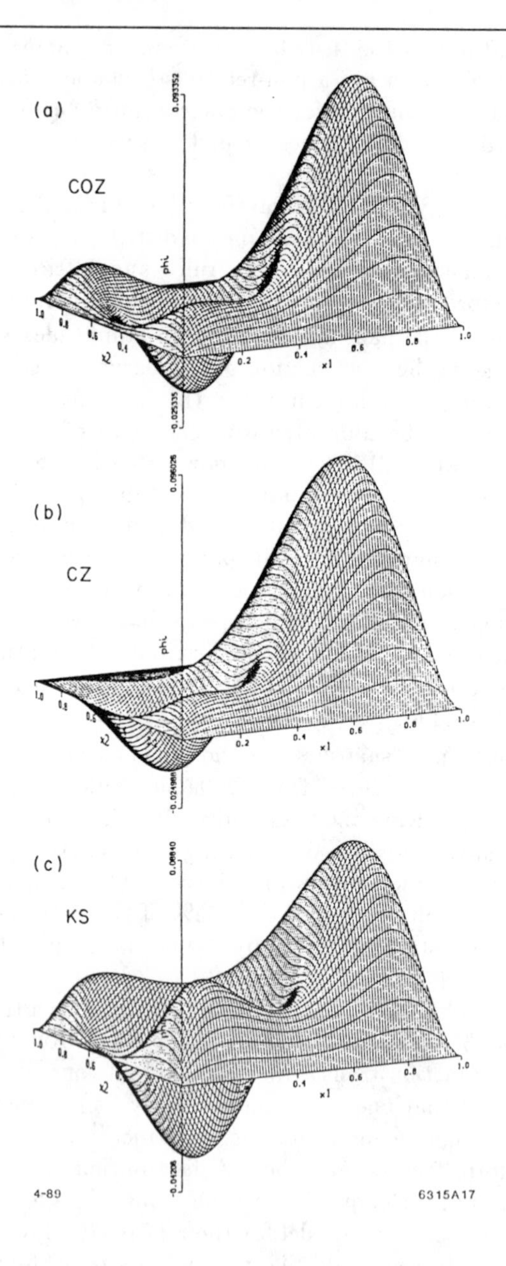

FIGURE 11.10
The proton distribution amplitude $\phi_p(x_i, \mu)$ determined at the scale $\mu \sim 1$ GeV from QCD sum rules.

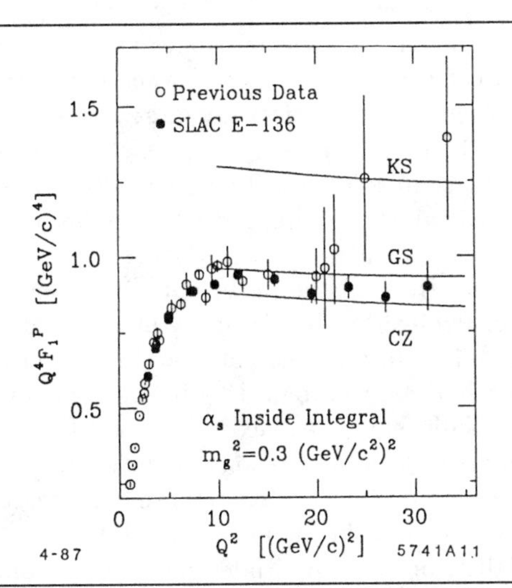

FIGURE 11.11
Predictions for the normalization and sign of the proton form factor at high Q^2 using perturbative QCD factorization and QCD sum rule predictions for the proton distribution amplitude (from Ref. [63]). The predictions use forms given by Chernyak and Zhitnitskii, King and Sachrajda [60] and Gari and Stefanis [62].

their model dominates over the soft contribution (obtained by convoluting the non-perturbative wave functions) at a scale $Q/N \approx 1$ GeV, where N is the number of valence constituents. (This criterion was also derived in Ref. [30].)

Gari and Stefanis [62] have developed a model for the nucleon form factors which incorporates the CZ distribution amplitude predictions at high Q^2 together with vector meson dominance (VMD) constraints at low Q^2. Their analysis predicts sizeable values for the neutron electric form factor at intermediate values of Q^2.

A detailed phenomenological analysis of the nucleon form factors for different shapes of the distribution amplitudes has been given by Ji, Sill, and Lombard-Nelsen [63]. Their results show that the CZ wave function is consistent with the sign and magnitude of the proton form factor at large Q^2 as recently measured by the American University/SLAC collaboration [47] (see Fig. 11.11).

It should be stressed that the magnitude of the proton form factor is sensitive to the $x \sim 1$ dependence of the proton distribution amplitude, where non-perturbative effects could be important [65]. The asymmetry of the distribution amplitude emphasizes contributions from the large x region. Since non-leading corrections are expected when the quark propagator scale $Q^2(1-x)$ is small, in principle relatively large momentum transfer is required to clearly test the perturbative QCD predictions. Chernyak *et al.* [59] have studied this effect in some detail and claim that their QCD sum rule predictions are not significantly changed when higher moments of the distribution amplitude are included.

It is important to notice that the perturbative scaling regime of the meson form factor is controlled by the virtuality of the quark propagator. When the quark is far off-shell, multiple gluon exchange contributions involving soft gluon insertions are suppressed by inverse powers of the quark propagator; there is not sufficient time to exchange soft gluons or gluonium. Thus the perturbative analysis is valid as long as the single gluon exchange propagator has inverse power behavior. There is thus no reason to require that the gluon be far off-shell, as in the analysis of Ref. [66].

The moments of distribution amplitudes can also be computed using lattice gauge theory [67]. In the case of the pion distribution amplitudes, there is good agreement of the lattice gauge theory computations of Martinelli and Sachrajda [68] with the QCD sum rule results. This check has strengthened confidence in the reliability of the QCD sum rule method, although the shape of the meson distribution amplitudes are unexpectedly structured: the pion distribution amplitude is broad and has a dip at $x=\frac{1}{2}$. The QCD sum rule meson distributions, combined with the perturbative QCD factorization predictions, account well for the scaling, normalization of the pion form factor and $\gamma\gamma \to M^+M^-$ cross sections.

In the case of the baryon, the asymmetric three-quark distributions are consistent with the normalization of the baryon form factor at large Q^2 and also the branching ratio for $J/\psi \to p\bar{p}$. The data for large angle Compton scattering $\gamma p \to \gamma p$ are also well described [69]. However, a very recent lattice calculation of the lowest two moments by Martinelli and Sachrajda [68] does not show skewing of the average fraction of momentum of the valence quarks in the proton. This lattice result is in contradiction to the predictions of the QCD sum rules and does cast some doubt on the validity of the model of the proton distribution proposed by Chernyak *et al.* [59]. The lattice calculation is performed in the quenched approximation with Wilson fermions and requires an extrapolation to the chiral limit.

The contribution of soft momentum exchange to the hadron form factors is a potentially serious complication when one uses the QCD sum rule model distribution amplitudes. In the analysis of Ref. [66] it was argued that only about 1% of the proton form factor comes from regions of integration in which all the propagators are hard. A new analysis by Dziembowski *et al.* [70] shows that the QCD sum rule [5] distribution amplitudes of Chernyak *et al.* [5] together with the perturbative QCD prediction gives contributions to the form factors which agree with the measured normalization of the pion form factor at $Q^2 > 4$ GeV^2 and proton form factor $Q^2 > 20$ GeV^2 to within a factor of two. In the calculation the virtuality of the exchanged gluon is restricted to $|k^2| > 0.25$ GeV^2. The authors assume $\alpha_s = 0.3$ and that the underlying wavefunctions fall off exponentially at the $x \simeq 1$ endpoints. Another model of the proton distribution amplitude with diquark clustering [71] chosen to satisfy the QCD sum rule moments come even closer. Considering the uncertainty in the magnitude of the higher order corrections, one really cannot expect better agreement between the QCD predictions and experiment.

The relative importance of non-perturbative contributions to form factors is also an issue. Unfortunately, there is little that can be said until we have a deeper understanding of the end-point behavior of hadronic wavefunctions, and of the role played by Sudakov form factors in the end-point region. Models have been constructed in which non-perturbative effects persist to high Q [66]. Other models have been constructed in which such effects vanish rapidly as Q increases [64, 72, 73].

If the QCD sum rule results are correct then, the light hadrons are highly structured oscillating momentum-space valence wavefunctions. In the case of mesons, the results from both the lattice calculations and QCD sum rules show that the light quarks are highly relativistic. This gives further indication that while nonrelativistic potential models are useful for enumerating the spectrum of hadrons (because they express the relevant degrees of freedom), they may not be reliable in predicting wavefunction structure.

11.7 The Pre-QCD Development of Exclusive Reactions

The study of exclusive processes in terms of underlying quark subprocesses in fact began before the discovery of QCD. The advent of the parton model and Bjorken scaling for deep inelastic structure functions in the late 1960's brought a new focus to the structure of form factors and exclusive processes at large momentum transfer. The underlying theme of the parton model was the concept that quarks carried the electromagnetic current within hadrons. The use of time-ordered perturbation theory in an "infinite momentum frame", or equivalently, quantization on the light cone, provided a natural language for hadrons as composites of relativistic partons, *i.e.*, point-like constituents [74]. Drell and Yan [36] showed how to compute current matrix elements in terms of a Fock state expansion at infinite momentum. (Later their result was shown to be exact in the light-cone quantization scheme.)

Drell and Yan suggested that hadron form factors are dominated by the end-point region $x \approx 1$. Then it is clear from the Drell-Yan formula that the form factor fall-off at large Q^2 is closely related to the $x \to 1$ behavior of the hadron structure function. The relation found by Drell and Yan was

$$F(Q^2) \sim \frac{1}{(Q^2)^n} \quad \text{if} \quad F_2(x, Q^2) \sim (1-x)^{2n-1} \,. \tag{11.18}$$

Gribov and Lipatov [75] extended this relationship to fragmentation functions $D(z, Q^2)$ at $z \to 1$, taking into account cancellations due to quark spin. Feynman [76] noted that the Drell-Yan relationship was also true in gauge theory models in which the endpoint behavior of structure functions is suppressed due to the emission of soft or "wee" partons by charged lines. The endpoint region is thus suppressed in QCD relative to the leading perturbative contributions.

The parton model was extended to exclusive processes such as hadron-hadron scattering and photoproduction by Blankenbecler, Brodsky, and Gunion [24] and by Landshoff and Polkinghorne [77]. It was recognized that independent of specific dynamics, hadrons could interact and scatter simply by exchanging their common constituents. These authors showed that the amplitude due to quark interchange (or rearrangement) could be written in closed form as an overlap of the light-cone wavefunctions of the incident and final hadrons. In order to make definite predictions, model wavefunctions were chosen to reproduce the fall-off of the form factors obtained from the Drell-Yan formula. Two-body exclusive amplitudes in the "constituent interchange model" then take the form of "fixed-angle" scaling laws

$$\frac{d\sigma}{dt}(AB \to CD) \sim \frac{f(\theta_{\rm cm})}{s^N} \,, \tag{11.19}$$

where the power N reflects the power-law fall-off of the elastic form factors of the scattered hadrons. The form of the angular dependence $f(\theta_{\rm cm})$ reflects the number of interchanged quarks.

Even though the constituent interchange model was motivated in part by the Drell-Yan endpoint analysis of form factors, many of the predictions and systematics of quark interchange remain applicable in the QCD analysis [24]. A comprehensive series of measurements of elastic meson nucleon scattering reactions has recently been carried out by Baller *et al.* [78] at BNL. Empirically, the quark interchange amplitudes gives a reasonable account of the scaling, angular dependence, and relative magnitudes of the various channels. For example, the strong differences between K^+p and K^-p scattering is accounted for by u quark interchange in the K^+p amplitude. It is inconsistent with gluon exchange as the dominant amplitude since this produces equal scattering for the two channels. The dominance of quark interchange over gluon exchange is a surprising result which eventually needs to be understood in the context of QCD.

The prediction of fixed angle scaling laws laid the groundwork for the derivation of the "dimensional counting rules." As discussed by Farrar and myself in Ref. [41], it is natural to assume that at large momentum transfer, an exclusive amplitude factorize as a convolution of hadron wavefunctions which couple the hadrons to their quark constituents with a hard scattering amplitude T_H which scatters the quarks from the initial to final direction. Since the hadron wavefunction is maximal when the quarks are nearly collinear with each parent hadron, the large momentum transfer occurs in T_H. The pre-QCD argument went as follows: the dimension of T_H is $[L^{n-4}]$ where $n = n_A + n_B + n_C + n_D$ is the total number of fields entering T_H. In a renormalizable theory where the coupling constant is dimensionless and masses can be neglected at large momentum transfer, all connected tree-graphs for T_H then scale as $[1/\sqrt{s}\,]^{n-4}$ at fixed t/s. This immediately gives the dimensional counting law [41]

$$\frac{d\sigma}{dt}(AB \to CD) \sim \frac{f(\theta_{\mathrm{cm}})}{s^{n_A+n_B+n_C+n_D-2}} \,. \tag{11.20}$$

In the case of incident or final photons or leptons $n = 1$. Specializing to elastic lepton-hadron scattering, this also implies $F(Q^2) \sim 1/(Q^2)^{n_H-1}$ for the spin averaged form factor, where n_H is the number of constituents in hadron H. These results were obtained independently by Matveev *et al.* [41] on the basis of an "automodality" principle, that the underlying constituent interactions are scale free.

As we have seen, the dimensional counting scaling laws will generally be modified by the accumulation of logarithms from higher loop corrections to the hard scattering amplitude T_H; the phenomenological success of the counting rules in their simplest form thus implies that the loop corrections be somewhat mild. As we have seen, it is the asymptotic freedom property of QCD which in fact makes higher order corrections an exponentiation of a $\log\log Q^2$ series, thus preserving the form of the dimensional counting rules modulo only logarithmic corrections.

11.8 Exclusive $\gamma\gamma$ Reactions

Two-photon reactions have a number of unique features which are especially important for testing QCD, especially in exclusive channels:

1. Any even charge conjugation hadronic state can be created in the annihilation of two photons—an initial state of minimum complexity.

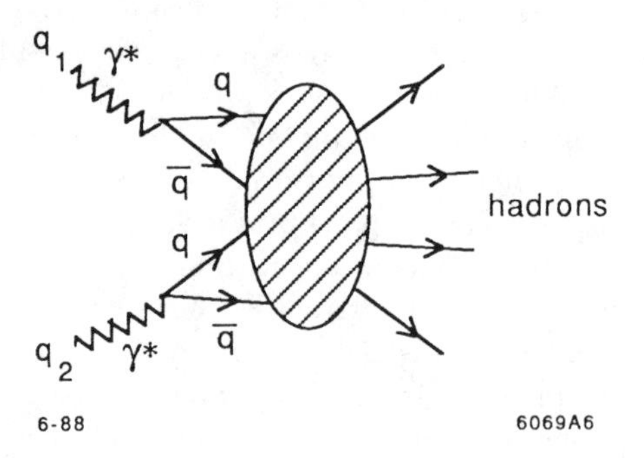

FIGURE 11.12
Photon-photon annihilation in QCD. The photons couple directly to one or two quark currents.

Because $\gamma\gamma$ annihilation is complete, there are no spectator hadrons to confuse resonance analyses. Thus, one has a clean environment for identifying the exotic color-singlet even C composites of quarks and gluons $|q\bar{q}\rangle$, $|gg\rangle$, $|ggg\rangle$, $|q\bar{q}g\rangle$, $|qq\bar{q}\bar{q}\rangle$, ... which are expected to be present in the few GeV mass range. (Because of mixing, the actual mass eigenstates of QCD may be complicated admixtures of the various Fock components.)

2. The mass and polarization of each of the incident virtual photons can be continuously varied, allowing highly detailed tests of theory. Because a spin-one state cannot couple to two on-shell photons, a $J = 1$ resonance can be uniquely identified by the onset of its production with increasing photon mass [80].

3. Two-photon physics plays an especially important role in probing dynamical mechanisms. In the low momentum transfer domain, $\gamma\gamma$ reactions such as the total annihilation cross section and exclusive vector meson pair production can give important insights into the nature of diffractive reactions in QCD. Photons in QCD couple directly to the quark currents at any resolution scale (see Fig. 11.12). Predictions for high momentum transfer $\gamma\gamma$ reactions, including the photon structure functions, $F_2^\gamma(x, Q^2)$ and $F_L^\gamma(x, Q^2)$, high p_T jet production, and exclusive channels are thus much more specific than corresponding hadron-induced reactions. The pointlike coupling of the annihilating photons leads to a host of special features which differ markedly with predictions based on vector meson dominance models.

4. Exclusive $\gamma\gamma$ processes provide a window for viewing the wavefunctions of hadrons in terms of their quark and gluon degrees of freedom. In the case of $\gamma\gamma$ annihilation into hadron pairs, the angular distribution of the production cross section directly reflects the shape of the distribution amplitude (valence wavefunction) of each hadron.

Thus far experiment has not been sufficiently precise to measure the logarithmic modification of dimensional counting rules predicted by QCD. Perturbative QCD predictions for $\gamma\gamma$ exclusive processes at high momentum transfer and high invariant pair mass provide some of the most severe tests of the theory [31]. A simple, but still very important example [38] is the Q^2-dependence of the reaction $\gamma^*\gamma \to M$ where M is a pseudoscalar meson such as the η. The invariant amplitude contains only one form factor:

$$M_{\mu\nu} = \epsilon_{\mu\nu\sigma\tau} p_\eta^\sigma q^\tau F_{\gamma\eta}(Q^2) \ . \tag{11.21}$$

It is easy to see from power counting at large Q^2 that the dominant amplitude (in light-cone gauge) gives $F_{\gamma\eta}(Q^2) \sim 1/Q^2$ and arises from diagrams (see Fig. 11.13) which have the minimum path carrying Q^2: *i.e.*, diagrams in which there is only a single quark propagator between the two photons. The coefficient of $1/Q^2$ involves only the two-particle $q\bar{q}$ distribution amplitude $\phi(x, Q)$, which evolves logarithmically on Q. Higher particle number Fock states give higher power-law falloff contributions to the exclusive amplitude.

The TPC/$\gamma\gamma$ data [81] shown in Fig. 11.14 are in striking agreement with the predicted QCD power: a fit to the data gives $F_{\gamma\eta}(Q^2) \sim (1/Q^2)^n$ with $n = 1.05 \pm 0.15$. Data for the η' from Pluto and the TPC/$\gamma\gamma$ experiments give similar results, consistent with scale-free behavior of the QCD quark propagator and the point coupling to the quark current for both the real and virtual photons. In the case of deep inelastic lepton scattering, the observation of Bjorken scaling tests these properties when both photons are virtual.

The QCD power law prediction, $F_{\gamma\eta}(Q^2) \sim 1/Q^2$, is consistent with dimensional counting [41] and also emerges from current algebra arguments (when both photons are very virtual) [82]. On the other hand, the $1/Q^2$ falloff is also expected in vector meson dominance models. The QCD and VMD predictions can be readily discriminated by studying $\gamma^*\gamma^* \to \eta$. In VMD one expects a product of form factors; in QCD the falloff of the amplitude is still $1/Q^2$ where Q^2 is a linear combination of Q_1^2 and Q_2^2. It is clearly very important to test this essential feature of QCD.

Exclusive two-body processes $\gamma\gamma \to H\overline{H}$ at large $s = W_{\gamma\gamma}^2 = (q_1 + q_2)^2$ and fixed $\theta_{\rm cm}^{\gamma\gamma}$ provide a particularly important laboratory for testing QCD, since the large momentum-transfer behavior, helicity structure, and often even the absolute normalization can be rigorously predicted [31, 69]. The

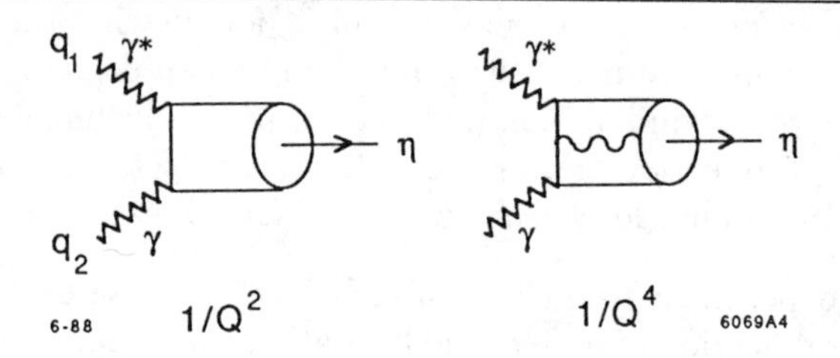

FIGURE 11.13
Calculation of the γ-η transition form factor in QCD from the valence $q\bar{q}$ and $q\bar{q}g$ Fock states.

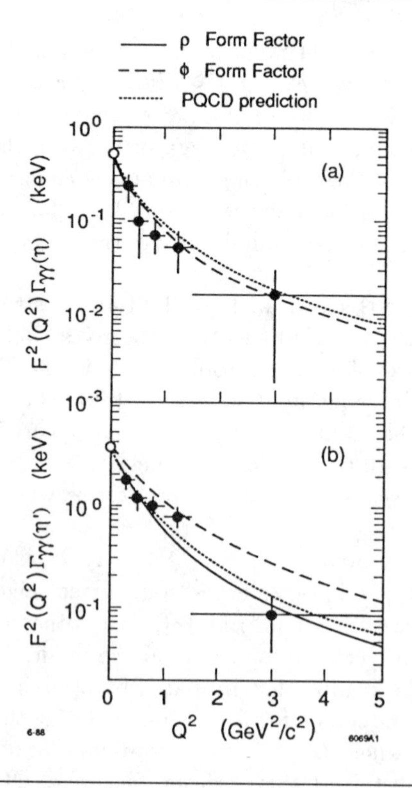

FIGURE 11.14
Comparison of TPC/$\gamma\gamma$ data [81] for the γ-η and γ-η' transition form factors with the QCD leading twist prediction of Ref. [31]. The VMD predictions are also shown.

angular dependence of some of the $\gamma\gamma \to H\overline{H}$ cross sections reflects the shape of the hadron distribution amplitudes $\phi_H(x_i, Q)$. The $\gamma_\lambda\gamma_{\lambda'} \to H\overline{H}$ amplitude can be written as a factorized form

$$\mathcal{M}_{\lambda\lambda'}(W_{\gamma\gamma}, \theta_{\rm cm}) = \int_0^1 [dy_i]\, \phi_H^*(x_i, Q)\, \phi_{\overline{H}}^*(y_i, Q)\, T_{\lambda\lambda'}(x, y; W_{\gamma\gamma}, \theta_{\rm cm}) \,, \tag{11.22}$$

where $T_{\lambda\lambda'}$ is the hard scattering helicity amplitude. To leading order $T \propto \alpha(\alpha_S/W_{\gamma\gamma}^2)^n$ and $d\sigma/dt \sim W_{\gamma\gamma}^{-(2n+2)} f(\theta_{\rm cm})$ where $n = 1$ for meson and $n = 2$ for baryon pairs.

Lowest order predictions for pseudo-scalar and vector-meson pairs for each helicity amplitude are given in Ref. [31]. In each case the helicities of the hadron pairs are equal and opposite to leading order in $1/W^2$. The normalization and angular dependence of the leading order predictions for $\gamma\gamma$ annihilation into charged meson pairs are almost model independent; *i.e.*, they are insensitive to the precise form of the meson distribution amplitude. If the meson distribution amplitude is symmetric in x and $(1-x)$, then the same quantity

$$\int_0^1 dx\, \frac{\phi_\pi(x, Q)}{(1-x)} \tag{11.23}$$

controls the x-integration for both $F_\pi(Q^2)$ and to high accuracy $M(\gamma\gamma \to \pi^+\pi^-)$. Thus for charged pion pairs one obtains the relation:

$$\frac{\frac{d\sigma}{dt}(\gamma\gamma \to \pi^+\pi^-)}{\frac{d\sigma}{dt}(\gamma\gamma \to \mu^+\mu^-)} \cong \frac{4|F_\pi(s)|^2}{1 - \cos^4\theta_{\rm cm}} \,. \tag{11.24}$$

Note that in the case of charged kaon pairs, the asymmetry of the distribution amplitude may give a small correction to this relation.

The scaling behavior, angular behavior, and normalization of the $\gamma\gamma$ exclusive pair production reactions are nontrivial predictions of QCD. Recent Mark II meson pair data and PEP4/PEP9 data [83] for separated $\pi^+\pi^-$ and K^+K^- production in the range $1.6 < W_{\gamma\gamma} < 3.2$ GeV near 90° are in satisfactory agreement with the normalization and energy dependence predicted by QCD (see Fig. 11.15). In the case of $\pi^0\pi^0$ production, the $\cos\theta_{\rm cm}$ dependence of the cross section can be inverted to determine the x-dependence of the pion distribution amplitude.

The wavefunction of hadrons containing light and heavy quarks such as the K, D-meson are likely to be asymmetric due to the disparity of the quark masses. In a gauge theory one expects that the wavefunction is maximum when the quarks have zero relative velocity; this corresponds to $x_i \propto m_{i\perp}$ where $m_\perp^2 = k_\perp^2 + m^2$. An explicit model for the skewing of the meson distribution amplitudes based on QCD sum rules is given

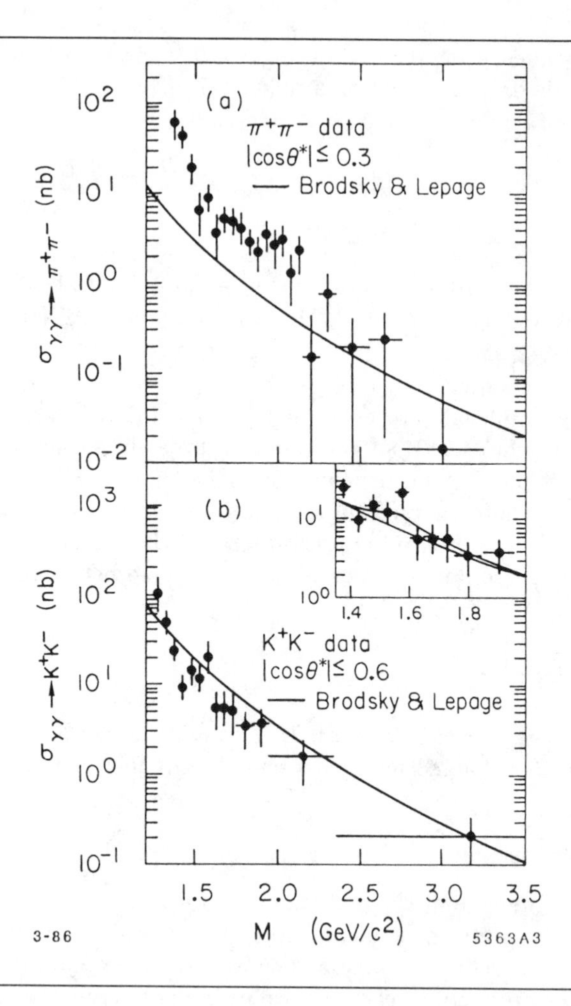

FIGURE 11.15
Comparison of $\gamma\gamma \to \pi^+\pi^-$ and $\gamma\gamma \to K^+K^-$ meson pair production data with the parameter-free perturbative QCD prediction of Ref. [31]. The theory predicts the normalization and scaling of the cross sections. The data are from the TPC/$\gamma\gamma$ collaboration [83].

by Benyayoun and Chernyak [84]. These authors also apply their model to two-photon exclusive processes such as $\gamma\gamma \to K^+K^-$ and obtain some modification compared to the strictly symmetric distribution amplitudes. If the same conventions are used to label the quark lines, the calculations of Benyayoun and Chernyak are in complete agreement with those of Ref. [31].

The one-loop corrections to the hard scattering amplitude for meson pairs have been calculated by Nizic [85]. The QCD predictions for mesons containing admixtures of the $|gg\rangle$ Fock state is given by Atkinson, Sucher, and Tsokos [69].

The perturbative QCD analysis has been extended to baryon-pair production in comprehensive analyses by Farrar *et al.* [86, 69] and by Gunion *et al.* [69]. Predictions are given for the "sideways" Compton process $\gamma\gamma \to p\bar{p}$, $\Delta\overline{\Delta}$ pair production, and the entire decuplet set of baryon pair states. The arduous calculation of 280 $\gamma\gamma \to qqq\bar{q}\bar{q}\bar{q}$ diagrams in T_H required for calculating $\gamma\gamma \to B\overline{B}$ is greatly simplified by using two-component spinor techniques. The doubly charged Δ pair is predicted to have a fairly small normalization. Experimentally such resonance pairs may be difficult to identify under the continuum background.

The normalization and angular distribution of the QCD predictions for proton-antiproton production shown in Fig. 11.16 depend in detail on the form of the nucleon distribution amplitude, and thus provide severe tests of the model form derived by Chernyak, Ogloblin, and Zhitnitskii [59] from QCD sum rules.

An important check of the QCD predictions can be obtained by combining data from $\gamma\gamma \to p\bar{p}$ and the annihilation reaction, $p\bar{p} \to \gamma\gamma$, with large angle Compton scattering $\gamma p \to \gamma p$. The available data [87] for large angle Compton scattering (see Fig. 11.17) for 5 $\mathrm{GeV}^2 < s < 10$ GeV^2 are consistent with the dimensional counting scaling prediction, $s^6 d\sigma/dt = f(\theta_{\mathrm{cm}})$. In general, comparisons between channels related by crossing of the Mandelstam variables place a severe constraint on the angular dependence and analytic form of the underlying QCD exclusive amplitude. Furthermore in $p\bar{p}$ collisions one can study timelike photon production into e^+e^- and examine the virtual photon mass dependence of the Compton amplitude. Predictions for the q^2 dependence of the $p\bar{p} \to \gamma\gamma^*$ amplitude can be obtained by crossing the results of Gunion and Millers [69].

The region of applicability of the leading power-law predictions for $\gamma\gamma \to p\bar{p}$ requires that one be beyond resonance or threshold effects. It presumably is set by the scale where $Q^4 G_M(Q^2)$ is roughly constant, *i.e.*, $Q^2 > 3$ GeV^2. Present measurements may thus be too close to threshold for meaningful tests [88]. It should be noted that unlike the case for charged meson pair production, the QCD predictions for baryons are sensitive to the form of the running coupling constant and the endpoint behavior of the wavefunctions.

The QCD predictions for $\gamma\gamma \to H\overline{H}$ can be extended to the case of one or two virtual photons, for measurements in which one or both electrons are tagged. Because of the direct coupling of the photons to the quarks, the Q_1^2 and Q_2^2 dependence of the $\gamma\gamma \to H\overline{H}$ amplitude for transversely

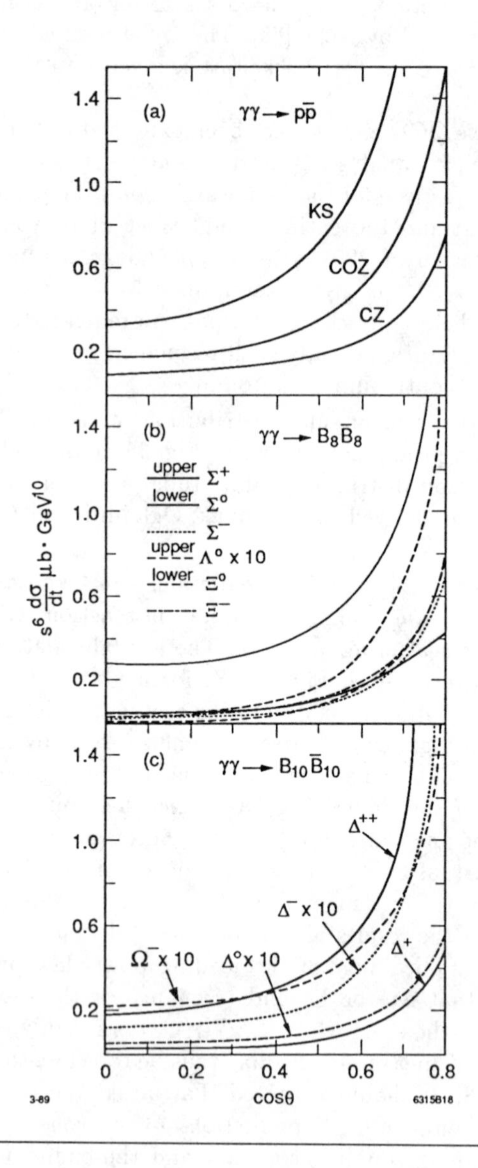

FIGURE 11.16
Perturbative QCD predictions by Farrar and Zhang for the $\cos(\theta_{\text{cm}})$ dependence of the $\gamma\gamma \to p\bar{p}$ cross section assuming the King-Sachrajda (KS), Chernyak, Ogloblin, and Zhitnitskii (COZ) [59] and original Chernyak and Zhitnitskii (CZ) [5] forms for the proton distribution amplitude, $\phi_p(x_i, Q)$.

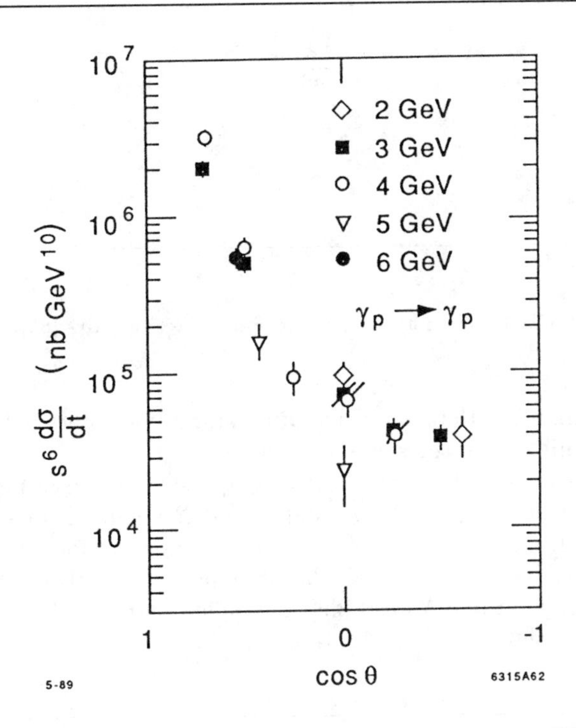

FIGURE 11.17
Test of dimensional counting for Compton scattering for $2 < E^{\gamma}_{\text{lab}} < 6$ GeV [87].

polarized photons is minimal at large W^2 and fixed θ_{cm}, since the off-shell quark and gluon propagators in T_H already transfer hard momenta; *i.e.*, the 2γ coupling is effectively local for Q_1^2, $Q_2^2 \ll p_T^2$. The $\gamma^*\gamma^* \to \overline{B}B$ and $M\overline{M}$ amplitudes for off-shell photons have been calculated by Millers and Gunion [69]. In each case, the predictions show strong sensitivity to the form of the respective baryon and meson distribution amplitudes.

We also note that photon-photon collisions provide a way to measure the running coupling constant in an exclusive channel, independent of the form of hadronic distribution amplitudes [31]. The photon-meson transition form factors $F_{\gamma\to M}(Q^2)$, $M = \pi^0, \eta^0, f$, *etc.*, are measurable in tagged $e\gamma \to e'M$ reactions. QCD predicts

$$\alpha_s(Q^2) = \frac{1}{4\pi} \frac{F_\pi(Q^2)}{Q^2 |F_{\pi\gamma}(Q^2)|^2}, \tag{11.25}$$

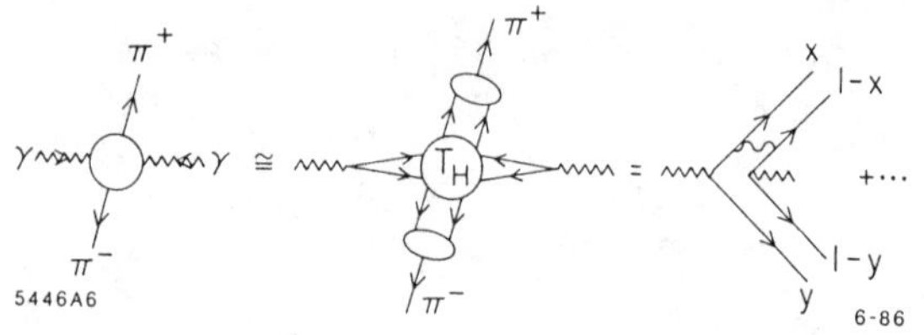

FIGURE 11.18
Application of QCD to two-photon production of meson pairs [89].

where to leading order the pion distribution amplitude enters both numerator and denominator in the same manner.

The complete calculations of the tree-graph structure (see Figs. 11.18, 11.19, 11.20) of both $\gamma\gamma \to M\overline{M}$ and $\gamma\gamma \to B\overline{B}$ amplitudes has now been completed. One can use crossing to compute T_H for $p\bar{p} \to \gamma\gamma$ to leading order in $\alpha_s(p_T^2)$ from the calculations reported by Farrar, Maina and Neri [69] and Gunion and Millers [69]. Examples of the predicted angular distributions are shown in Figs. 11.21 and 11.22.

11.9 Exclusive Nuclear Reactions— Reduced Amplitudes

The nucleus is itself an interesting QCD structure. At short distances nuclear wavefunctions and nuclear interactions necessarily involve *hidden color*, degrees of freedom orthogonal to the channels described by the usual nucleon or isobar degrees of freedom. At asymptotic momentum transfer, the deuteron form factor and distribution amplitude are rigorously calculable. One can also derive new types of testable scaling laws for exclusive nuclear amplitudes in terms of the reduced amplitude formalism.

An ultimate goal of QCD phenomenology is to describe the nuclear force and the structure of nuclei in terms of quark and gluon degrees of freedom. Explicit signals of QCD in nuclei have been elusive, in part because of the fact that an effective Lagrangian containing meson and nucleon degrees of freedom must be in some sense equivalent to QCD if one is limited to low-energy probes. On the other hand, an effective local field theory of nucleon and meson fields cannot correctly describe the observed off-shell falloff of form factors, vertex amplitudes, Z-graph diagrams, *etc.*, because hadron compositeness is not taken into account.

We have already mentioned the prediction $F_d(Q^2) \sim 1/Q^{10}$ which

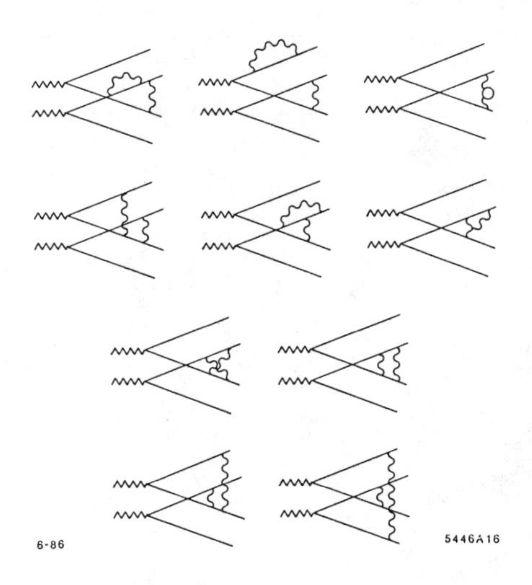

FIGURE 11.19
Next-to-leading perturbative contribution to T_H for the process $\gamma\gamma \to M\overline{M}$. The calculation has been done by Nizic [89].

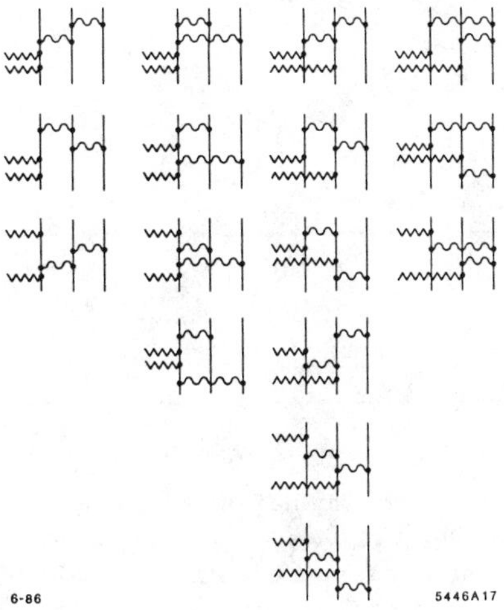

6-86 5446A17

FIGURE 11.20
Leading diagrams for $\gamma + \gamma \to \bar{p} + p$ calculated in Ref. [69].

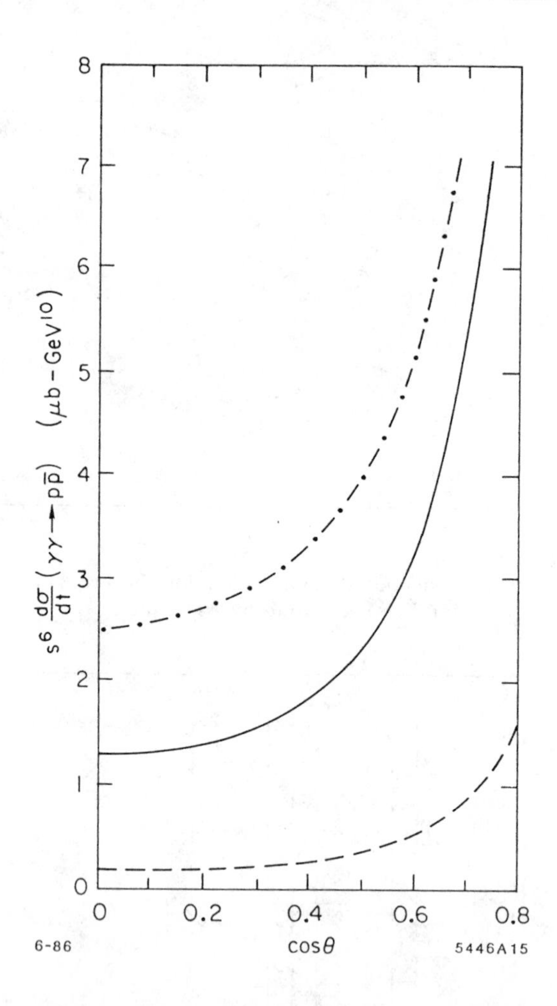

FIGURE 11.21
QCD prediction for the scaling and angular distribution for $\gamma + \gamma \to \bar{p} + p$ calculated by Farrar *et al.* [69]. The dashed-dot curve corresponds to $4\Lambda^2/s = 0.0016$ and a maximum running coupling constant $\alpha_s^{max} = 0.8$. The solid curve corresponds to $4\Lambda^2/s = 0.016$ and a maximum running coupling constant $\alpha_s^{max} = 0.5$. The dashed curve corresponds to a fixed $\alpha_s = 0.3$. The results are very sensitive to the endpoint behavior of the proton distribution amplitude. The CZ form is assumed.

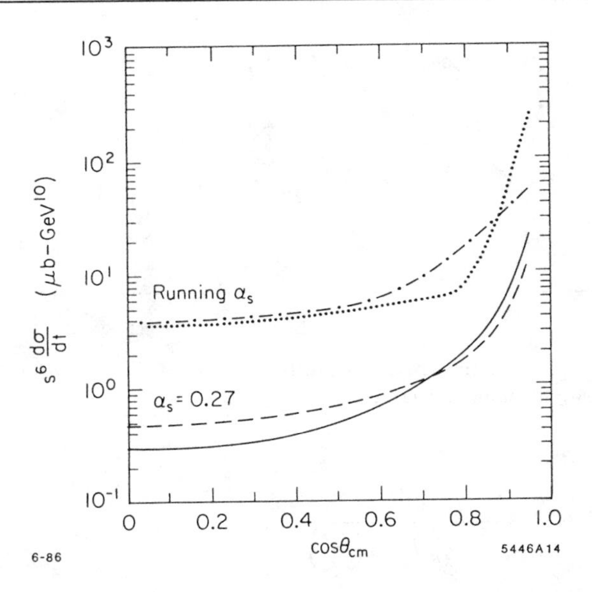

FIGURE 11.22
QCD prediction for the scaling and angular distribution for $\gamma + \gamma \rightarrow \bar{p} + p$ calculated by Gunion, Sparks and Millers [69]. CZ distribution amplitudes are assumed. The dashed and dot-dashed curves correspond to one photon space-like, with $Q_b^2/s = 0.1$. The solid and running curves are for real photon annihilation.

comes from simple quark counting rules, as well as perturbative QCD. One cannot expect this asymptotic prediction to become accurate until very large Q^2 is reached since the momentum transfer has to be shared by at least six constituents. However there is a simple way to isolate the QCD physics due to the compositeness of the nucleus, not the nucleons. The deuteron form factor is the probability amplitude for the deuteron to scatter from p to $p + q$ but remain intact. Note that for vanishing nuclear binding energy $\epsilon_d \rightarrow 0$, the deuteron can be regarded as two nucleons sharing the deuteron four-momentum (see Fig. 11.23). The momentum ℓ is limited by the binding and can thus be neglected. To first approximation the proton and neutron share the deuteron's momentum equally. Since the deuteron form factor contains the probability amplitudes for the proton and neutron to scatter from $p/2$ to $p/2 + q/2$; it is natural to define the reduced deuteron form factor [90, 91]

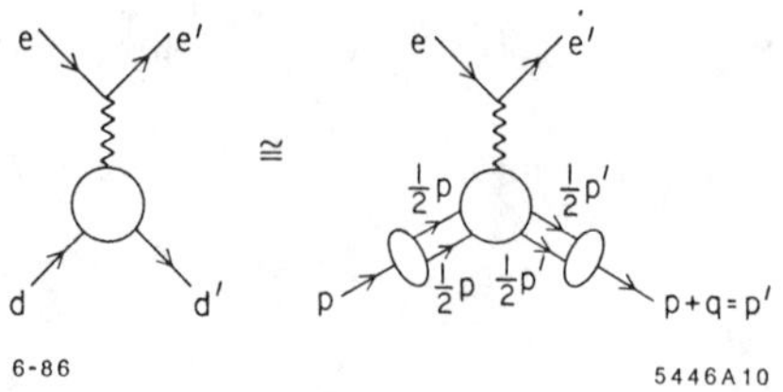

FIGURE 11.23
Application of the reduced amplitude formalism to the deuteron form factor at large momentum transfer.

$$f_d(Q^2) \equiv \frac{F_d(Q^2)}{F_{1N}\left(\frac{Q^2}{4}\right) F_{1N}\left(\frac{Q^2}{4}\right)} . \tag{11.26}$$

The effect of nucleon compositeness is removed from the reduced form factor. QCD then predicts the scaling

$$f_d(Q^2) \sim \frac{1}{Q^2} , \tag{11.27}$$

i.e., the same scaling law as a meson form factor. Diagrammatically, the extra power of $1/Q^2$ comes from the propagator of the struck quark line, the one propagator not contained in the nucleon form factors. Because of hadron helicity conservation, the prediction is for the leading helicity-conserving deuteron form factor ($\lambda = \lambda' = 0$). As shown in Fig. 11.24, this scaling is consistent with experiment for $Q = p_T \gtrsim 1$ GeV [92].

The distinction between the QCD and other treatments of nuclear amplitudes is particularly clear in the reaction $\gamma d \to np$; *i.e.*, photodisintegration of the deuteron at fixed center of mass angle. Using dimensional counting, the leading power-law prediction from QCD is simply $\frac{d\sigma}{dt}(\gamma d \to np) \sim \frac{1}{s^{11}} F(\theta_{\rm cm})$. Again we note that the virtual momenta are partitioned among many quarks and gluons, so that finite mass corrections will be significant at low to medium energies. Nevertheless, one can test the basic QCD dynamics in these reactions taking into account much of the finite-mass, higher-twist corrections by using the "reduced amplitude" formalism [90, 91]. Thus the photodisintegration amplitude contains the probability amplitude (*i.e.*, nucleon form factors) for the proton and neutron to each remain intact after absorbing momentum transfers $p_p - \frac{1}{2}p_d$ and $p_n - \frac{1}{2}p_d$, respectively (see Fig. 11.25). After the form factors are removed, the remaining "reduced" amplitude should scale as $F(\theta_{\rm cm})/p_T$. The

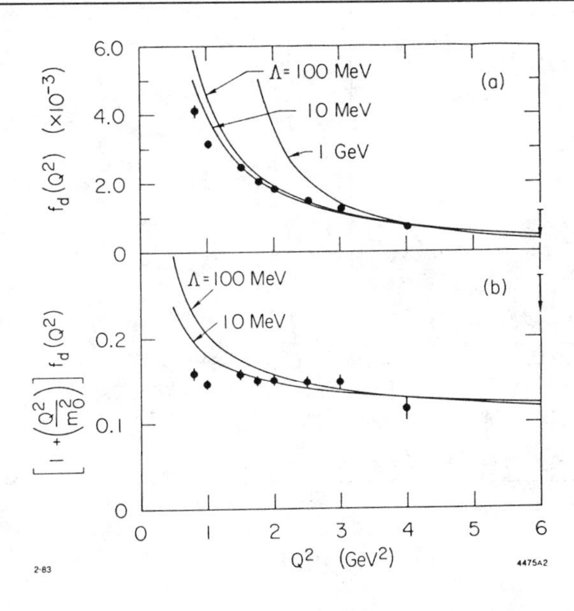

FIGURE 11.24
Scaling of the deuteron reduced form factor. The data are summarized in Ref. [90].

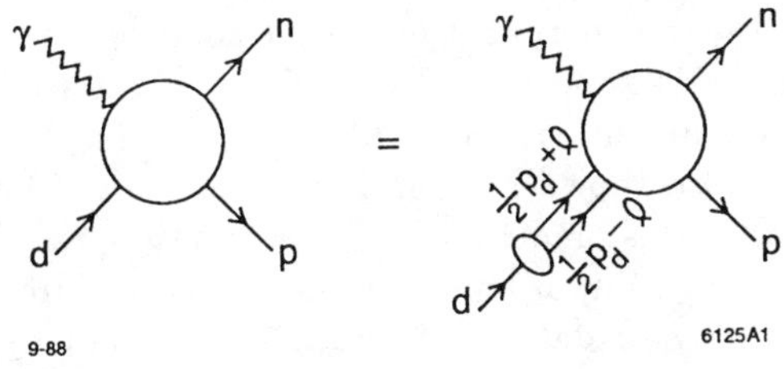

FIGURE 11.25
Construction of the reduced nuclear amplitude for two-body inelastic deuteron reactions [90].

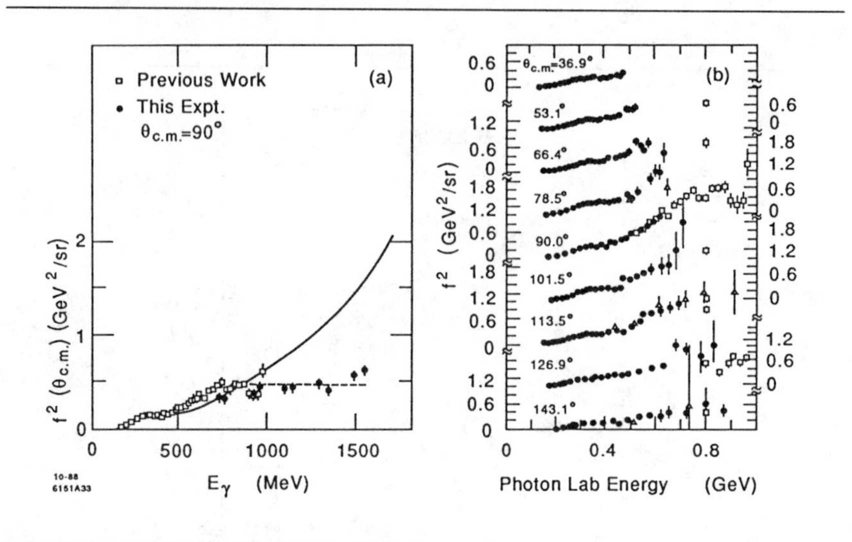

FIGURE 11.26
Comparison of deuteron photodisintegration data with the scaling prediction which requires $f^2(\theta_{\rm cm})$ to be at most logarithmically dependent on energy at large momentum transfer. The data in (a) are from the recent experiment of Ref. [93]. The nuclear physics prediction shown in (a) is from Ref. [94]. The data in (b) are from Ref. [95].

single inverse power of transverse momentum p_T is the slowest conceivable in any theory, but it is the unique power predicted by PQCD.

The prediction that $f(\theta_{\rm cm})$ is energy dependent at high-momentum transfer is compared with experiment in Fig. 11.26. It is particularly striking to see the QCD prediction verified at incident photon lab energies as low as 1 GeV. A comparison with a standard nuclear physics model with exchange currents is also shown for comparison as the solid curve in Fig. 11.26 (a). The fact that this prediction falls less fast than the data suggests that meson and nucleon compositeness are not taken to into account correctly. An extension of these data to other angles and higher energy would clearly be very valuable.

An important question is whether the normalization of the $\gamma d \to pn$ amplitude is correctly predicted by perturbative QCD. A recent analysis by Fujita [96] shows that mass corrections to the leading QCD prediction are not significant in the region in which the data show scaling. However Fujita also finds that in a model based on simple one-gluon plus quark-interchange mechanism, normalized to the nucleon-nucleon scattering amplitude, gives a photo-disintegration amplitude with a normalization an order of magnitude below the data. However this model only allows for diagrams in which the photon insertion acts only on the quark lines which couple to the exchanged gluon. It is expected that including other diagrams in which the photon couples to the current of the other four quarks will increase the photo-disintegration amplitude by a large factor.

The derivation of the evolution equation for the deuteron and other multi-quark states is given in Refs. [97] and [91]. In the case of the deuteron, the evolution equation couples five different color singlet states composed of the six quarks. The leading anomalous dimension for the deuteron distribution amplitude and the helicity-conserving deuteron form factor at asymptotic Q^2 is given in Ref. [97].

There are a number of related tests of QCD and reduced amplitudes which require $\bar{p}$ beams [91] such as $\bar{p}d \to \gamma n$ and $\bar{p}d \to \pi^- p$ in the fixed $\theta_{\rm cm}$ region. These reactions are particularly interesting tests of QCD in nuclei. Dimensional counting rules predict the asymptotic behavior $\frac{d\sigma}{dt}(\bar{p}d \to \pi^- p) \sim \frac{1}{(p_T^2)^{12}} f(\theta_{\rm cm})$ since there are 14 initial and final quanta involved. Again one notes that the $\bar{p}d \to \pi^- p$ amplitude contains a factor representing the probability amplitude (*i.e.*, form factor) for the proton to remain intact after absorbing momentum transfer squared $\hat{t} = (p - \frac{1}{2}p_d)^2$ and the $\bar{N}N$ time-like form factor at $\hat{s} = (\bar{p} + \frac{1}{2}p_d)^2$. Thus $\mathcal{M}_{\bar{p}d \to \pi^- p} \sim F_{1N}(\hat{t})\, F_{1N}(\hat{s})\, \mathcal{M}_r$, where $\mathcal{M}_r$ has the same QCD scaling properties as quark meson scattering. One thus predicts

$$\frac{\frac{d\sigma}{d\Omega}(\bar{p}d \to \pi^- p)}{F_{1N}^2(\hat{t})\, F_{1N}^2(\hat{s})} \sim \frac{f(\Omega)}{p_T^2} . \tag{11.28}$$

The reduced amplitude scaling for $\gamma d \to pn$ at large angles and $p_T \gtrsim 1$ GeV is shown in Fig. 11.26. One thus expects similar precocious scaling behavior to hold for $\bar{p}d \to \pi^- p$ and other $\bar{p}d$ exclusive reduced amplitudes. Recent analyses by Kondratyuk and Sapozhnikov [98] show that standard nuclear physics wavefunctions and interactions cannot explain the magnitude of the data for two-body anti-proton annihilation reactions such as $\bar{p}d \to \pi^- p$.

11.10 A Test of Color Transparency

A striking feature of the QCD description of exclusive processes is "color transparency": the only part of the hadronic wavefunction that scatters at large momentum transfer is its valence Fock state where the quarks are at small relative impact separation. Such a fluctuation has a small color-dipole moment and thus has negligible interactions with other hadrons. Since such a state stays small over a distance proportional to its energy, this implies

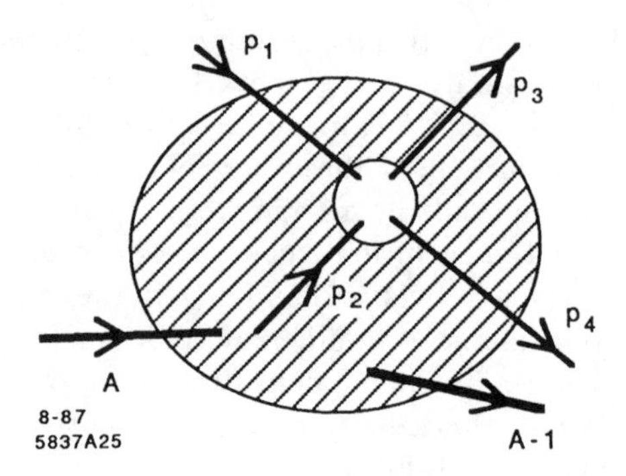

FIGURE 11.27
Quasi-elastic pp scattering inside a nuclear target. Normally one expects such processes to be attenuated by elastic and inelastic interactions of the incident proton and the final state interaction of the scattered proton. Perturbative QCD predicts minimal attenuation; *i.e.*, "color transparency" at large momentum transfer [14].

that quasi-elastic hadron-nucleon scattering at large momentum transfer as illustrated in Fig. 11.27 can occur additively on all of the nucleons in a nucleus with minimal attenuation due to elastic or inelastic final state interactions in the nucleus, *i.e.*, the nucleus becomes "transparent". By contrast, in conventional Glauber scattering, one predicts strong, nearly energy-independent initial and final state attenuation. A detailed discussion of the time and energy scales required for the validity of the PQCD prediction is given by Farrar *et al.* and Mueller in Ref. [14].

A recent experiment [99] at BNL measuring quasi-elastic $pp \to pp$ scattering at $\theta_{\rm cm} = 90°$ in various nuclei appears to confirm the color transparency prediction— at least for $p_{\rm lab}$ up to 10 GeV/c (see Fig. 11.28). Descriptions of elastic scattering which involve soft hadronic wavefunctions cannot account for the data. However, at higher energies, $p_{\rm lab} \sim 12$ GeV/c, normal attenuation is observed in the BNL experiment. This is the same kinematical region $E_{\rm cm} \sim 5$ GeV where the large spin correlation in A_{NN} are observed [100]. I shall argue that both features may be signaling new s-channel physics associated with the onset of charmed hadron production [101]. Clearly, much more testing of the color transparency phenomena is required, particularly in quasi-elastic lepton-proton scattering, Compton scattering, antiproton-proton scattering, *etc.* The cleanest test of the PQCD prediction is to check for minimal attenuation in large momentum

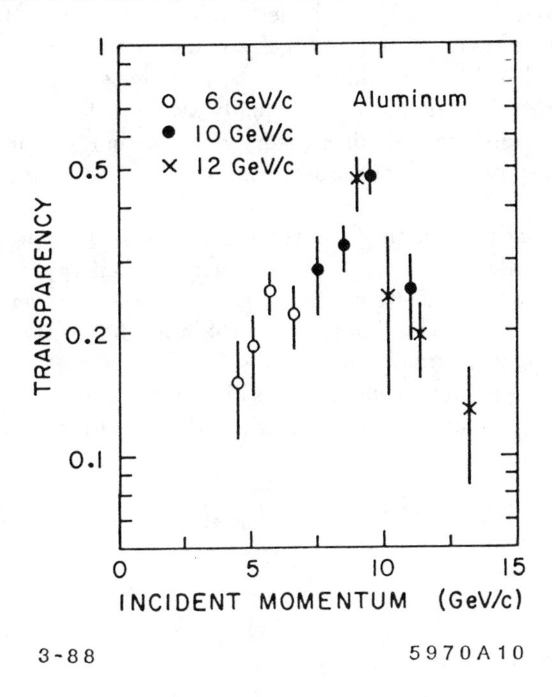

FIGURE 11.28
Measurements of the transparency ratio

$$T = \frac{Z_{eff}}{Z} = \frac{d\sigma}{dt}[pA \to p(A-1)] / \frac{d\sigma}{dt}[pA \to pp]$$

near 90° on Aluminum [99]. Conventional theory predicts that T should be small and roughly constant in energy. Perturbative QCD [14] predicts a monotonic rise to $T = 1$.

transfer lepton-proton scattering in nuclei since there are no complications from pinch singularities or resonance interference effects.

One can also understand the origin of color transparency as a consequence of the PQCD prediction that soft initial-state corrections to reactions such as $\bar{p}p \to \bar{\ell}\ell$ are suppressed at high lepton pair mass. This is a remarkable consequence of gauge theory and is quite contrary to normal treatments of initial interactions based on Glauber theory. This novel effect can be studied in quasielastic $\bar{p}A \to \bar{\ell}\ell\ (A-1)$ reaction, in which there are no extra hadrons produced and the produced leptons are coplanar with the beam. (The nucleus $(A-1)$ can be left excited). Since PQCD predicts the absence of initial-state elastic and inelastic interactions, the number of such events should be strictly additive in the number Z of protons in the nucleus, every proton in the nucleus is equally available for short-distance annihilation. In traditional Glauber theory only the surface protons can participate because of the strong absorption of the $\bar{p}$ as it traverses the nucleus.

The above description is the ideal result for large s. QCD predicts that additivity is approached monotonically with increasing energy, corresponding to two effects: a) the effective transverse size of the $\bar{p}$ wavefunction is $b_\perp \sim 1/\sqrt{s}$, and b) the formation time for the $\bar{p}$ is sufficiently long, such that the Fock state stays small during transit of the nucleus.

The color transparency phenomenon is also important to test in purely hadronic quasiexclusive antiproton-nuclear reactions. For large p_T one predicts

$$\frac{d\sigma}{dt\,dy}\left(\bar{p}A \to \pi^+\pi^- + (A-1)\right) \simeq \sum_{p\epsilon A} G_{p/A}(y)\,\frac{d\sigma}{dt}\left(\bar{p}p \to \pi^+\pi^-\right), \tag{11.29}$$

where $G_{p/A}(y)$ is the probability distribution to find the proton in the nucleus with light-cone momentum fraction $y = (p^0 + p^z)/(p^0_A + p^z_A)$, and

$$\frac{d\sigma}{dt}(\bar{p}p \to \pi^+\pi^-) \simeq \left(\frac{1}{p_T^2}\right)^8 f(\cos\theta_{\rm cm}) . \tag{11.30}$$

The distribution $G_{p/A}(y)$ can also be measured in $eA \to ep(A-1)$ quasiexclusive reactions. A remarkable feature of the above prediction is that there are no corrections required from initial-state absorption of the $\bar{p}$ as it traverses the nucleus, nor final-state interactions of the outgoing pions. Again the basic point is that the only part of hadron wavefunctions which is involved in the large p_T reaction is $\psi_H(b_\perp \sim \mathcal{O}(1/p_T))$, *i.e.*, the amplitude where all the valence quarks are at small relative impact parameter. These configurations correspond to small color singlet states which, because of color cancellations, have negligible hadronic interactions in the target. Measurements of these reactions thus test a fundamental feature of the Fock state description of large p_T exclusive reactions.

Another interesting feature which can be probed in such reactions is the behavior of $G_{p/A}(y)$ for y well away from the Fermi distribution peak at $y \sim m_N/M_A$. For $y \to 1$ spectator counting rules [102] predict $G_{p/A}(y) \sim (1-y)^{2N_s-1} = (1-y)^{6A-7}$ where $N_s = 3(A-1)$ is the number of quark spectators required to "stop" $(y_i \to 0)$ as $y \to 1$. This simple formula has been quite successful in accounting for distributions measured in the forward fragmentation of nuclei at the BEVALAC [28]. Color transparency

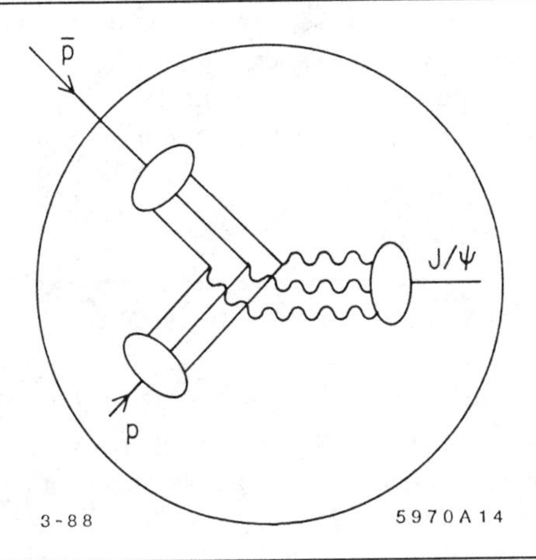

FIGURE 11.29
Schematic representation of quasielastic charmonium production in $\bar{p}A$ reactions.

can also be studied by measuring quasiexclusive J/ψ production by antiprotons in a nuclear target $\bar{p}A \to J/\psi(A-1)$ where the nucleus is left in a ground or excited state, but extra hadrons are not created (see Fig. 11.29). The cross section involves a convolution of the $\bar{p}p \to J/\psi$ subprocess cross section with the distribution $G_{p/A}(y)$ where $y = (p^0 + p^3)/(p_A^0 + p_A^3)$ is the boost-invariant light-cone fraction for protons in the nucleus. This distribution can be determined from quasiexclusive lepton-nucleon scattering $\ell A \to \ell p(A-1)$.

In first approximation $\bar{p}p \to J/\psi$ involves $qqq + \bar{q}\bar{q}\bar{q}$ annihilation into three charmed quarks. The transverse momentum integrations are controlled by the charm mass scale and thus only the Fock state of the incident antiproton which contains three antiquarks at small impact separation can annihilate. Again it follows that this state has a relatively small color dipole moment, and thus it should have a longer than usual mean-free path in nuclear matter; *i.e.*, color transparency. Unlike traditional expectations, QCD predicts that the $\bar{p}p$ annihilation into charmonium is not restricted to the front surface of the nucleus. The exact nuclear dependence depends on the formation time for the physical $\bar{p}$ to couple to the small $\bar{q}\bar{q}\bar{q}$ configuration, $\tau_F \propto E_p$. It may be possible to study the effect of finite formation time by varying the beam energy, E_p, and using the Fermi-motion of the nucleon to stay at the J/ψ resonance. Since the J/ψ is produced at nonrelativistic velocities in this low energy experiment, it is formed inside the nucleus. The A-dependence of the quasiexclusive reaction can thus be used to determine the J/ψ-nucleon cross section at low energies. For a normal hadronic reaction $\bar{p}A \to HX$, we expect $A_{\text{eff}} \sim A^{1/3}$, corresponding to absorption in the initial and final state. In the case of $\bar{p}A \to J/\psi\ X$ one expects A_{eff} much closer to A^1 if color transparency is fully effective and $\sigma(J/\psi\ N)$ is small.

11.11 Spin Correlations in Proton-Proton Scattering

One of the most serious challenges to quantum chromodynamics is the behavior of the spin-spin correlation asymmetry $A_{NN} = \frac{[d\sigma(\uparrow\uparrow)-d\sigma(\uparrow\downarrow)]}{[d\sigma(\uparrow\uparrow)+d\sigma(\uparrow\downarrow)]}$ measured in large momentum transfer pp elastic scattering (see Fig. 11.30). At $p_{\text{lab}} = 11.75$ GeV/c and $\theta_{\text{cm}} = \pi/2$, A_{NN} rises to $\simeq 60\%$, corresponding to four times more probability for protons to scatter with their incident spins both normal to the scattering plane and parallel, rather than normal and opposite.

The polarized cross section shows a striking energy and angular dependence not expected from the slowly-changing perturbative QCD predictions. However, the unpolarized data is in first approximation consistent with the fixed angle scaling law $s^{10}d\sigma/dt(pp \to pp) = f(\theta_{\text{cm}})$ expected from the perturbative analysis (see Fig. 11.9). The onset of new structure [104] at $s \simeq 23$ GeV^2 is a sign of new degrees of freedom in the two-baryon system. In this section, I will discuss a possible explanation [101] for (1) the observed spin correlations, (2) the deviations from fixed-angle scaling laws, and (3) the anomalous energy dependence of absorptive corrections to quasielastic pp scattering in nuclear targets, in terms of a simple model based on two $J = L = S = 1$ broad resonances (or threshold enhancements) interfering with a perturbative QCD quark-interchange background amplitude. The structures in the $pp \to pp$ amplitude may be associated with the onset of strange and charmed thresholds. The fact that the produced quark and anti-quark have opposite parity explains why the $L = 1$ channel is involved. If the charm threshold explanation is correct, large angle pp elastic scattering would have been virtually featureless for $p_{\text{lab}} \geq 5$ GeV/c, had it not been for the onset of heavy flavor production. As a further illustration of the threshold effect, one can see the effect in A_{NN} due to a narrow $^3F_3$ pp resonance at $\sqrt{s} = 2.17$ GeV ($p_{\text{lab}} = 1.26$ GeV/c) associated with the $p\Delta$ threshold.

The perturbative QCD analysis [34] of exclusive amplitudes assumes that large momentum transfer exclusive scattering reactions are controlled by short distance quark-gluon subprocesses, and that corrections from quark masses and intrinsic transverse momenta can be ignored. The main predictions are fixed-angle scaling laws [41] (with small corrections due to

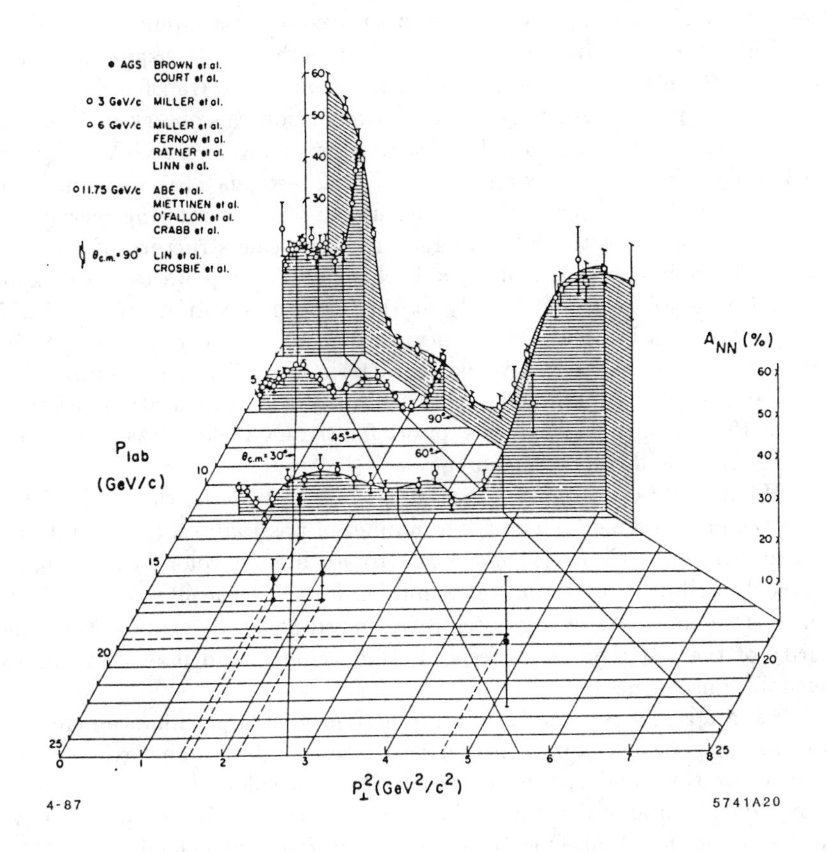

FIGURE 11.30
The spin-spin correlation A_{NN} for elastic pp scattering with beam and target protons polarized normal to the scattering plane [103]. $A_{NN} = 60\%$ implies that it is four times more probable for the protons to scatter with spins parallel rather than antiparallel.

evolution of the distribution amplitudes, the running coupling constant, and pinch singularities), hadron helicity conservation, [39] and the novel phenomenon, "color transparency."

As discussed in Sec. 11.9, a test of color transparency in large momentum transfer quasielastic pp scattering at $\theta_{\rm cm} \simeq \pi/2$ has recently been carried out at BNL using several nuclear targets (C, Al, Pb) [99]. The attenuation at $p_{\rm lab} = 10$ GeV/c in the various nuclear targets was observed to be in fact much less than that predicted by traditional Glauber theory (see Fig. 11.28). This appears to support the color transparency prediction.

The expectation from perturbative QCD is that the transparency effect should become even more apparent as the momentum transfer rises. Nevertheless, at $p_{\text{lab}} = 12$ GeV/c, normal attenuation was observed. One can explain this surprising result if the scattering at $p_{\text{lab}} = 12$ GeV/c ($\sqrt{s} = 4.93$ GeV), is dominated by an s-channel B=2 resonance (or resonance-like structure) with mass near 5 GeV, since unlike a hard-scattering reaction, a resonance couples to the fully-interacting large-scale structure of the proton. If the resonance has spin $S = 1$, this can also explain the large spin correlation A_{NN} measured nearly at the same momentum, $p_{\text{lab}} = 11.75$ GeV/c. Conversely, in the momentum range $p_{\text{lab}} = 5$ to 10 GeV/c one predicts that the perturbative hard-scattering amplitude is dominant at large angles. The experimental observation of diminished attenuation at $p_{\text{lab}} = 10$ GeV/c thus provides support for the QCD description of exclusive reactions and color transparency.

What could cause a resonance at $\sqrt{s} = 5$ GeV, more than 3 GeV beyond the pp threshold? There are a number of possibilities: (a) a multigluonic excitation such as $|qqqqqqggg\rangle$, (b) a "hidden color" color singlet $|qqqqqq\rangle$ excitation [105], or (c) a "hidden flavor" $|qqqqqqQ\overline{Q}\rangle$ excitation, which is the most interesting possibility, since it naturally explains the spin-parity of the resonance or threshold enhancement, and it leads to many testable consequences.

As in QED, where final state interactions give large enhancement factors for attractive channels in which $Z\alpha/v_{rel}$ is large, one expects resonances or threshold enhancements in QCD in color-singlet channels at heavy quark production thresholds since all the produced quarks have similar velocities [106]. One thus can expect resonant behavior at $M^* = 2.55$ GeV and $M^* = 5.08$ GeV, corresponding to the threshold values for open strangeness: $pp \to \Lambda K^+ p$, and open charm: $pp \to \Lambda_c D^0 p$, respectively. In any case, the structure at 5 GeV is highly inelastic: its branching ratio to the proton-proton channel is $B^{pp} \simeq 1.5\%$.

A model for this phenomenon is given in Ref. [101]. In order not to overcomplicate the phenomenology, the simplest Breit–Wigner parameterization of the resonances was used. There has not been an attempt to optimize the parameters of the model to obtain a best fit. It is possible that what is identified a single resonance is actually a cluster of resonances.

The background component of the model is the perturbative QCD amplitude. Although complete calculations are not yet available, many features of the QCD predictions are understood, including the approximate s^{-4} scaling of the $pp \to pp$ amplitude at fixed θ_{cm} and the dominance of those amplitudes that conserve hadron helicity [39]. Furthermore, recent data comparing different exclusive two-body scattering channels from BNL

[42] show that quark interchange amplitudes [107] dominate quark annihilation or gluon exchange contributions. Assuming the usual symmetries, there are five independent pp helicity amplitudes: $\phi_1 = M(++,++)$, $\phi_2 = M(--,++)$, $\phi_3 = M(+-,+-)$, $\phi_4 = M(-+,+-)$, $\phi_5 = M(++,+-)$. The helicity amplitudes for quark interchange have a definite relationship [51]:

$$\begin{aligned}\phi_1(\text{PQCD}) &= 2\phi_3(\text{PQCD}) = -2\phi_4(\text{PQCD}) \\ &= 4\pi C F(t) F(u) \left[\frac{t - m_d^2}{u - m_d^2} + (u \leftrightarrow t)\right] e^{i\delta} \, . \end{aligned} \tag{11.31}$$

The hadron helicity non-conserving amplitudes, $\phi_2(\text{PQCD})$ and $\phi_5(\text{PQCD})$ are zero. This form is consistent with the nominal power-law dependence predicted by perturbative QCD and also gives a good representation of the angular distribution over a broad range of energies [108]. Here $F(t)$ is the helicity conserving proton form factor, taken as the standard dipole form: $F(t) = (1 - t/m_d^2)^{-2}$, with $m_d^2 = 0.71\ \text{GeV}^2$. As shown in Ref. [51], the PQCD-quark-interchange structure alone predicts $A_{NN} \simeq \frac{1}{3}$, nearly independent of energy and angle.

Because of the rapid fixed-angle s^{-4} falloff of the perturbative QCD amplitude, even a very weakly-coupled resonance can have a sizeable effect at large momentum transfer. The large empirical values for A_{NN} suggest a resonant $pp \to pp$ amplitude with $J = L = S = 1$ since this gives $A_{NN} = 1$ (in absence of background) and a smooth angular distribution. Because of the Pauli principle, an $S = 1$ di-proton resonances must have odd parity and thus odd orbital angular momentum. The two non-zero helicity amplitudes for a $J = L = S = 1$ resonance can be parameterized in Breit–Wigner form:

$$\begin{aligned}\phi_3(\text{resonance}) &= 12\pi \frac{\sqrt{s}}{p_{\text{cm}}} d^1_{1,1}(\theta_{\text{cm}}) \frac{\frac{1}{2}\Gamma^{pp}(s)}{M^* - E_{\text{cm}} - \frac{i}{2}\Gamma} \, , \\ \phi_4(\text{resonance}) &= -12\pi \frac{\sqrt{s}}{p_{\text{cm}}} d^1_{-1,1}(\theta_{\text{cm}}) \frac{\frac{1}{2}\Gamma^{pp}(s)}{M^* - E_{\text{cm}} - \frac{i}{2}\Gamma} \, . \end{aligned} \tag{11.32}$$

(The ${}^3F_3$ resonance amplitudes have the same form with $d^3_{\pm1,1}$ replacing $d^1_{\pm1,1}$.) As in the case of a narrow resonance like the Z^0, the partial width into nucleon pairs is proportional to the square of the time-like proton form factor: $\Gamma^{pp}(s)/\Gamma = B^{pp}|F(s)|^2/|F(M^{*2})|^2$, corresponding to the formation of two protons at this invariant energy. The resonant amplitudes then die away by one inverse power of $(E_{\text{cm}} - M^*)$ relative to the dominant PQCD amplitudes. (In this sense, they are higher twist contributions relative to the leading twist perturbative QCD amplitudes.) The model is thus very simple: each pp helicity amplitude ϕ_i is the coherent sum of PQCD plus resonance components: $\phi = \phi(\text{PQCD}) + \Sigma\phi(\text{resonance})$. Because of pinch

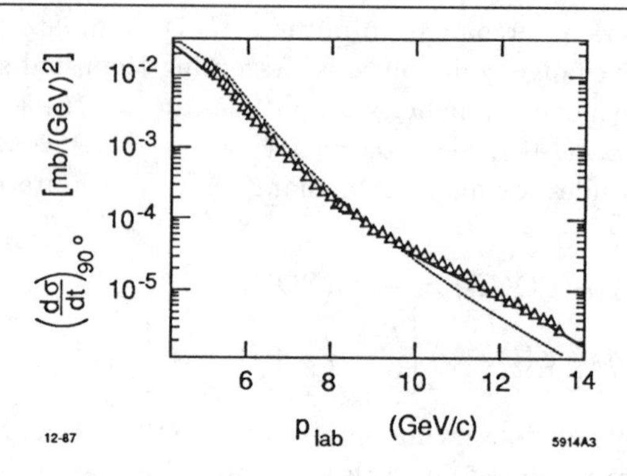

FIGURE 11.31
Prediction (solid curve) for $d\sigma/dt(pp \to pp)$ at $\theta_{\rm cm} = \pi/2$ compared with the data of Akerlof *et al.* [110]. The dotted line is the background PQCD prediction.

singularities and higher-order corrections, the hard QCD amplitudes are expected to have a nontrivial phase [15]; the model allows for a constant phase δ in ϕ(PQCD). Because of the absence of the ϕ_5 helicity-flip amplitude, the model predicts zero single spin asymmetry A_N. This is consistent with the large angle data at $p_{\rm lab} = 11.75$ GeV/c [109].

At low transverse momentum, $p_T \le 1.5$ GeV, the power-law fall-off of ϕ(PQCD) in s disagrees with the more slowly falling large-angle data, and one has little guidance from basic theory. The main interest in this low-energy region is to illustrate the effects of resonances and threshold effects on A_{NN}. In order to keep the model tractable, one can extend the background quark interchange and the resonance amplitudes at low energies using the same forms as above but replacing the dipole form factor by a phenomenological form $F(t) \propto e^{-1/2\beta\sqrt{|t|}}$. A kinematic factor of $\sqrt{s}/2p_{\rm cm}$ is included in the background amplitude. The value $\beta = 0.85$ GeV^{-1} then gives a good fit to $d\sigma/dt$ at $\theta_{\rm cm} = \pi/2$ for $p_{\rm lab} \le 5.5$ GeV/c [110]. The normalizations are chosen to maintain continuity of the amplitudes.

The predictions of the model and comparison with experiment are shown in Figs. 11.31–11.36. The following parameters are chosen: $C = 2.9 \times 10^3$, $\delta = -1$ for the normalization and phase of ϕ(PQCD). The mass, width and pp branching ratio for the three resonances are $M_d^* = 2.17$ GeV, $\Gamma_d = 0.04$ GeV, $B_d^{pp} = 1$; $M_s^* = 2.55$ GeV, $\Gamma_s = 1.6$ GeV, $B_s^{pp} = 0.65$; and $M_c^* = 5.08$ GeV, $\Gamma_c = 1.0$ GeV, $B_c^{pp} = 0.0155$, respectively. As shown in Figs. 11.31 and 11.32, the deviations from the simple scaling predicted by

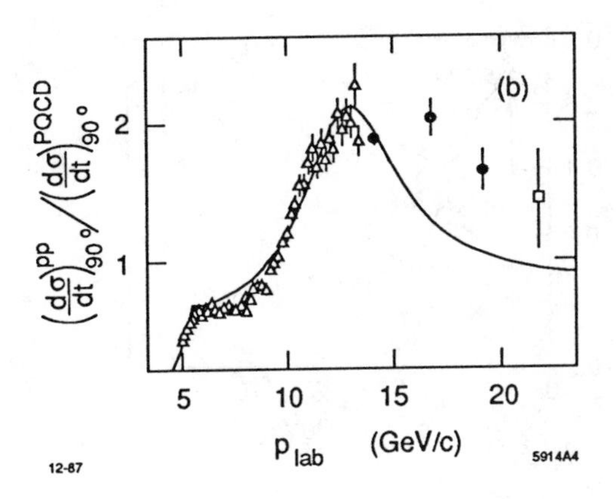

FIGURE 11.32
Ratio of $d\sigma/dt(pp \to pp)$ at $\theta_{\rm cm} = \pi/2$ to the PQCD prediction. The data [110] are from Akerlof *et al.* (open triangles), Allaby *et al.* (solid dots) and Cocconi *et al.* (open square). The cusp at $p_{\rm lab} = 5.5$ GeV/c indicates the change of regime from PQCD.

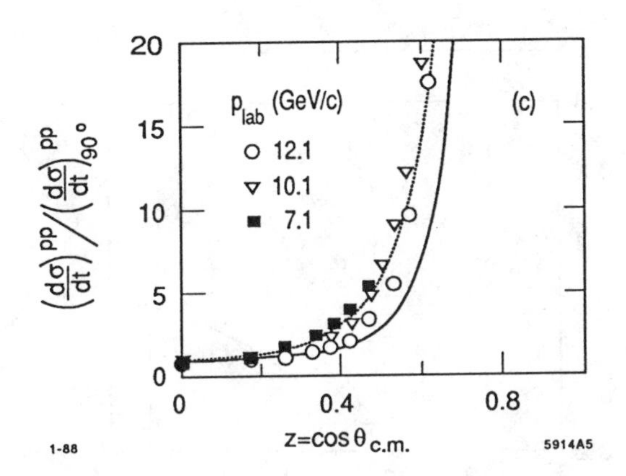

FIGURE 11.33
The $pp \to pp$ angular distribution normalized at $\theta_{\rm cm} = \pi/2$. The data are from the compilation given in Sivers *et al.*, Ref. [24]. The solid and dotted lines are predictions for $p_{\rm lab} = 12.1$ and 7.1 GeV/c, respectively, showing the broadening near resonance.

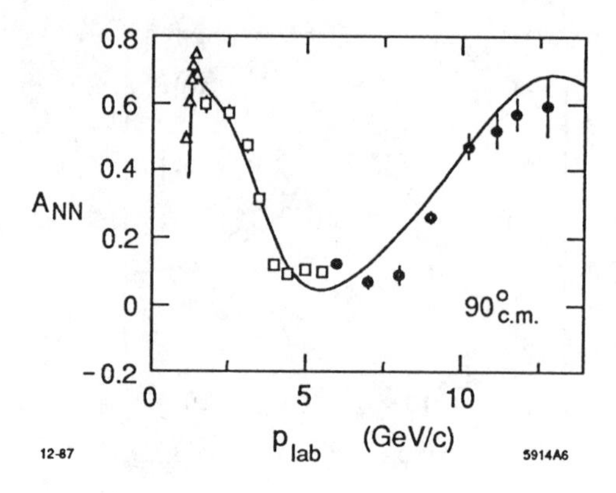

FIGURE 11.34
A_{NN} as a function of p_{lab} at $\theta_{\text{cm}} = \pi/2$. The data [110] are from Crosbie *et al.* (solid dots), Lin *et al.* (open squares) and Bhatia *et al.* (open triangles). The peak at $p_{\text{lab}} = 1.26$ GeV/c corresponds to the $p\Delta$ threshold. The data are well reproduced by the interference of the broad resonant structures at the strange ($p_{\text{lab}} = 2.35$ GeV/c) and charm ($p_{\text{lab}} = 12.8$ GeV/c) thresholds, interfering with a PQCD background. The value of A_{NN} from PQCD alone is $\frac{1}{3}$.

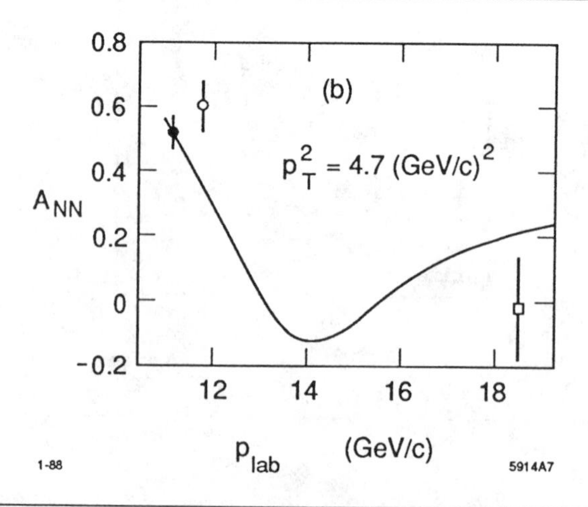

FIGURE 11.35
A_{NN} at fixed $p_T^2 = (4.7 \text{ GeV/c})^2$. The data point [110] at $p_{\text{lab}} = 18.5$ GeV/c is from Court *et al.*

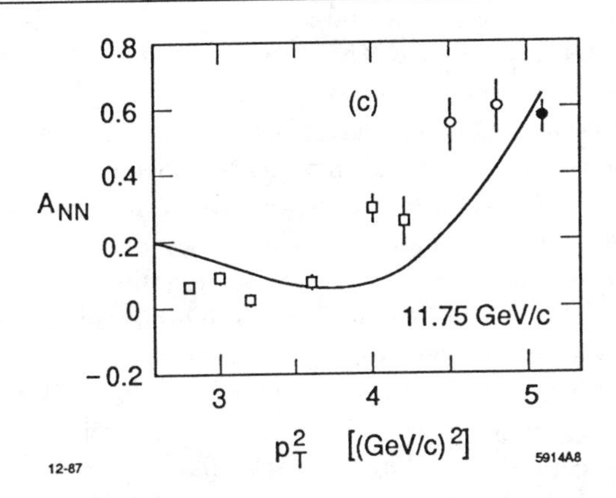

FIGURE 11.36
A_{NN} as a function of transverse momentum. The data [103] are from Crabb *et al.* (open circles) and O'Fallon *et al.* (open squares). Diffractive contributions should be included for $p_T^2 \lesssim 3$ GeV^2.

the PQCD amplitudes are readily accounted for by the resonance structures. The cusp which appears in Fig. 11.32 marks the change in regime below $p_{\rm lab} = 5.5$ GeV/c where PQCD becomes inapplicable. It is interesting to note that in this energy region normal attenuation of quasielastic pp scattering is observed [99]. The angular distribution (normalized to the data at $\theta_{\rm cm} = \pi/2$) is predicted to broaden relative to the steeper perturbative QCD form, when the resonance dominates. As shown in Fig. 11.33 this is consistent with experiment, comparing data at $p_{\rm lab} = 7.1$ and 12.1 GeV/c.

The most striking test of the model is its prediction for the spin correlation A_{NN} shown in Fig. 11.34. The rise of A_{NN} to $\simeq 60\%$ at $p_{\rm lab} = 11.75$ GeV/c is correctly reproduced by the high energy J=1 resonance interfering with ϕ(PQCD). The narrow peak which appears in the data of Fig. 11.34 corresponds to the onset of the $pp \to p\Delta(1232)$ channel which can be interpreted as a $uuuuddq\bar{q}$ resonant state. Because of spin-color statistics one expects in this case a higher orbital angular momentum state, such as a $pp\ {}^3F_3$ resonance. The model is also consistent with the recent high-energy data point for A_{NN} at $p_{\rm lab} = 18.5$ GeV/c and $p_T^2 = 4.7$ GeV^2 (see Fig. 11.35). The data show a dramatic decrease of A_{NN} to zero or negative values. This is explained in the model by the destructive interference effects above the resonance region. The same effect accounts for the depression of A_{NN} for $p_{\rm lab} \approx 6$ GeV/c shown in Fig. 11.34. The comparison of the angular dependence of A_{NN} with data at $p_{\rm lab} = 11.75$ GeV/c is shown in Fig. 11.36. The agreement with the data [111] for the longitudinal spin correlation A_{LL} at the same $p_{\rm lab}$ is somewhat worse.

The simple model discussed here shows that many features can be naturally explained with only a few ingredients: a perturbative QCD background plus resonant amplitudes associated with rapid changes of the inelastic pp cross section. The model provides a good description of the s and t dependence of the differential cross section, including its "oscillatory" dependence [112] in s at fixed $\theta_{\rm cm}$, and the broadening of the angular distribution near the resonances. Most important, it gives a consistent explanation for the striking behavior of both the spin-spin correlations and the anomalous energy dependence of the attenuation of quasielastic pp scattering in nuclei. It is predicted that color transparency should reappear at higher energies ($p_{\rm lab} \geq 16$ GeV/c), and also at smaller angles ($\theta_{\rm cm} \approx 60^\circ$) at $p_{\rm lab} = 12$ GeV/c where the perturbative QCD amplitude dominates. If the J=1 resonance structures in A_{NN} are indeed associated with heavy quark degrees of freedom, then the model predicts inelastic pp cross sections of the order of 1 mb and 1μb for the production of strange and charmed hadrons near their respective thresholds [113]. Thus a crucial test of the heavy quark hypothesis for explaining A_{NN}, rather than hidden color or gluonic excitations, is the observation of significant charm hadron production at $p_{\rm lab} \geq 12$ GeV/c.

Recently Ralston and Pire [15] have proposed that the oscillations of the pp elastic cross section and the apparent breakdown of color transparency are associated with the dominance of the Landshoff pinch contributions at $\sqrt{s} \sim 5$ GeV. The oscillating behavior of $d\sigma/dt$ is due to the energy dependence of the relative phase between the pinch and hard-scattering contributions. They assume color transparency will disappear whenever the pinch contributions are dominant since such contributions could couple to wavefunctions of large transverse size. The large spin correlation in A_{NN} is not readily explained in the Ralston-Pire model. Furthermore, the recent analysis by Botts and Sterman [58] suggests that the pinch contributions should satisfy color transparency. In any event, more data and analysis are needed to discriminate between models.

11.11.1 Nuclear-Bound Quarkonia

The above analysis also has implications for the production of hidden charm near threshold in hadronic and nuclear collisions. For example, consider the reaction $dd \to \alpha(c\bar{c})$ where the charmonium state is produced nearly at rest. At the threshold for charm production, the incident nuclei will be stopped (in the center of mass frame) and will evidently fuse into a compound nucleus (the α) because of the strong attractive nuclear force.

The charmonium state will be attracted to the nucleus by the QCD gluonic Van der Waals force. It is thus likely that a new type of nuclear bound state will be formed: charmonium bound to nuclear matter. Such a state should be observable at a distinct dd energy, and it will decay to unique signatures such as $dd \to \alpha\mu^+\mu^-$. The binding energy in the nucleus gives a precision measure of the charmonium's interactions with ordinary hadrons and nuclei; its decays will measure hadron-nucleus interactions and test color transparency starting from a unique initial state condition.

11.12 Discretized Light-Cone Quantization

Only a small fraction of strong interaction and nuclear physics can be addressed by perturbative QCD analyses. The solution to the mass and wavefunction of the proton requires a solution to the QCD bound-state problem. Even with the simplicity of the e^+e^- and $\gamma\gamma$ initial state, the full complexity of hadron dynamics is involved in understanding resonance production, exclusive channels near threshold, jet hadronization, the hadronic contribution to the photon structure function, and the total e^+e^- or $\gamma\gamma$ annihilation cross section. A primary question is whether we can ever hope to confront QCD directly in its nonperturbative domain. Lattice gauge theory and effective Lagrangian methods such as the Skyrme model offer some hope in understanding the low-lying hadron spectrum but dynamical computations relevant to $\gamma\gamma$ annihilation appear intractable. Considerable information [5] on the spectrum and the moments of hadron valence wavefunctions has been obtained using the ITEP QCD sum rule method, but the region of applicability of this method to dynamical problems appears limited.

Recently a new method for analysing QCD in the nonperturbative domain has been developed: discretized light-cone quantization (DLCQ) [114]. The method has the potential for providing detailed information on all the hadron's Fock light-cone components. DLCQ has been used to obtain the complete spectrum of neutral states in QED [9] and QCD [115] in one space and one time for any mass and coupling constant. The QED results agree with the Schwinger solution at infinite coupling. We will review the QCD[1+1] results below. Studies of QED in 3+1 dimensions are now underway [116]. Thus one can envision a nonperturbative method which in principle could allow a quantitative confrontation of QCD with the data even at low energies and momentum transfer.

The basic idea of DLCQ is as follows: QCD dynamics takes a rather simple form when quantized at equal light-cone "time" $\tau = t + z/c$. In light-cone gauge $A^+ = A^0 + A^z = 0$, the QCD light-cone Hamiltonian

$$H_{\rm QCD} = H_0 + gH_1 + g^2 H_2 \tag{11.33}$$

contains the usual 3-point and 4-point interactions plus induced terms from instantaneous gluon exchange and instantaneous quark exchange diagrams. The perturbative vacuum is an eigenstate of $H_{\rm QCD}$ and serves as the lowest state in constructing a complete basis set of color singlet Fock states of H_0 in momentum space. Solving QCD is then equivalent to solving the eigenvalue problem:

$$H_{\rm QCD}|\Psi\rangle = M^2|\Psi\rangle \tag{11.34}$$

as a matrix equation on the free Fock basis. The set of eigenvalues $\{M^2\}$ represents the spectrum of the color-singlet states in QCD. The Fock projections of the eigenfunction corresponding to each hadron eigenvalue gives the quark and gluon Fock state wavefunctions $\psi_n(x_i, k_{\perp i}, \lambda_i)$ required to compute structure functions, distribution amplitudes, decay amplitudes, *etc.* For example, as shown by Drell and Yan [36], the form-factor of a hadron can be computed at any momentum transfer Q from an overlap integral of the ψ_n summed over particle number n. The e^+e^- annihilation cross section into a given $J = 1$ hadronic channel can be computed directly from its $\psi_{q\bar{q}}$ Fock state wavefunction.

The light-cone momentum space Fock basis becomes discrete and amenable to computer representation if one chooses (anti-)periodic boundary conditions for the quark and gluon fields along the $z^- = z - ct$ and $z_\perp$ directions. In the case of renormalizable theories, a covariant ultraviolet cutoff Λ is introduced which limits the maximum invariant mass of the particles in any Fock state. One thus obtains a finite matrix representation of $H^{(\Lambda)}_{\rm QCD}$ which has a straightforward continuum limit. The entire analysis is frame independent, and fermions present no special difficulties.

Since H_{LC}, P^+, $\vec{P}_\perp$, and the conserved charges all commute, H_{LC} is block diagonal. By choosing periodic (or anti-periodic) boundary conditions for the basis states along the negative light-cone $\psi(z^- = +L) = \pm\psi(z^- = -L)$, the Fock basis becomes restricted to finite dimensional representations. The eigenvalue problem thus reduces to the diagonalization of a finite Hermitian matrix. To see this, note that periodicity in z^- requires $P^+ = \frac{2\pi}{L}K$, $k_i^+ = \frac{2\pi}{L}\, n_i$, $\sum_{i=1}^n n_i = K$. The dimension of the representation corresponds to the number of partitions of the integer K as a sum of positive integers n. For a finite resolution K, the wavefunction is sampled at the discrete points

$$x_i = \frac{k_i^+}{P+} = \frac{n_i}{K} = \left\{\frac{1}{K}, \frac{2}{K}, \ldots, \frac{K-1}{K}\right\}. \tag{11.35}$$

The continuum limit is clearly $K \to \infty$.

One can easily show that P^- scales as L. One thus defines $P^- \equiv \frac{L}{2\pi}H$.

The eigenstates with $P^2 = M^2$ at fixed P^+ and $\vec{P}_\perp = 0$ thus satisfy $H_{LC}|\Psi\rangle = KH|\Psi\rangle = M^2|\Psi\rangle$, independent of L (which corresponds to a Lorentz boost factor).

The basis of the DLCQ method is thus conceptually simple: one quantizes the independent fields at equal light-cone time τ and requires them to be periodic or anti-periodic in light-cone space with period $2L$. The commuting operators, the light-cone momentum $P^+ = \frac{2\pi}{L}K$ and the light cone energy $P^- = \frac{L}{2\pi}H$ are constructed explicitly in a Fock space representation and diagonalized simultaneously. The eigenvalues give the physical spectrum: the invariant mass squared $M^2 = P^\nu P_\nu$. The eigenfunctions give the wavefunctions at equal τ and allow one to compute the current matrix elements, structure functions, and distribution amplitudes required for physical processes. All of these quantities are manifestly independent of L, since $M^2 = P^+P^- = HK$. Lorentz-invariance is violated by periodicity, but re-established at the end of the calculation by going to the continuum limit: $L \to \infty$, $K \to \infty$ with P^+ finite. In the case of gauge theory, the use of the light-cone gauge $A^+ = 0$ eliminates negative metric states in both Abelian and non–Abelian theories.

Since continuum as well as single hadron color singlet hadronic wavefunctions are obtained by the diagonalization of H_{LC}, one can also calculate scattering amplitudes as well as decay rates from overlap matrix elements of the interaction Hamiltonian for the weak or electromagnetic interactions. An important point is that all higher Fock amplitudes including spectator gluons are kept in the light-cone quantization approach; such contributions cannot generally be neglected in decay amplitudes involving light quarks.

The simplest application of DLCQ to local gauge theory is QED in one-space and one-time dimensions. Since $A^+ = 0$ is a physical gauge there are no photon degrees of freedom. Explicit forms for the matrix representation of H_{QED} are given in Ref. [9].

The basic interactions which occur in H_{LC}(QCD) are illustrated in Fig. 11.37. Recently Hornbostel [115] has used DLCQ to obtain the complete color-singlet spectrum of QCD in one space and one time dimension for $N_C = 2, 3, 4$. The hadronic spectra are obtained as a function of quark mass and QCD coupling constant (see Fig. 11.38).

Where they are available, the spectra agree with results obtained earlier; in particular, the lowest meson mass in SU(2) agrees within errors with lattice Hamiltonian results [117]. The meson mass at $N_C = 4$ is close to the value obtained in the large N_C limit. The method also provides the first results for the baryon spectrum in a non–Abelian gauge theory. The lowest baryon mass is shown in Fig. 11.38 as a function of coupling constant. The ratio of meson to baryon mass as a function of N_C also agrees at strong coupling with results obtained by Frishman and Sonnenschein [118].

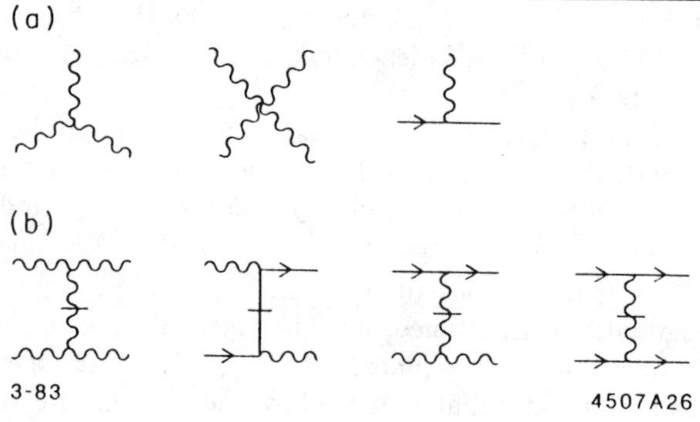

FIGURE 11.37
Diagrams appearing in the interaction Hamiltonian for QCD on the light cone. The propagators with horizontal bars represent "instantaneous" gluon and quark exchange which arise from reduction of the dependent fields in $A^+ = 0$ gauge. (a) Basic interaction vertices in QCD. (b) "Instantaneous" contributions.

Precise values for the mass eigenvalue can be obtained by extrapolation to large K since the functional dependence in $1/K$ is understood.

As emphasized above, when the light-cone Hamiltonian is diagonalized for a finite resolution K, one gets a complete set of eigenvalues corresponding to the total dimension of the Fock state basis. A representative example of the spectrum is shown in Fig. 11.39 for baryon states ($B = 1$) as a function of the dimensionless variable $\lambda = 1/(1+\pi m^2/g^2)$. Note that spectrum automatically includes continuum states with $B = 1$.

The structure functions for the lowest meson and baryon states in SU(3) at two different coupling strengths $m/g = 1.6$ and $m/g = 0.1$ are shown in Figs. 11.40 and 11.41. Higher Fock states have a very small probability; representative contributions to the baryon structure functions are shown in Figs. 11.42 and 11.43. For comparison, the valence wavefunction of a higher mass state which can be identified as a composite of meson pairs (analogous to a nucleus) is shown in Fig. 11.44. The interactions of the quarks in the pair state produce Fermi motion beyond $x = 0.5$. Although these results are for one time one space theory they do suggest that the sea quark distributions in physical hadrons may be highly structured.

In the case of gauge theory in 3+1 dimensions, one also takes the $k^i_\perp = (2\pi/L_\perp)n^i_\perp$ as discrete variables on a finite cartesian basis. The theory is covariantly regulated if one restricts states by the condition

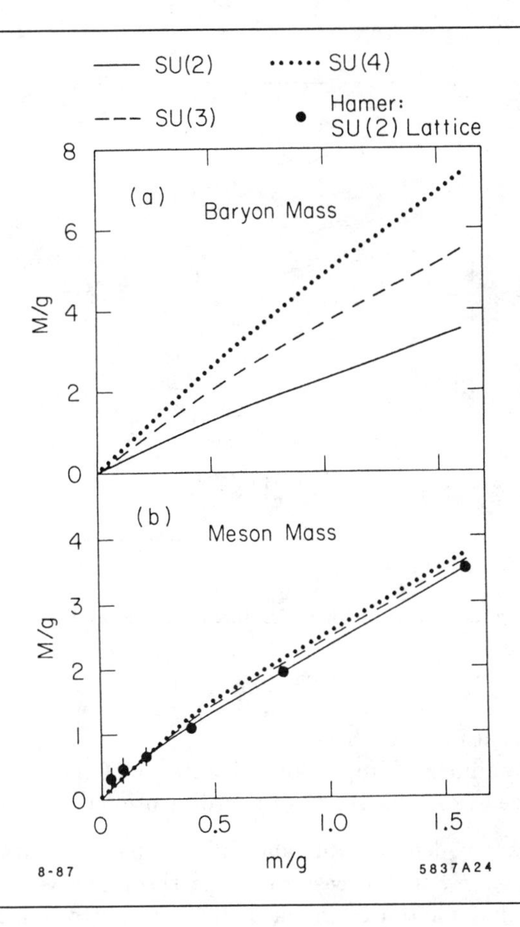

FIGURE 11.38
The baryon and meson spectrum in QCD [1+1] computed in DLCQ for $N_C = 2, 3, 4$ as a function of quark mass and coupling constant [115].

$$\sum_i \frac{k_{\perp i}^2 + m_i^2}{x_i} \leq \Lambda^2 \quad , \tag{11.36}$$

where Λ is the ultraviolet cutoff. In effect, states with total light-cone kinetic energy beyond Λ^2 are cut off. In a renormalizable theory physical quantities are independent of physics beyond the ultraviolet regulator; the only dependence on Λ appears in the coupling constant and mass parameters of the Hamiltonian, consistent with the renormalization group [119]. The resolution parameters need to be taken sufficiently large such that the theory is controlled by the continuum regulator Λ, rather than the discrete

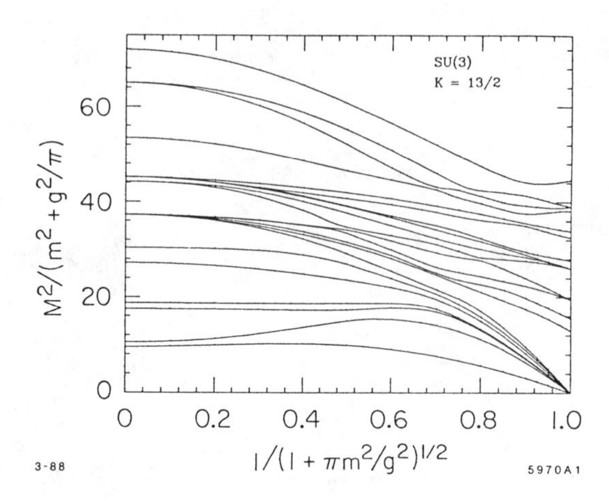

FIGURE 11.39
Representative baryon spectrum for QCD in one-space and one-time dimension [115].

scales of the momentum space basis.

There are a number of important advantages of the DLCQ method which have emerged from this study of two-dimensional field theories:

1. The Fock space is denumerable and finite in particle number for any fixed resolution K. In the case of gauge theory in 3+1 dimensions, one expects that photon or gluon quanta with zero four-momentum decouple from neutral or color-singlet bound states, and thus need not be included in the Fock basis.

2. Because one is using a discrete momentum space representation, rather than a space-time lattice, there are no special difficulties with fermions: *e.g.*, no fermion doubling, fermion determinants, or necessity for a quenched approximation. Furthermore, the discretized theory has basically the same ultraviolet structure as the continuum theory. It should be emphasized that unlike lattice calculations, there is no constraint or relationship between the physical size of the bound state and the length scale L.

3. The DLCQ method has the remarkable feature of generating the complete spectrum of the theory; bound states and continuum states alike. These can be separated by tracing their minimum Fock state content down to small coupling constant since the continuum states have higher

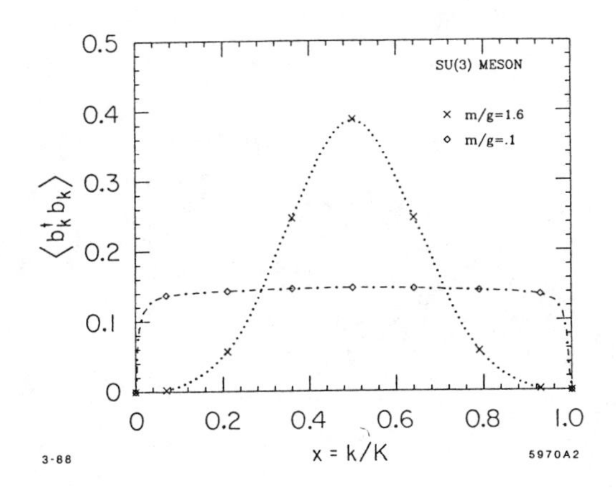

FIGURE 11.40
The meson quark momentum distribution in QCD[1+1] computed using DLCQ [115].

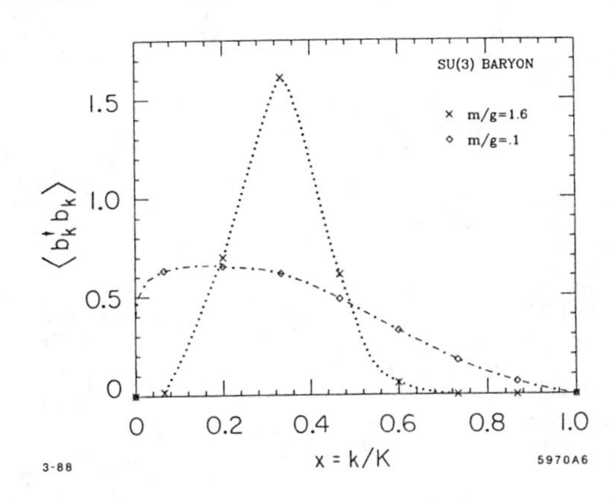

FIGURE 11.41
The baryon quark momentum distribution in QCD[1+1] computed using DLCQ [115].

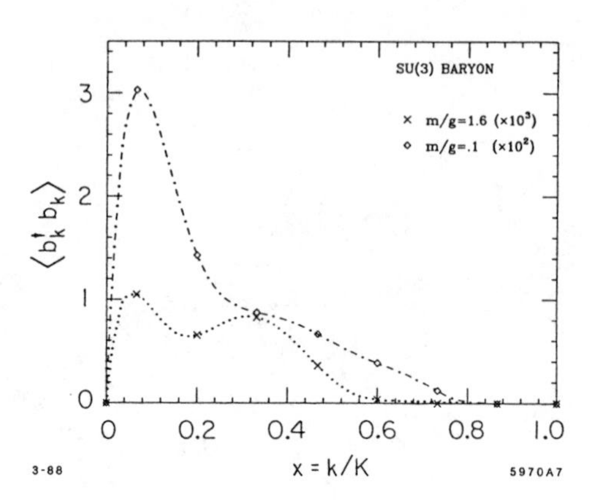

FIGURE 11.42
Contribution to the baryon quark momentum distribution from $qqq\bar{q}\bar{q}$ states for QCD[1+1] [115].

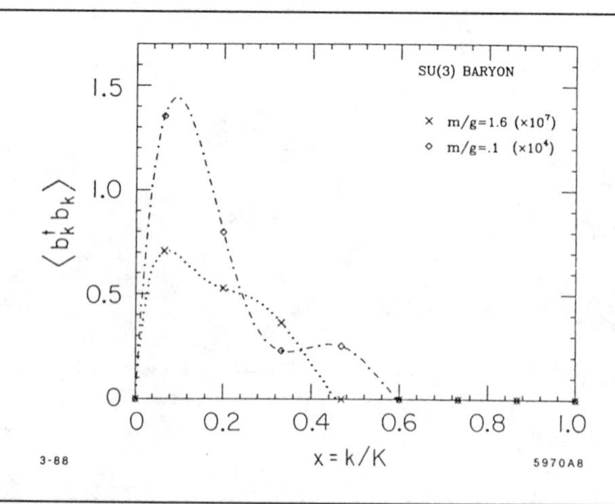

FIGURE 11.43
Contribution to the baryon quark momentum distribution from $qqq\bar{q}\bar{q}\bar{q}\bar{q}$ states for QCD[1+1] [115].

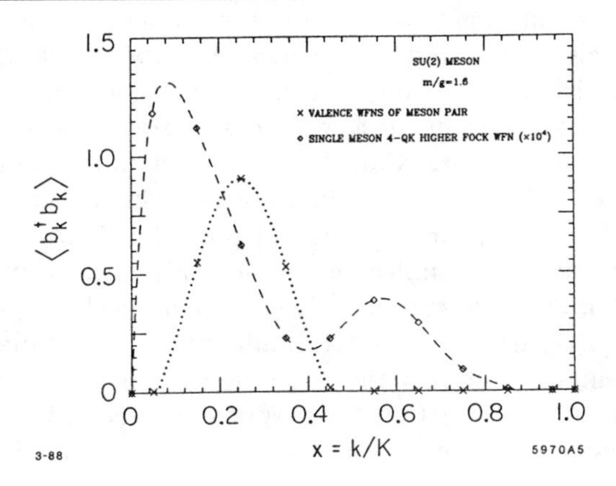

FIGURE 11.44
Comparison of the meson quark distributions in the $qq\bar{q}\bar{q}$ Fock state with that of a continuum meson pair state. The structure in the former may be due to the fact that these four-particle wavefunctions are orthogonal [115].

particle number content. In lattice gauge theory it appears intractable to obtain information on excited or scattering states or their correlations. The wavefunctions generated at equal light cone time have the immediate form required for relativistic scattering problems. In particular one can calculate the relativistic form factor from the matrix element of currents.

4. DLCQ is basically a relativistic many-body theory, including particle number creation and destruction, and is thus a basis for relativistic nuclear and atomic problems. In the nonrelativistic limit the theory is equivalent to the many-body Schrödinger theory.

Whether QCD can be solved using DLCQ — considering its large number of degrees of freedom is unclear. The studies for Abelian and non–Abelian gauge theory carried out so far in 1+1 dimensions give grounds for optimism.

11.12.1 Other Applications of Light-Cone Quantization

In the discretized light-cone quantization method, one can construct an explicit matrix representation of the QCD Hamiltonian on the light-cone momentum space Fock representation. The kinetic energy operator in this representation is diagonal. In principle one can diagonalize the total Hamiltonian on this representation to obtain not only the discrete and continuum eigenvalues, but also the corresponding light-cone wavefunctions required to compute intrinsic structure functions and distribution amplitudes. Since we are primarily interested in the lowest mass eigenstates of the hadron spectrum, we can use the variational method and simply minimize the expectation value of the light-cone Hamiltonian. This is currently being carried out by Tang [116] for the study of positronium at large α. The evaluation of the Fock state sum can be made highly efficient by using vectorized code and importance sampling algorithms such as Lepage's program *VEGAS.* On the other hand if the total Hamiltonian could be diagonalized, one could immediately construct the resolvent, and thus the T-matrix for scattering problems. The fractional experimental resolution in center of mass energy squared $\delta s/s$ can be matched to the corresponding resolution $1/K$.

The light-cone Fock state representation can also be used advantageously in perturbation theory. For example, one can calculate any scattering amplitude in terms of the usual Lippman-Schwinger series:

$$T = H_I + H_I \frac{1}{P^- - H_0 + i\epsilon} H_I + \ldots \quad . \tag{11.37}$$

Langnau and I are currently applying this method to the higher order calculation of the electron's anomalous magnetic moment in quantum electrodynamics. The sum over intermediate Fock states is equivalent to summing all τ-ordered diagrams and integrating over the transverse momentum and light-cone fractions x. Because of the restriction to positive x, diagrams corresponding to vacuum fluctuations or those containing backward-moving lines are eliminated. The amplitudes are regulated in the infrared and ultraviolet by cutting off the invariant mass. The ultraviolet regularization and renormalization of the perturbative contributions may be carried out by using the "alternating denominator method" [120] which yields an automatic construction of mass renormalization counter-terms.

The same method can also be used to compute perturbative contributions to the annihilation ratio $R_{e^+e^-} = \sigma(e^+e^- \to hadrons)/\sigma(e^+e^- \to \mu^+\mu^-)$ as well as the quark and gluon jet distribution. The results are obtained in the light-cone variables, x, $k_\perp$, λ, which are the natural covariant variables for this problem. Since there are no Fadeev-Popov or Gupta-Bleuler ghost fields in the light-cone gauge $A^+ = 0$, the calculations are explicitly unitary. It is hoped that one can in this way check the three-loop calculation of Gorishny, et al. [121] who found a surprisingly large value of 64.9 for the coefficient of $(\alpha_s/\pi)^3$ of $R_{e^+e^-}$ in the $\overline{MS}$ scheme.

11.13 Conclusions

In this colloquium I have emphasized several novel features of quantum chromodynamics, features which lead to new insights into the structure of the hadrons and their interactions. Among the highlights:

1. The structure of the proton now appears both theoretically and experimentally to be surprisingly complex, very much at variance with intuition based on non-relativistic quark model. The most convenient covariant representation of the hadron in QCD is given by the light-cone Fock basis. According to QCD sum rules, the valence Fock state wavefunction of the proton turns out to be highly structured and asymmetric between the valence u and d quarks. Polarized deep inelastic structure function measurements by the SLAC-Yale and CERN-EMC collaborations show that the gluons and strange quarks have strong spin correlations with the proton spin. There is even the possibility of a small admixture of hidden charm in the nucleon wavefunction. I have also discussed the distinctions between intrinsic (bound state) versus extrinsic (collision-induced) contributions to the proton structure functions, and a new approach to understanding the non-additive shadowing and anti-shadowing features of the leading twist nuclear structure functions.

2. The perturbative QCD analysis of exclusive amplitudes has now become a highly-developed field, based on all-orders factorization theorems, evolution equations, Sudakov-regulated pinch contributions, *etc.* The application to experiment has been highly successful; the recent confirmation by the TPC–$\gamma\gamma$ experiment of the PQCD predictions for the photon-η transition form factor is an important verification of the theory, as significant as Bjorken scaling in deep inelastic inclusive reactions. The recent observation at SLAC of reduced-amplitude scaling for large angle photo-disintegration provides a striking demonstration of the dominance of simple quark-gluon degrees of freedom in nuclear amplitudes at the few GeV scale. The observation at BNL of increasing color transparency of quasi-elastic pp scattering in nuclei has confirmed perhaps the most novel feature of perturbative QCD. The experimental results contradict the standard Glauber treatment of initial and final state interactions but support the PQCD prediction that large-angle pp scattering involves only the small color-dipole moment configurations of the proton Fock state. The observation of color transparency rules against a description of large momentum exclusive amplitudes in terms of the convolution of soft hadronic wavefunctions. It is clearly essential that color transparency be tested in other chan-

nels, particularly, quasi-elastic $e - p$ scattering.

3. It should be emphasized that experimental and theoretical studies of exclusive amplitudes are necessary for the fundamental understanding of the structure of the hadronic wavefunctions. Exclusive amplitudes provide a testing ground for hadronization in the simplest, most controlled amplitudes. These tests are essential if we are ever able to understand coherence and coalescence phenomena in the hadronization of QCD jets. The calculation of weak decay matrix elements and the extraction of quark mixing parameters of electroweak theory also require a detailed understanding of hadronic wavefunctions.
4. I have described a new approach to the problem of solving QCD in the non-perturbative domain— discretized light-cone quantization. The application of the method to QCD in one-space and one-time has been very encouraging. The challenge now is to apply this method to obtain the mass spectrum and light-cone Fock wave functions of the hadrons in QCD[3+1]. A very interesting feature of the DLCQ results for QCD[1+1] are the oscillations which emerge in the higher Fock state contributions to the hadron structure functions. The DLCQ method also leads naturally to a perturbative method for computing $R_{e^+e^-}$ as well as coherent contributions to jet observables at the amplitude rather than probabilistic level.
5. One of the most important challenges to the PQCD analysis of exclusive reactions is the striking behavior observed in the spin-spin correlation A_{NN} in large-angle pp scattering at $E_{\rm cm} \sim 5\ GeV$. As I have discussed in this lecture, this phenomena can be interpreted as due to a threshold enhancement or resonance due to open charm production in the intermediate state. This explanation also naturally accounts for the observed diminishing of color transparency seen in the BNL experiment at the same kinematic domain. A corollary of this explanation is the prediction of new bound states of charmonium with nucleons or nuclei, just below the production threshold for open charm.

Quantum Chromodynamics has now emerged as a science in itself, unifying hadron and nuclear physics in terms of a common set of fundamental degrees of freedom. It is clear that we have only begun the study of its novel perturbative and non-perturbative features.

Acknowledgements

Some of the work discussed in this lecture was based on collaboration with others, including: G. P. Lepage, J. F. Gunion, E. L. Berger, P. Hoyer, I. Schmidt, H. Lu, T. Huang, A. Langnau, G. Farrar, G. de Teramond, J. R. Hiller, K. Hornbostel, C. R. Ji, A. H. Mueller, H. C. Pauli, D. E. Soper, and A. Tang. I also wish to thank Ed Tang, Paul Ellis and their colleagues at the University of Minnesota for their outstanding hospitality. This work was supported by the Department of Energy under contract DE-AC03-76SF00515.

Bibliography

[1] For general reviews of QCD see J. C. Collins and D. E. Soper, Ann. Rev. Nucl. Part. Sci. **37** (1987) 383; E. Reya, Phys. Rep. **69** (1981) 195; A.H. Mueller, Lectures on perturbative QCD given at the Theoretical Advanced Study Inst. (New Haven, 1985); *Quarks and Gluons in Particle and Nuclei*, Proc. of the UCSB Institute for Theoretical Physics Workshop on Nuclear Chromodynamics, ed. S.J. Brodsky and E. Moniz (World Scientific, 1985).
For a more detailed discussion of exclusive processes in QCD, see S. J. Brodsky and G. P. Lepage, SLAC-PUB-4947 (1989).

[2] C. Klopfenstein *et al.*, CUSB 83-07 (1983).

[3] K. Varvell *et al.*, CERN/EP 87-46 (1987).

[4] S.J. Brodsky, G.P. Lepage and P.B. Mackenzie, Phys. Rev. **D28** (1983) 228. The problem of setting the scale of the argument of the running coupling constant is discussed in this paper.

[5] V.L. Chernyak and A.R. Zhitnitskii, Phys. Rep. **112** (1984) 173; see also Xiao-Duang Xiang, Wang Xin-Nian, and Huang Tao, BIHEP-TH-84, 23 and 29 (1984).

[6] S. Bethke, LBL-26958 (1989).

[7] I.D. King and C.T. Sachrajda, SHEP-85/86-15 (1986), p. 36.

[8] G. Martinelli and C.T. Sachrajda, CERN-TH-4637/87 (1987). The results are based on the method of S. Gottlieb and A.S. Kronfeld, Phys. Rev. **D33** (1986) 227; A.S. Kronfeld and D.M. Photiadis, Phys. Rev. **D31** (1985) 2939.

[9] T. Eller, H. C. Pauli and S. J. Brodsky, Phys. Rev. **D35** (1987) 1493.

[10] P.A.M. Dirac, Rev. Mod. Phys. **21** (1949) 392. Further references to light-cone quantization are given in Ref. [9].

[11] A. V. Crewe, Science **154** (1966) 729.

[12] S.J. Brodsky, J.F. Gunion and R.L. Jaffe, Phys. Rev. **D6** (1972) 2487.

[13] P. Hoodbhoy and R.L. Jaffe, Phys. Rev. **D35** (1987) 113; R.L. Jaffe, CTP #1315 (1985).

[14] A.H. Mueller, *Proc. XVII Recontre de Moriond* (1982); S.J. Brodsky, *Proc. XIII Int. Symp. on Multiparticle Dynamics* (Volendam, 1982). See also G. Bertsch, A.S. Goldhaber, and J.F. Gunion, Phys. Rev. Lett. **47** (1981) 297; G.R. Farrar, H. Liu, L.L. Frankfurt and M.J. Strikmann, Phys. Rev. Lett. **61** (1988) 686; A.H. Mueller, CU-TP-415, talk given at the DPF meeting (Stoors, Conn, 1988), and CU-TP-412 talk given at the *Workshop on Nuclear and Particle Physics on the Light-Cone* (Los Alamos, 1988); B. Pire and J.P. Ralston [15].

[15] J.P. Ralston and B. Pire, Phys. Rev. Lett. **57** (1986) 2330; Phys. Lett. **B117** (1982) 233.

[16] S.J. Brodsky, G.T. Bodwin and G.P. Lepage, in *Proc. of the Volendam Multipart. Dyn. Conf.* (1982) p. 841; *Proc. of the Banff Summer Inst.* (1981) p. 513. This effect is related to the formation zone principle of L. Landau and I. Pomeranchuk, Dok. Akademii Nauk SSSR **92** (1953) 535,735.

[17] G.T. Bodwin, Phys. Rev. **D31** (1985) 2616; G.T. Bodwin, S.J. Brodsky and G.P. Lepage, ANL-HEP-CP-85-32-mc (1985), presented at *20th Rencontre de Moriond* (Les Arcs, France, March 10-17, 1985).

[18] A. Sommerfeld, *Atombau and Spektallinen* (Vieweg, Braunschweig, 1939).

[19] G. T. Bodwin, S. J. Brodsky and G. P. Lepage, Phys. Rev. **D39** (1989) 3287.

[20] S. J. Brodsky and Hung Lu, in preparation.

[21] See *e.g.*, S.J. Brodsky, J. Ellis and M. Karliner, Phys. Lett. **B206** (1988) 309.

[22] Further discussion will appear in S. J. Brodsky and I. Schmidt, to be pub. For a corresponding example in atomic physics see M.L. Goldberger and F.E. Low, Phys. Rev. **176** (1968) 1778.

[23] J.F. Gunion, P. Nason and R. Blankenbecler, Phys. Rev. **D29** (1984) 2491; Phys. Lett. **B117** (1982) 353.

[24] J.F. Gunion, S.J. Brodsky and R. Blankenbecler, Phys. Rev. **D8** (1973) 287; Phys. Lett. **B39** (1972) 649; D. Sivers, S.J. Brodsky and R. Blankenbecler, Phys. Rep. **23C** (1976) 1. Extensive references to fixed angle scattering are given in this review.

[25] E.L. Berger and S.J. Brodsky, Phys. Rev. **D24** (1981) 2428.

[26] E.L. Berger and F. Coester, ANL-HEP-PR-87-13 (1987).

[27] J.J. Aubert *et al.*, Phys. Lett. **B123** (1983) 275; For a recent review see E.L. Berger and F. Coester, ANL-HEP-PR-87-13 (to be pub. in Ann. Rev. Nucl. Part. Sci.).

[28] I.A. Schmidt and R. Blankenbecler, Phys. Rev. **D15** (1977) 3321.

[29] S.J. Brodsky and G.R. Farrar, Phys. Rev. Lett. **31** (1973) 1153; Phys. Rev. **D11** (1975) 1309.

[30] S.J. Brodsky and B.T. Chertok, Phys. Rev. Lett. **37** (1976) 269; Phys. Rev. **D114** (1976) 3003.

[31] S.J. Brodsky and G.P. Lepage, Phys. Rev. **D24** (1981) 1808. The next to leading order evaluation of T_H for these processes is given by B. Nizic, Ph.D. Thesis, Cornell Univ. (1985).

[32] S.J. Brodsky, J.F. Gunion and D.E. Soper, Phys. Rev. **D36** (1987) 2710.

[33] S.J. Brodsky, G. Kopp, and P. Zerwas, Phys. Rev. Lett. **58** (1987) 443.

[34] General QCD analyses of exclusive processes are given in Ref. [38], S.J. Brodsky and G.P. Lepage, SLAC-PUB-2294, presented at the *Workshop on Current Topics in High Energy Physics* (Cal Tech, Feb. 1979); S.J. Brodsky, in *Proc. of the La Jolla Inst. Summer Workshop on QCD* (La Jolla, 1978); A.V. Efremov and A.V. Radyushkin, Phys. Lett. **B94** (1980) 245; V.L. Chernyak, V.G. Serbo and A.R. Zhitnitskii, Yad. Fiz. **31** (1980) 1069; S.J. Brodsky, Y. Frishman, G.P. Lepage and C. Sachrajda, Phys. Lett. **B91** (1980) 239; A. Duncan and A.H. Mueller, Phys. Rev. **D21** (1980) 1636.

[35] J. Ashman *et al.*, Phys. Lett. **B206** (1988) 384.

[36] S.D. Drell and T.M. Yan, Phys. Rev. Lett. **24** (1970) 181.

[37] S.J. Brodsky, SLAC–PUB–4342 and in *Proceedings of the VIIIth Nuclear and Particle Physics Summer School* (Launceston, Australia, 1987).

[38] G.P. Lepage and S.J. Brodsky, Phys. Rev. **D22** (1980) 2157; Phys. Lett. **B87** (1979) 359; Phys. Rev. Lett. **43** (1979) 545, 1625(E).

[39] S.J. Brodsky and G.P. Lepage, Phys. Rev. **D24** (1981) 2848.

[40] Arguments for the conservation of baryon chirality in large-momentum transfer processes have been given by B.L. Ioffe, Phys. Lett. **B63** (1976) 425. For some processes this rule leads to predictions which differ from the QCD results given here. The QCD helicity conservation rule also differs from the electroproduction helicity rules given in O. Nachtmann, Nucl. Phys. **B115** (1976) 61.

[41] See Ref. [29] and also V.A. Matveev, R.M. Muradyan and A.V. Tavkheldize, Lett. Nuovo Cimento **7** (1973) 719.

[42] G.C. Blazey *et al.*, Phys. Rev. Lett. **55** (1985) 1820; G.C. Blazey, Ph.D. Thesis, University of Minnesota (1987); B.R. Baller, Ph.D. Thesis, University of Minnesota (1987); D.S. Barton *et al.*, J. de Phys. **46** (1985) C2, Supp. 2. For a review, see D. Sivers, Ref. [24].

[43] V. D. Burkert, CEBAF-PR-87-006.

[44] M. Peskin, Phys. Lett. **B88** (1979) 128; A. Duncan and A.H. Mueller, Phys. Lett. **B90** (1980) 159; Phys. Rev. **D21** (1980) 1636.

[45] M.D. Mestayer, SLAC-Report 214 (1978); F. Martin *et al.*, Phys. Rev. Lett. **38** (1977) 1320; W.P. Schultz *et al*, Phys. Rev. Lett. **38** (1977) 259; R.G. Arnold *et al.*, Phys. Rev. Lett. **40** (1978) 1429, SLAC-PUB-2373 (1979); B.T. Chertok, Phys. Lett. **41** (1978) 1155; D. Day *et al.*, Phys. Rev. Lett. **43** (1979) 1143. Summaries of the data for nucleon and nuclear form factors at large Q^2 are given in B.T. Chertok, in *Progress in Particle and Nuclear Physics, Proceeding of the International School of Nuclear Physics*, 5th Course (Erice, 1978) and *Proceedings of the XVI Rencontre de Moriond* (Les Arcs, Savoie, France, 1981).

[46] S.J. Brodsky, Y. Frishman, G.P. Lepage and C. Sachrajda, Phys. Lett. **B91** (1980) 239.

[47] R. G. Arnold *et al.*, Phys. Rev. Lett. **57** (1986) 174.

[48] R. L. Anderson *et al.*, Phys. Rev. Lett. **30** (1973) 627.

[49] A.W. Hendry, Phys. Rev. **D10** (1974) 2300.

[50] G.R. Court *et al.*, UM–HE–86–03 (April, 1986) 14 pp.

[51] S.J. Brodsky, C.E. Carlson and H.J. Lipkin, Phys. Rev. **D20** (1979) 2278; G.R. Farrar, S. Gottlieb, D. Sivers and G. Thomas, Phys. Rev. **D20** (1979) 202.

[52] G.R. Farrar, RU–85–46, (1986).

[53] S.J. Brodsky, C.E. Carlson and H.J. Lipkin, Ref. [51]; H.J. Lipkin, priv. comm.

[54] A.H. Mueller, Phys. Rep. **73** (1981) 237. See also S.S. Kanwal, Phys. Lett. **142B** (1984) 294.

[55] S. Gupta, Phys. Rev. **D24** (1981) 1169.

[56] P.H. Damgaard, Nucl. Phys. **B211** (1983) 435.

[57] G.R. Farrar, G. Sterman and H. Zhang, Rutgers Preprint 89-07 (1989).

[58] J. Botts and G. Sterman, ITP-SB-89-7,8,44 (1989).

[59] V.L. Chernyak, A.A. Ogloblin and I.R. Zhitnitskii, Novosibirsk preprints INP 87–135,136 and references therein. See also Xiao-Duang Xiang, Wang Xin-Nian and Huang Tao, BIHEP-TH-84, 23 and 29 (1984) and M.J. Lavelle, ICTP-84-85-12; Nucl. Phys. **B260** (1985) 323.

[60] I.D. King and C.T. Sachrajda, Nucl. Phys. **B279** (1987) 785.

[61] G.P. Lepage, S.J. Brodsky, Tao Huang and P.B. Mackenzie, in *Proceedings of the Banff Summer Institute* (1981).

[62] M. Gari and N. Stefanis, Phys. Lett. **B175** (1986) 462; M. Gari and N. Stefanis, Phys. Lett. **B187** (1987) 401.

[63] C-R Ji, A.F. Sill and R.M. Lombard-Nelsen, Phys. Rev. **D36** (1987) 165.

[64] Z. Dziembowski and L. Mankiewicz, Phys. Rev. Lett. **58** (1987) 2175; Z. Dziembowski, Phys. Rev. **D37** (1988) 768, 778, 2030.

[65] See also G. R. Farrar, presented at the *Workshop on Quantum Chromodynamics at Santa Barbara* (1988).

[66] N. Isgur and C.H. Llewellyn Smith, Phys. Rev. Lett. **52** (1984) 1080. G.P. Korchemskii and A.V. Radyushkin, Sov. J. Nucl. Phys. **45** (1987) 910 and references therein.

[67] S. Gottlieb and A.S. Kronfeld, Ref. [8]; CLNS–85/646 (June 1985).

[68] G. Martinelli and C.T. Sachrajda, Phys. Lett. **B190** (1987) 151; *ibid.* **B196** (1987) 184; *ibid.* **B217** (1989) 319.

[69] G.W. Atkinson, J. Sucher, and K. Tsokos, Phys. Lett. **B137** (1984) 407; G.R. Farrar, E. Maina, and F. Neri, Nucl. Phys. **B259** (1985) 702 (Err. *ibid.* **B263** (1986) 746); E. Maina, Rutgers Ph.D. Thesis (1985); J.F. Gunion, D. Millers and K. Sparks, Phys. Rev. **D33** (1986) 689; P.H. Damgaard, Ref. [56]; B. Nizic, Ph.D. Thesis, Cornell University (1985); D. Millers and J.F. Gunion, Phys. Rev. **D34** (1986) 2657.

[70] Z. Dziembowski, G.R. Farrar, H. Zhang and L. Mankiewicz, contribution to the *12th Int. Conf. on Few Body Problems in Physics* (Vancouver, 1989).

[71] Z. Dziembowski and J. Franklin, contribution to the *12th Int. Conf. on Few Body Problems in Physics* (Vancouver, 1989).

[72] C. Carlson and F. Gross, Phys. Rev. Lett. **53** (1984) 127; Phys. Rev. **D36** (1987) 2060.

[73] O.C. Jacob and L.S. Kisslinger, Phys. Rev. Lett. **56** (1986) 225.

[74] The connection of the parton model to QCD is discussed in G. Altarelli, Phys. Rep. **81** (1982) No. 1.

[75] V.N. Gribov and L.V. Lipatov, Sov. Jour. Nucl Phys. **15** (1972) 438, 675.

[76] R.P. Feynman, *Photon-Hadron Interactions* (W. A. Benjamin, Reading, MA, 1972).

[77] P.V. Landshoff and J.C. Polkinghorne Phys. Rev. **D10** (1974) 891.

[78] B.R. Baller *et al.*, Phys. Rev. Lett. **60** (1988) 1118.

[79] For general discussions of $\gamma\gamma$ annihilation in $e^+e^- \to e^+e^-X$ reactions, see S.J. Brodsky, T. Kinoshita, and H. Terazawa, Phys. Rev. Lett. **25** (1970) 972; Phys. Rev. **D4** (1971) 1532; V.E. Balakin, V.M. Budnev, and I.F. Ginzburg, JETP Lett. **11** (1970) 388; N. Arteaga-Romero, A. Jaccarini and P. Kessler, Phys. Rev. **D3** (1971) 1569; R.W. Brown and I.J. Muzinich, Phys. Rev. **D4** (1971) 1496; C.E. Carlson and W.-K. Tung, Phys. Rev. **D4** (1971) 2873. Reviews and further references are given in H. Kolanoski and P.M. Zerwas, DESY 87-175 (1987); H. Kolanoski, *Two-Photon Physics in e^+e^- Storage Rings* (Springer–Verlag, 1984); Ch. Berger and W. Wagner, Phys. Rep. **136** (1987); J.H. Field, University of Paris preprint LPNHE 84-04 (1984).

[80] G. Köpp, T.F. Walsh and P. Zerwas, Nucl. Phys. **B70** (1974) 461; F. M. Renard, *Proc. of the Vth International Workshop on $\gamma\gamma$ Interactions*, and Nuovo Cim. **80** (1984) 1. Backgrounds to the $C = +$, $J = 1$ signal can occur from tagged $e^+e^- \to e^+e^-X$ events which produce $C = -$ resonances.

[81] H. Aihara *et al.*, Phys. Rev. Lett. **57** (1986) 51, 404. Mark II data for combined charged meson pair production are also in good agreement with the PQCD predictions. See J. Boyer *et al.*, Phys. Rev. Lett. **56** (1986) 207.

[82] H. Suura, T.F. Walsh and B.L. Young, Lett. Nuovo Cimento **4** (1972) 505. See also M.K. Chase, Nucl. Phys. **B167** (1980) 125.

[83] J. Boyer *et al.*, Ref. [81]; TPC/Two Gamma Collaboration (H. Aihara *et al.*), Phys. Rev. Lett. **57** (1986) 404.

[84] M. Benyayoun and V.L. Chernyak, College de France preprint LPC 89 10 (1989).

[85] B. Nizic, Phys. Rev. **D35** (1987) 80.

[86] G.R. Farrar RU-88-47, Invited talk given at *Workshop on Particle and Nuclear Physics on the Light Cone* (Los Alamos, New Mexico, 1988); G.R. Farrar, H. Zhang, A.A. Globlin and I.R. Zhitnitskii, Nucl. Phys. **B311** (1989) 585; G.R. Farrar, E. Maina and F. Neri, Phys. Rev. Lett. **53** (1984) 28, 742.

[87] M.A. Shupe *et al.*, Phys. Rev. **D19** (1979) 1921.

[88] A simple method for estimating hadron pair production cross sections near threshold in $\gamma\gamma$ collisions is given in Ref. [33].

[89] See Ref. [31]. The next-to-leading order evaluation of T_H for these processes is given by B. Nizic, Ph.D. Thesis, Cornell University (1985).

[90] See Ref. [30], and S.J. Brodsky and J.R. Hiller, Phys. Rev. **C28** (1983) 475.

[91] C.R. Ji and S.J. Brodsky, Phys. Rev. **D34** (1986) 1460; *ibid.* **D33** (1986) 1951, 1406, 2653. For a review of multi-quark evolution, see S.J. Brodsky and C.-R. Ji, SLAC-PUB-3747 (1985).

[92] The data are compiled in Brodsky and Hiller, Ref. [90].

[93] J. Napolitano *et al.*, ANL preprint PHY–5265–ME–88 (1988).

[94] T.S.-H. Lee, ANL preprint (1988).

[95] H. Myers *et al.*, Phys. Rev. **121** (1961) 630; R. Ching and C. Schaerf, Phys. Rev. **141** (1966) 1320; P. Dougan *et al.*, Z. Phys. **A276** (1976) 55.

[96] T. Fujita, MPI-Heidelberg preprint (1989).

[97] S.J. Brodsky, C.-R. Ji and G.P. Lepage, Phys. Rev. Lett. **51** (1983) 83.

[98] L.A. Kondratyuk and M.G. Sapozhnikov, Dubna preprint E4-88-808.

[99] A. S. Carroll *et al.*, Phys. Rev. Lett. **61** (1988) 1698.

[100] G.R. Court *et al.*, Phys. Rev. Lett. **57** (1986) 507.

[101] S.J. Brodsky and G. de Teramond, Phys. Rev. Lett. **60** (1988) 1924.

[102] R. Blankenbecler and S.J. Brodsky, Phys. Rev. **D10** (1974) 2973.

[103] See Ref. [100]; T.S. Bhatia *et al.*, Phys. Rev. Lett. **49** (1982) 1135; E.A. Crosbie *et al.*, Phys. Rev. **D23** (1981) 600; A. Lin *et al.*, Phys. Lett. **B74** (1978) 273; D.G. Crabb *et al.*, Phys. Rev. Lett. **41** (1978) 1257; J.R. O'Fallon *et al.*, Phys. Rev. Lett. **39** (1977) 733. For a review, see A.D. Krisch, UM–HE–86–39 (1987).

[104] For other attempts to explain the spin correlation data, see C. Avilez, G. Cocho and M. Moreno, Phys. Rev. **D24** (1981) 634; G.R. Farrar, Phys. Rev. Lett. **56** (1986) 1643 (Err *ibid.* **56** (1986) 2771); H.J. Lipkin, Nature **324** (1986) 14; S.M. Troshin and N.E. Tyurin, JETP Lett. **44** (1986) 149 (Pisma Zh. Eksp. Teor. Fiz. **44** (1986) 117); G. Preparata and J. Soffer, Phys. Lett. **B180** (1986) 281; S.V. Goloskokov, S.P. Kuleshov and O.V. Seljugin, *Proceedings of the VII International Symposium on High Energy Spin Physics*, Protvino (1986); C. Bourrely and J. Soffer, Phys. Rev. **D35** (1987) 145.

[105] There are five different combinations of six quarks which yield a color singlet B=2 state. It is expected that these QCD degrees of freedom should be expressed as B=2 resonances. See, *e.g.*, S.J. Brodsky and C.R. Ji, Ref. [91].

[106] For other examples of threshold enhancements in QCD, see S. J. Brodsky, J.F. Gunion and D.E. Soper, Ref. [32] and also Ref. [88]. Resonances are often associated with the onset of a new threshold. For a discussion, see D. Bugg, presented at the *IV LEAR Workshop* (Villars-Sur-Ollon, Switzerland, September 6–13, 1987).

[107] J.F. Gunion, R. Blankenbecler and S.J. Brodsky, Phys. Rev. **D6** (1972) 2652.

[108] With the above normalization, the unpolarized pp elastic cross section is $d\sigma/dt = \Sigma_{i=1,2,\ldots 5} \mid \phi_i^2 \mid /(128\pi s p_{\rm cm}^2)$.

[109] At low momentum transfers one expects the presence of both helicity-conserving and helicity non-conserving pomeron amplitudes. It is possible that the data for A_N at $p_{\rm lab} = 11.75$ GeV/c can be understood over the full angular range in these terms. The large value of $A_N = 24 \pm 8\%$ at $p_{\rm lab} = 28$ GeV/c and $p_T^2 = 6.5$ GeV^2 remains an open problem. See P.R. Cameron *et al.*, Phys. Rev. **D32** (1985) 3070.

[110] K. Abe *et al.*, Phys. Rev. **D12** (1975) 1, and references therein. The high energy data for $d\sigma/dt$ at $\theta_{\rm cm} = \pi/2$ are from C.W. Akerlof *et al.*, Phys. Rev. **159** (1967) 1138; G. Cocconi *et al.*, Phys. Rev. Lett. **11** (1963) 499; J.V. Allaby *et al.*, Phys. Lett. **23** (1966) 389.

[111] I.P. Auer *et al.*, Phys. Rev. Lett. **52** (1984) 808. Comparison with the low energy data for A_{LL} at $\theta_{\rm cm} = \pi/2$ suggests that the resonant amplitude below $p_{\rm lab} = 5.5$ GeV/c has more structure than the single resonance form adopted here. See I.P. Auer *et al.*, Phys. Rev. Lett. **48** (1982) 1150.

[112] See Ref. [49], and N. Jahren and J. Hiller, University of Minnesota preprint (1987).

[113] The neutral strange inclusive pp cross section measured at $p_{\rm lab} = 5.5$ GeV/c is 0.45 ± 0.04 mb; see G. Alexander *et al.*, Phys. Rev. **154** (1967) 1284.

[114] H.C. Pauli and S.J. Brodsky, Phys. Rev. **D32** (1985) 1993, 2001 and Ref. [9].

[115] K. Hornbostel, SLAC-0333 (Dec 1988); K. Hornbostel, S.J. Brodsky, and H.C. Pauli, SLAC-PUB-4678, talk presented to *Workshop on Relativistic Many Body Physics* (Columbus, Ohio, June, 1988).

[116] S.J. Brodsky, H.C. Pauli and A. Tang, in preparation.

[117] C.J. Burden and C.J. Hamer, Phys. Rev. **D37** (1988) 479 and references therein.

[118] Y. Frishman and J. Sonnenschein, Nucl. Phys. **B294** (1987) 801 and Nucl. Phys. **B301** (1988) 346.

[119] For a discussion of renormalization in light-cone perturbation theory, see Ref. [120] and also Ref. [38].

[120] S.J. Brodsky, R. Suaya and R. Roskies, Phys. Rev. **D8** (1973) 4574.

[121] S.G. Gorishny, A.L. Kataev and S.A. Larin, Phys. Lett. **B212** (1988) 238.

12

Aspects of the Heavy Fermion Problem

C. M. Varma
AT&T Bell Laboratories
600 Mountain Avenue
Murray Hill, NJ 07974

12.1 Introduction and Scope of this Review

In this overview of the heavy fermion problem, I will discuss the physical basis of the problem, what progress has been achieved and what problems remain unsolved. A large part of what I discuss is drawn from lectures given at Tsukuba two years ago [1]. The first part, on the different kinds of masses that occur in condensed matter physics, is drawn from the Gerry Brown festschrift [2].

The enormous activity, for over a decade, on the heavy fermion and mixed-valence problem, has abated considerably in the last two years. The reason is the discovery of the high temperature superconductors which has drawn the efforts of most of the theorists and experimentalists working on the difficult and fascinating problems of correlated fermions. (However many of the current ideas on high T_c superconductors are strongly influenced by the understanding that has been achieved in the heavy fermion problem.) The principal new result has been the discovery of multiple superconducting/antiferromagnetic phase transitions in the heavy fermion compound UPt_3 and the elucidation of this phase diagram. This has effectively sealed the case for heavy fermion superconductivity as a new symmetry and one driven microscopically by electron-electron interactions rather than electron-phonon interactions. However, in this review I will not discuss superconductivity in the heavy fermions.

The physics of the heavy fermion compounds has three well springs: (1) The electronic structure of rare-earth atoms and compounds, (2) The theorems on Fermi-liquid theory [3] suggested by Landau and his followers; it is a remarkable triumph of the ideas of analyticity and continuity enshrined in Fermi-liquid theory, and (3) Aspects of infrared divergences of the Fermi sea. Some aspects of this, the Kondo problem, are solved. But although I expect it is so, I cannot assert confidently that the heavy fermion problem can be understood, in principle, with just this knowledge.

I shall review all three aspects. But first I want to talk in general about the concept of mass in condensed matter physics.

12.2 Masses in Condensed Matter Physics

In mundane terms certain solids are said to display heavy-fermion behavior because they exhibit specific heats or entropies which vary at low temperatures as γT with the coefficient $\gamma \approx \mathcal{O}(10^3)$ times that of ordinary metals. At finite temperature quasi-particles and quasi-holes of relative density $N(E_F)T$ are excited across the Fermi surface in a Fermi liquid, contributing an entropy proportional to $N(E_F)k_BT$. Here $N(E_F)$ is the density of states of quasi-particles at the Fermi energy, which simply comes from counting phase space and is proportional to the average of the inverse Fermi velocity $v_F^{-1} \equiv m^*/k_F$, where k_F is an average Fermi momentum. One of the fundamental tenets of Fermi-liquid theory is that the number of quasi-particles is equal to the number of particles, which implies that k_F is unaffected by the interactions. Therefore, a γ that is a thousand times larger than that in ordinary metals implies a mass renormalization of $\mathcal{O}(10^3)$. This is a spectacularly large many-body effect and raises several important questions.

I would like to digress briefly to say that, to a condensed-matter theorist there are many kinds of masses:

1. Bare mass, m.
2. Band mass, m_b, which arises because the energy-momentum relation, $E(k)$, of electrons is changed in a solid from its free-electron value due to the periodic potential;
$$\langle m_b^{-1}(k)\rangle = \left\langle \frac{d^2E}{dk^2} \right\rangle .$$
3. Dynamical mass, m_d, which is the ratio of the momentum carried by the quasi-particle to the current carried by it. If the problem being

studied is Galilean invariant, for instance liquid ^{3}He, so that the current operator commutes with the Hamiltonian, the current carried is the same with or without interactions and therefore the dynamical mass is equal to the bare mass. In more physical terms, since the renormalizations are provided by ^{3}He $-^3$He interactions and since all of them move together in response to the potential coupling to the current, the effect of the interactions is not felt in the response. For the same reason, if we weigh a solid, say on a spring balance, neither the interactions nor the periodic potential play a role and we observe the bare mass.

4. Entropic mass m^*, which, being a thermodynamically defined mass, includes the effects of all the fluctuations and usually has the highest value of any of the above. For liquid ^{3}He near the melting line, wherein the strongest renormalization prior to the discovery of heavy fermions was observed, $m^* \approx 6m$.

The dynamical mass of the heavy fermions can be deduced, for example, in an experiment measuring the London penetration depth in the superconducting state. This and many other experiments strongly indicate that the dynamical mass of the heavy fermions is about the same as the entropic mass. This gives a clue to the nature of renormalizations in heavy fermions which I will dwell upon in the section on phenomenology.

12.3 Electronic Structure of the Rare-earths and the Actinides

The heavy fermion phenomena are observed in rare-earths and actinide compounds. The genesis of the special physics of the rare-earth and the actinide compounds lies in the atomic physics of the f-shell. Because of the large centrifugal repulsion $\ell(\ell+1)/r^2$ with $\ell = 3$ for the f-shell, the $4f$ orbital is unoccupied, while the $6s, 5p$ and $5d$ shells are filled. Consider Ba which contains 2 electrons in the $6s$ shell outside a Xe core. The $4f$ orbital has almost zero binding energy and a large radial extent with a peak at about 14 atomic units from the nucleus (see Fig. 12.1). Let us now increase the nuclear charge by two, arriving at Ce. Because of the incomplete screening of the nuclear charge by the two extra added electrons, all the orbitals $5d, 6s, 6p$, *etc.* are pulled in a little bit. But this effect is small since, for example, the $6s$ orbital has to be orthogonal to the $5s$, which has to be orthogonal to the $4s$ orbital and so on. On the other hand the $4f$ orbital, being the first orbital of its annular symmetry, has no such restriction. The relatively small nuclear charge increase pulls this orbital all the way in, so that the maximum in its charge density is at about 0.7

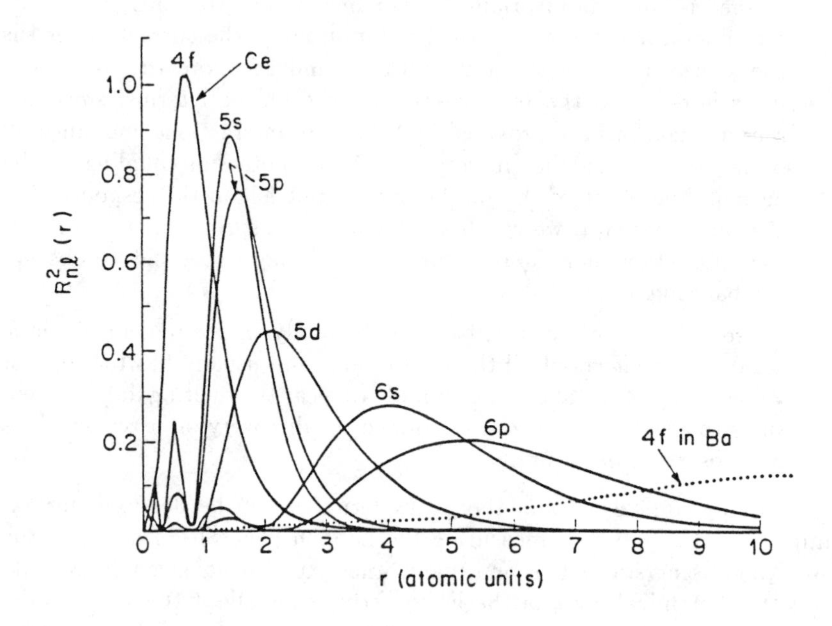

FIGURE 12.1
The charge density of the $4f$ orbital of Ce in relation to the $4f$ orbital in Ba and to the charge density in the other orbitals which is almost the same for Ce and Ba.

a.u. inside even the core-like $5s$ and $5p$ orbitals. But its ionization energy is still small— close to that of the widely spread out $5d, 6s, 6p$ orbitals. The latter make the wide valence and conduction bands in the solid state. Being core-like the $4f$ orbitals have large correlation energies— the difference U of the ionization energy I and the affinity energy A for a given charge state in the rare-earths (even after screening effects are taken into account) is large:

$$U = I - A\,. \tag{12.1}$$

This near degeneracy (on the atomic scale) of highly correlated core-like orbitals with the well spread out, weakly-correlated orbitals which form bands in the solid state is at the root of the special physics of the rare-earths and the actinides.

The nature of the phenomena actually observed in the rare-earth and actinide compounds depends very sensitively on the difference between the ionization energy of the f-orbitals and the chemical potential set by the

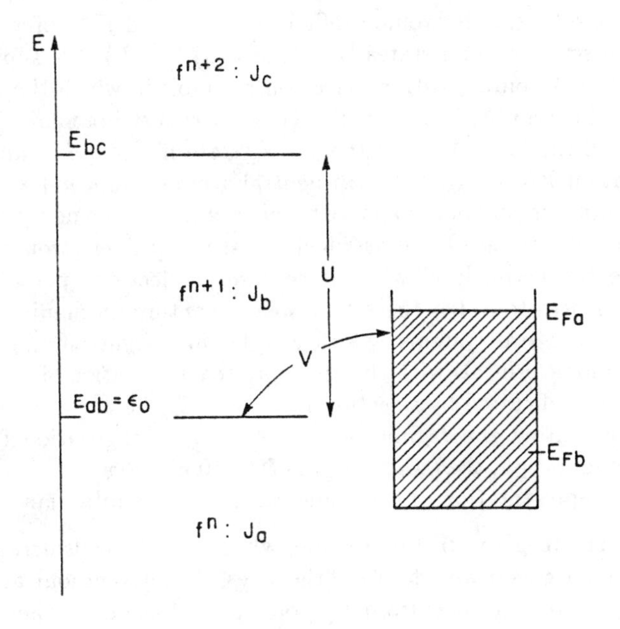

FIGURE 12.2
Pictorial representation of the Anderson model to illustrate the three different regimes of the physics of the rare earths and the actinides.

Fermi energy. The problem is best discussed through the Anderson lattice Hamiltonian first introduced for such problems by Varma and Yafet [4]. This Hamiltonian is depicted pictorially in Fig. 12.2.

$$H = \epsilon_0 \sum_{i\sigma} n_{fi\sigma} + \sum_{k\sigma} \epsilon_k c^\dagger_{k\sigma} c_{k\sigma} + U \sum_{i\sigma} n_{fi\sigma} n_{fi-\sigma} + \sum_{ki\sigma} V_k e^{i\vec{k}\cdot\vec{R}_i} c^\dagger_{fi\sigma} c_{k\sigma} , \qquad (12.2)$$

where ϵ_0 is the ionization energy of the f-orbitals which have large correlation energies U and weak hybridization V with conduction orbitals moving in a wide band of dispersion ϵ_k. For understanding the properties of actual materials quantitatively, it is undoubtedly necessary to include the details of multiple orbitals, spin-orbit coupling, crystal fields, *etc.* But the essential physics is believed to be already contained in Eq. (12.2).

There appear to be three different regimes in the physics of these materials, which are usefully discussed with reference to Eq. (12.2) and Fig. 12.2. Consider the electronic configurations f^n and f^{n+1} of an isolated ion with respective ground states $|f^n : J_a\rangle$ and $|f^{n+1} : J_b\rangle$. We suppose the ion to be in weak contact with an electron reservoir in which the chemical potential is at energy E_F. There will exist some energy boundary E_{ab} such that at $T = 0$ the ion will be in the $f^n : J_a$ state if $E_F < E_{ab}$ and in the $f^{n+1} : J_b$ state if $E_F > E_{ab}$. Let each neutral atom contain $n + r$ electrons and let us form a crystal out of a large number of them. The non-f electrons go into a set of (mostly) d-bands which constitute the electron reservoir. Let E_{Fa} be the Fermi level when it contains r electrons per atom and E_{Fb} for $r - 1$ per atom. Let there be a weak quantum mechanical contact $\{V\}$ between the two systems, which couples in a tight-binding manner the localized ionic orbitals with the appropriate combination of conduction band orbitals as dictated by selection rules. Let E_{bc} be another ionization boundary separating the regime of the $f^{n+1} : J_b$ configuration from the $f^{n+2} : J_c$ configuration, then $E_{bc} - E_{ab} \equiv U \approx 10$ eV. Fig. 12.2 is of course the pictorial representation of the Anderson lattice Hamiltonian.

a. **Magnetic Regime** In this regime, which may be considered understood and occurs in about 90% of the cases, the relevant ionization level of the f-state E_{ab} is far from E_{Fa} or E_{Fb} and therefore the chemical potential is at, say, E_{Fa}. The f-orbitals have no charge fluctuations (integral valence) and may be thought to be in a Mott insulating state. They obey atomic spectral rules supplemented by crystal fields. There is weak residual spin polarization of the conduction electrons, RKKY interactions between the local moments, magnetic transition at low temperatures and spin-wave excitations. The conduction electrons at low temperatures scatter off these spin waves. The degrees of freedom of the local moments are described by spin Hamiltonians and their coupling to conduction electrons can be treated perturbatively.

b. **Heavy Fermions** Now suppose we adjust just one parameter so that the ionization boundary, E_{ab}, comes closer to the chemical potential but is still at a separation much larger than $\mathcal{O}(\Gamma \approx \pi V^2 \rho)$ where V is the hybridization matrix element (taking into account orbital degeneracy) for f-orbitals with the conducting electrons and ρ the relevant partial density of states of the conduction electrons in the f-channel. In this regime also, at low temperatures, real quantum-mechanical valence fluctuations may be considered negligible. The properties of such materials are radically different from the first regime. Below a certain cross-over temperature T_x, the degrees of freedom of the local moments appear in the form of entropy characteristic of a Fermi liquid. The same entropy in regime **a** occurs, of course, in the spin waves plus paramagnetic fluctuations above the magnetic transition temperature.

Whereas in regime **a** these are distinguishable and have only a weak coupling between them, in this regime they are so strongly coupled that one cannot speak of them separately. At low temperatures sometimes these degrees of freedom have instability to the antiferromagnetic state with weak moments (as in $CeAl_2$, U_2Zn_{17}, *etc.*) or as superconductivity (as in $CeCu_2Si_2$, UBe_{13}). Some cases are known ($CeAl_3$, $CeCu_6$) where down to the lowest temperature investigated ($\sim$ 10 mK) the Fermi-liquid behavior persists.

The specific heat of these materials [5] in the Fermi-liquid regime in terms of the coefficient γ of ($C = \gamma T$) ranges from about 50 mJ/mole K in $CeAl_2$ to about 1.6 J/mole K for $CeCu_6$. If as is customary, the specific heat is expressed in terms of an effective mass m^*, then $m^* \approx$ 500 times the free electron mass for UBe_{13}. The magnetic susceptibility at low temperatures tends to a constant, as in a Fermi liquid, and its value is enhanced with respect to non-interacting electrons by about the same ratio as the specific heat, *i.e.*, by about m^*/m.

c. **Mixed- or Fluctuating-Valence Regime** The third regime of the rare earths and actinides [6] occurs if E_{ab} should happen to lie between E_{Fa} and E_{Fb}; then the mixed-valence situation arises (Fig. 12.2). It requires no remarkable accident of nature for this to happen. The state of the system in which all ions have n f-electrons is impossible because E_{Fa} is too high to be consistent with the $f^n : J_a$ ionic states, and similarly having all ions in $f^{n+1} : J_b$ is impossible because the corresponding E_{Fb} lies below E_{ab}. Thus the Fermi level in the reservoir will lie at the boundary E_{ab} with some fraction x of ions in the f^{n+1} and the rest in the f^n states, as demanded by having the correct total number of electrons. This consideration also reveals that about 10% of rare-earth compounds should belong to the intermediate valence category, since typically $E_{Fa} - E_{Fb} \approx 1$ eV, while $U \approx 10$ eV.

The third regime occurs frequently in cerium, thulium and ytterbium compounds as well as in samarium and europium. In the former, the underlying closed shell makes the $4f$ and the $5d$ atomic energy levels closed, and, in the latter two, the preference of Hund's rule for the f^7 configuration makes different configurations close in energy.

The demarcation between regimes **b** and **c** is not always clear from a phenomenological point of view. The various experimental methods of valence determination can only be trusted, at best, to about 10% and the uncertainty in the difference $E_{ab} - E_{Fa}$ is several times Γ. The Fermi-liquid behavior and the high temperature regime of free moments are qualitatively quite similar, although I do not know of any case in a well-established intermediate-valence case that $m^*/m \gtrsim 10^2$ or of a magnetic transition (except in TmSe) or a superconducting transition in them. From a theoretical point of view, the intermediate-valence regime must have f-charge and f-moment fluctuations to about equal degree, while in regime **b** real (as opposed to virtual) charge fluctuation may be considered negligible at low temperatures.

In this review, I will deal exclusively with the heavy fermion regime in which real or resonant charge fluctuations can be neglected. One can then work with a simpler version of the Hamiltonian (12.2) obtained by a canonical transformation [7] to eliminate charge fluctuations.

$$H = \sum_{k\sigma} \epsilon_k c^\dagger_{k\sigma} c_{k\sigma} + J \sum_i c^\dagger_i \vec{\sigma}\, c_i \cdot \vec{S}_i \,, \tag{12.3}$$

where

$$J \approx \sum_k \frac{2|V_k|^2\, U}{\epsilon_0(\epsilon_0 + U)} \tag{12.4}$$

with the Fermi level taken at zero energy. In Eq. (12.3) $\vec{S}_i$ is the local moment at site i in the f-orbital. If there were a local moment only at one site, any $i = 0$, we would have just the Kondo problem characterized by the temperature

$$T_k \approx D\,(\rho J)^{\frac{1}{2}} \exp(-\tfrac{1}{2}\rho J) \tag{12.5}$$

at which the local moment begins to be renormalized. In Eq. (12.5), D is the bandwidth of the conduction electrons and ρ the density of states near the Fermi level.

To second order in J an interaction develops between two moments at sites i and j. This is the well known RKKY interaction

$$H_{\text{int}} = \sum_{i<j} K_{ij}\, \vec{S}_i \cdot \vec{S}_j \,, \tag{12.6}$$

$$K_{ij} = K_0 f(R_{ij}) \,, \qquad K_0 \equiv J^2 \rho \,. \tag{12.7}$$

For $K_0 \gg T_K$, magnetic order develops and the Kondo effect is moot. The heavy fermion behavior arises for $T_K < K_0$. This condition can be met in our simplified model only for unrealistically large J of order ρ^{-1}. But for realistic starting points, with orbital degeneracy, crystal field splitting *etc.*, it is achieved [8] for reasonable values of $J\rho$. Note that the condition $T_K > K_0$ requires, in terms of the Hamiltonian (12.2), that the ionization potential (ϵ_0) not be too far below the Fermi level and the hybridization parameter V be large enough.

12.4 Phenomenology

The Hamiltonian (12.3) is suitable for describing the ordinary rare-earth magnetic metals and compounds as well as the heavy fermion compounds. All that is involved is a change in the value of the parameter $J\rho$. In fact at high temperatures the heavy fermion compounds exhibit magnetic susceptibility of the Curie-form, $\chi \sim \mu^2/T$ with μ nearly equal to the full localized magnetic moment. As the temperature decreases, there is a cross-over to a Fermi-liquid magnetic susceptibility and a concomitant specific heat C_V linearly increasing with temperature.

I will now give a simple argument for the conditions under which a very heavy fermion mass of $\mathcal{O}(10^2 - 10^3)$ is mandated [9]. Consider the contributions to the entropy of an ordinary rare-earth metal or compound. There is a small linear contribution γT from the fermions corresponding to a mass of $\mathcal{O}(1)$, and a contribution from the local moments which, at room temperature, is nearly temperature independent and two to three orders of magnitude larger than the γT contribution. The spin contribution to the entropy decreases as temperature is decreased, and, as illustrated in Fig. 12.3, has a change in slope at the magnetic ordering temperature T_N which is typically of $\mathcal{O}(10\text{ K})$. Below T_N, one gets the spin-wave contribution to the entropy. The heavy fermions are described by the same Hamiltonian as the typical rare-earth solid albeit with somewhat different parameters. Suppose that we demand that there be no magnetic phase transition and that the low-energy excitations in these solids be purely fermionic, *i.e.*, there is a gap of $\sim T_F$ which is the same order of magnitude as T_N for spin-fluctuation excitations. Then, as illustrated in Fig. 12.3, a heavy-fermion mass of $\mathcal{O}(10^2 - 10^3)$ is necessary, since total entropy in the present problem is the same as for the ordinary rare earths and the high-temperature entropies must be identical.

The discussion above may be summarized by the statement that heavy fermions arise from ordinary conduction electrons by exchange of spin fluctuations with total spectral weight similar to that carried by the local moments and that if these spin fluctuations must have a gap of $\mathcal{O}(T_F)$, the specific heat of the fermions will be $\mathcal{O}(T/T_F)$. One really observes only the spin-fluctuation entropy, but in a fermionic form, *i.e.*, $\sim T$.

This point of view is reinforced and at the same time made more precise by the observation that although the thermodynamic properties like the specific heat and the magnetic susceptibility are severely renormalized, many transport properties (the coefficient of the leading low temperature terms in the thermal conductivity, ultrasonic attenuation, nuclear and impurity spin-lattice relaxation rate, the residual resistivity, *etc.*) appear un-

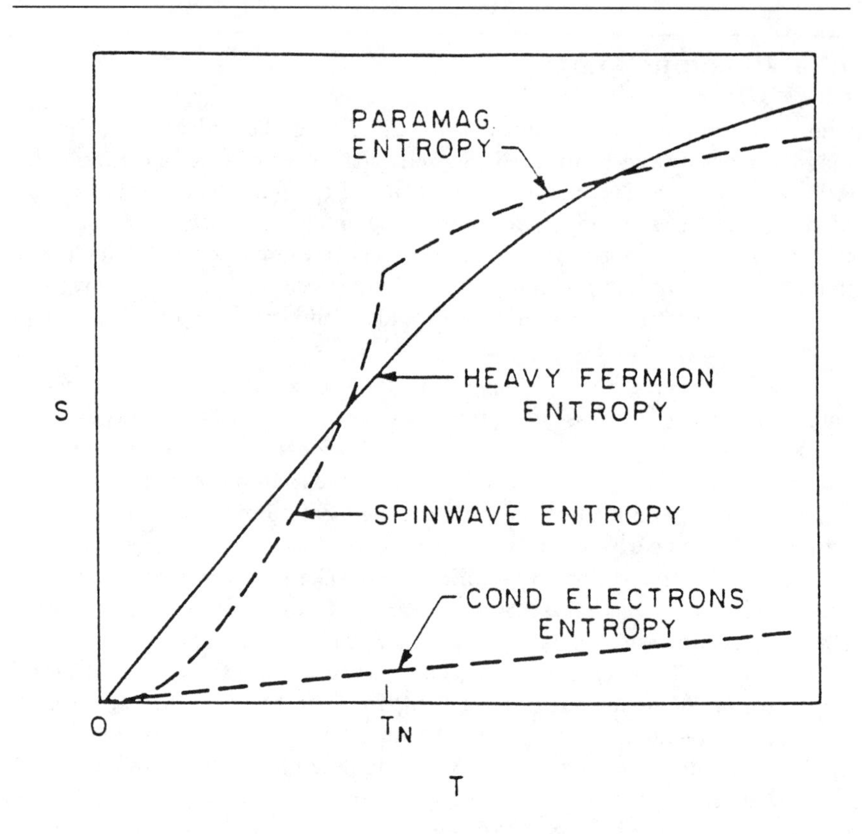

FIGURE 12.3
The magnetic and the conduction electron contribution to the entropy of an ordinary rare-earth magnetic metal (dashed lines) compared with the entropy for a heavy fermion material

renormalized. Naively all these quantities depend on $(m^*/m)^2$ and one would expect enormous renormalizations; none are seen. Such a behavior is understood by asserting that in the single particle fermion Green's function

$$G(\vec{k},\omega) = (\omega - \epsilon(\vec{k}) - \Sigma(\vec{k},\omega))^{-1} , \tag{12.8}$$

$\epsilon(\vec{k})$ is the dispersion obtained from some self-consistent one-electron theory and $\Sigma(\vec{k},\omega)$, the self-energy function, is required to have the property

$$\frac{\partial \Sigma}{\partial \omega} \gg \vec{v}_0^{\,-1} \cdot \frac{\partial \Sigma}{\partial \vec{k}} \quad , \tag{12.9}$$

where $\vec{v}_0$ is defined by $\epsilon(\vec{k}) = \vec{v}_0 \cdot (\vec{k} - \vec{k}_f)$ near the Fermi surface. Then near the Fermi surface

$$G(\vec{k},\omega) \approx z(\omega - z\vec{v}_0 \cdot (\vec{k} - \vec{k}_F))^{-1} , \tag{12.10}$$

where z, the quasi-particle renormalization amplitude, is

$$z = (1 - \frac{\partial \Sigma}{\partial \omega})^{-1} . \tag{12.11}$$

The coefficient of the linear specific heat is then proportional to the renormalized density of states $\sim z^{-1} \int d\vec{S}_k |\vec{v}_0(\vec{k})|^{-1}$. Therefore

$$\frac{m^*}{m} \approx z^{-1} . \tag{12.12}$$

On the other hand, with the condition (12.9), the self-energy and the vertex renormalization cancel in most transport properties. This is usually shown through the use of a Ward identity. One may motivate it physically by noting that most transport properties involve a quantity τ^*/m^*, where τ^* is the renormalized scattering time. Now in the Born approximation:

$$(\tau^*)^{-1} \sim |\langle\psi_i^*|M|\psi_f^*\rangle|^2 n_f^*(E) , \tag{12.13}$$

where ψ_i^* and ψ_f^* are quasi-particle wave functions, M is the appropriate operator for the transport process under consideration and n_f^* is the renormalized final density of states. If the coherent part of the Green's function, which is proportional to the product of two quasi-particle wave functions, is proportional to z, each quasi-particle wave function has a weight $z^{\frac{1}{2}}$. Further $m^* \sim n_f \sim z^{-1}$. We then see that τ^*/m^* is independent of the renormalizations.

These arguments as well as the more formal proof using the Ward-identity work for strong scattering as well. They also survive into the superconducting state provided only that the condition (12.9) is met.

The condition (12.9) is equivalent to renormalizations through fluctuations whose energy scale is much smaller than the Fermi energy of conduction electrons, but whose momentum scale is comparable to the Fermi wave vector. Physically the cancellation of the renormalization in transport properties arises because the condition (12.9) is equivalent to the statement that the self-energy rides with the local chemical potential, so that in any perturbation, where the latter is altered and relaxes towards equilibrium, the renormalizations are always the same and their derivatives with respect to time are not felt.

This situation is very familiar from the case of electron-phonon interactions. Quite apart from the two orders-of-magnitude difference between heavy fermions and the electron-phonon problem, there is another important difference— magnetic susceptibility is not renormalized in the

TABLE 12.1
Comparison of renormalizations in the heavy fermion problem with those in liquid ^{3}He.

	^{3}He	Heavy Fermions
Compressibility $\frac{dn}{d\mu} = \frac{m^*/m}{1+F_0}$	Renormalized	Almost unrenormalized
Entropic mass m^*/m	$1 + F_1^S/3$	$1 + F_0^S$
Dynamic mass m_d	m	$\approx m^*$
Magnetic susceptibility χ	$\frac{m^*/m}{1+F_0^a}$	$\frac{m^*/m}{1+F_0^a}$
Transport properties	Usually renormalized	Usually unrenormalized

electron-phonon problem. For the renormalization of susceptibility, as in the heavy fermions, the fluctuations exchanged by the conduction electrons must carry spin.

In Table 12.1, we contrast the renormalizations in heavy fermions with those in liquid ^{3}He in terms of the Landau parameters. The result $m^*/m = 1 + \frac{1}{3}F_1^S$ is derived in one-component Fermi liquids solely using Galilean invariance. In a two component situation (we will soon discuss that microscopically, that is how we have to start thinking of the heavy fermions) in which only one compound gives a linear specific heat and the effective mass is defined only in relation to it, but the momentum is shared between the two components, Galilean invariance cannot be used. This is true even if there is no lattice and both the components are continuums.

The dynamical mass m_d in Table 12.1 is defined through the operator relation

$$\vec{p} = \frac{\vec{J}}{m_d} \, . \tag{12.14}$$

The result that $m_d \approx m^*$ has important consequence in the superconducting state where it is measured through the London penetration depth.

It is important to note that the condition (12.9) implies that the shape of the Fermi surface is not renormalized by the many-body effects. This is consistent with the de Haas-van Alphen measurements.

The energy scales of renormalizations in the heavy fermions are characteristic of those in the Kondo effect. If we simply assert that all the magnetic moments of the periodic array of rare-earth atoms in the heavy fermion solids undergo a Kondo effect and renormalize the mass of the conduction electrons interacting with them, the specific heat predicted per

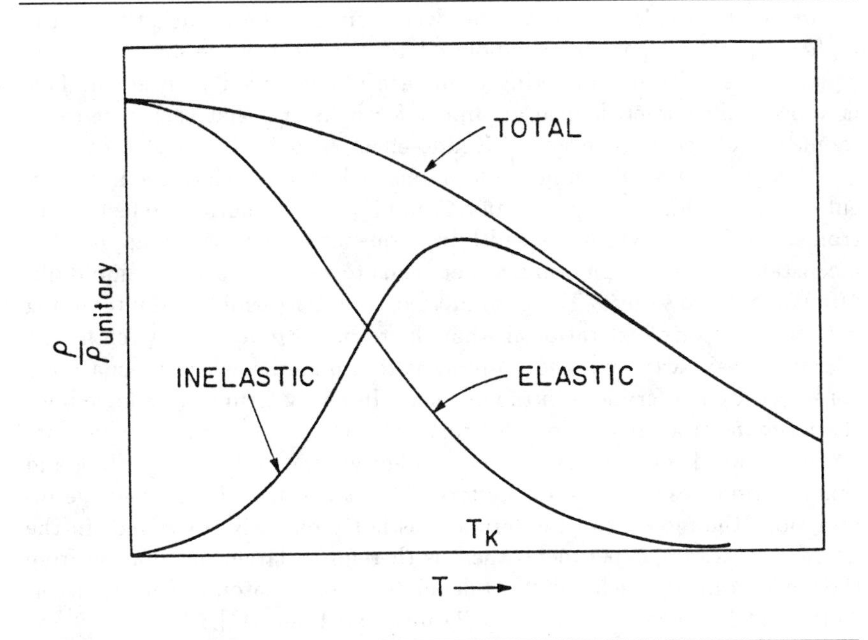

FIGURE 12.4
Decomposition of the total resistivity of a Kondo impurity into an elastic and an inelastic part.

rare-earth atom at low temperature would be $\mathcal{O}(T/T_K)$. If we take T_K as deduced for similar isolated magnetic impurities in conduction electrons, the experimentally observed magnitude is obtained.

This phenomenological statement raises a number of important questions which we shall touch on. Before we do so, an interesting relation between the single impurity Kondo effect resistivity and the resistivity commonly observed in the heavy fermion solids is discussed.

Fig. 12.4 shows the resistivity for a single impurity, the high temperature $\ln(T/T_K)$ dependence, and the asymptotic form $\rho_0\left[1-(T/T_K)^2\right]$, where ρ_0 is the unitarity value for an impurity. This resistivity can be strictly separated [10] into the inelastic part, which contributes all of the logarithmic term at high temperatures and which goes to zero as $(T/T_K)^2$ at low temperatures, and the elastic part. In a periodic array of scatterers the elastic part simply leads to a new electronic band and only the inelastic part can lead to electrical resistance. It is indeed remarkable that the resistivity of heavy fermions is qualitatively similar to the *inelastic* part of the single Kondo impurity. The overall magnitude is such that if the high temperature part is extrapolated to $T = 0$, a value approximately equal to the unitarity scattering from each magnetic ion is obtained. This is another phenomenological argument for believing that each rare-earth moment undergoes some sort of Kondo-effect renormalization.

Many experiments on heavy fermions yield results which are in apparent conflict with those expected of a Fermi liquid. One such is inelastic neutron scattering for which the width in frequency of the spectra approaches a constant as the momentum transfer tends to zero. For a Fermi liquid like ^{3}He it should go to zero. The general theorem for a Fermi liquid interacting with other types of excitations is that the response function is given by the (Fermi) quasi-particles if the coupling is to a quantity which is separately conserved by the fermions in the problem. In heavy fermions we have both the fermions (the conduction electrons) and the local moments which are not fermions. They interact with each other via the exchange coupling and this renormalizes the mass of the former. But because of the exchange interactions the moment of the fermions is not separately conserved. In the neutron scattering experiment, there is therefore a large contribution from the local moment which has the q-dependence of the atomic f-form factor. This point has also been made by Pethick and Pines [11].

12.5 Aspects of the Kondo Problem

Kondo discovered that the perturbation theory in J (*cf.* Eqs. (12.3) and (12.4)) for the problem of conduction electrons exchange-scattering off a local moment diverges for antiferromagnetic J. The second order contribution to the T-matrix (see Fig. 12.5) consists of two terms— one with an intermediate electron line and the other with an intermediate hole line. The quantum mechanical nature of the spins is essential here; the two processes add because of the anti-commutative nature of the spin operators. It was soon realized that the problem cannot be solved by perturbation theory. An important development towards the eventual solution of the problem was the work of Anderson and collaborators [12] who cast it in a path-integral representation eliminating the local moment fluctuations. The conduction electron spins scatters up and down (see Fig. 12.6) at successive times $t_1, t_2, \ldots$. The solution of the problem consists in solving for electrons near the Fermi sea moving in such a time dependent potential. One conclusion of their work is that T_K is only a cross-over temperature and a singlet state of the local moment and conduction electrons is arrived at only at $T = 0$. Through the work of various people it was realized qualitatively that at low temperatures the Kondo problem is equivalent to a resonance at the Fermi energy of width T_K.

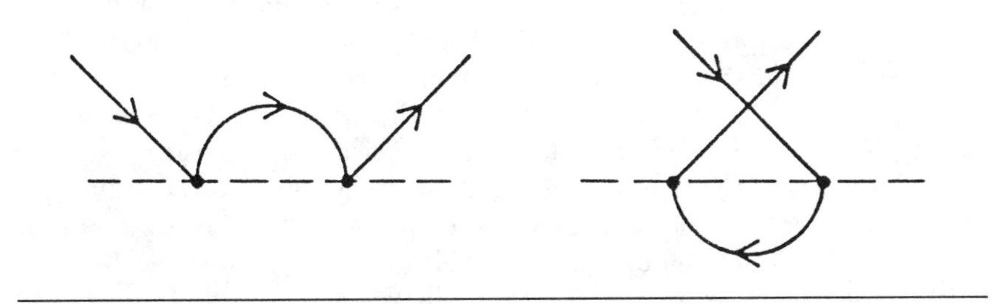

FIGURE 12.5
The leading contributions to the T-matrix for conduction electron-magnetic impurity scattering.

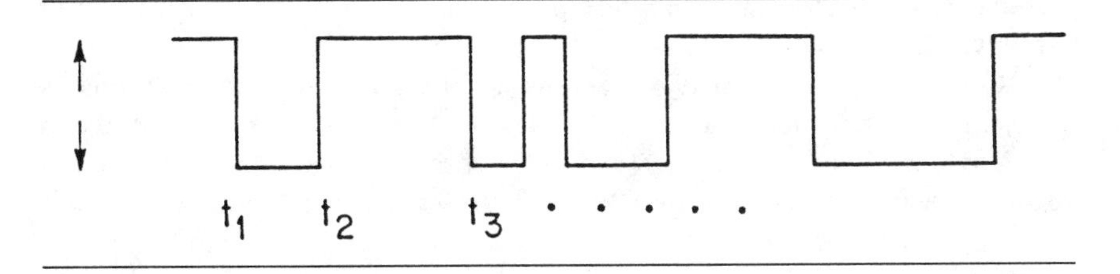

FIGURE 12.6
The path-integral representation of the single Kondo impurity problem.

The complete and enlightening solution was achieved by Wilson [13] with a numerical renormalization group method. An analytic solution was obtained by Andrei [14] and by Weigmann [15]. Wilson's solution was interpreted in a phenomenological fashion by Nozières [10]. For our subsequent discussion on heavy fermions it is necessary to review these. For the same reason we review first the important work by Yamada and Yosida [16].

Yamada and Yosida's work is an interesting application of the principle of analyticity and continuity which underlies the derivation, for instance, of Fermi-liquid theory. Consider the Hartree-Fock solution [17] of the Anderson model for the single magnetic impurity problem. In general there are two solutions: (i) a magnetic solution with a local resonance for a given spin direction below the Fermi energy, and a local resonance with the opposite spin direction above the Fermi energy. Each resonance has a width $\sim V^2\rho$ and the two are separated by U. (ii) a non-magnetic solution with a half- occupied resonance also of width $\sim V^2\rho$ straddling the Fermi energy. For the range of parameters of interest the magnetic solution has the lower energy in the Hartree-Fock approximation. But we know that the correct ground state is non-magnetic. Most of the brilliant (and difficult) methods used in the Kondo problem start from the broken symmetry magnetic solution and work hard to restore symmetry and obtain the non-magnetic solution. Yamada and Yosida's point is that even though U may be a large parameter, it is legitimate to calculate in perturbation theory about the *non-magnetic* Hartree-Fock solution. The solution will analytically and continuously give several features of the right answers if the starting point has the same symmetry as the correct solution. The low temperature "Kondo resonance" is in this picture just a modification in width and strength of the Hartree-Fock non-magnetic resonance. Yamada and Yosida were able to show that at each order in perturbation theory in U/V, the Wilson ratio, the dimensionless ratio of the magnetic susceptibility to the coefficient of the linear term in the specific heat, is 2. It is a reward for adopting the correct point of view that the absolute answers for the magnetic susceptibility, to within a few percent, are obtained already in fourth order perturbation theory in U/V which may be as large as 10^2 [18]!

Certain aspects of heavy fermions can be obtained from just such a priniciple of analyticity. As mentioned, the shape of the Fermi surface is given by self-consistent one-electron theory provided one treats the f-states as propagating states. Also the Fermi surface for rare-earth magnetic metals is correctly given by self-consistent one-electron theory provided one treats the f-states as local core-like states with frozen charges. The analyticity principle sometimes does not go too far. For instance, one may have certain global symmetries which are the same as those in a simpler problem but in addition have some additional local constraints which are not present in the simpler problem. The additional constraints may then not appear by doing perturbation theory about the simpler problem. For example, we will see in the problem of two Kondo impurities an asymptotic Fermi-liquid behavior but with a correlation between the magnetic moments. So far it has not been possible to see such correlations emerging in perturbation theory about the non-magnetic Hartree-Fock resonances.

12.5.1 Wilson's Method for the Kondo Problem

Wilson's renormalization group [13] generates a sequence of Hamiltonians suitable for describing accurately the properties of successively lower and lower temperatures. Consider the Kondo problem

$$H_K = J\vec{S}_0 \cdot \psi_0^\dagger \vec{\sigma} \psi_0 + t \sum_{i\sigma} \psi_{i\sigma}^\dagger \psi_{i+n,\sigma} \, , \tag{12.15}$$

where $i+n$ are neighbors of i. The effective Hamiltonians at intermediate temperatures are very complicated indeed, but for $T \to 0$ they approach the fixed point Hamiltonian

$$H^* = t \sum_{\substack{i \neq 0 \\ \sigma}} \psi^\dagger_{i\sigma} \psi_{i+n,\sigma} \,, \tag{12.16}$$

where the impurity is simply blocked out and one electron is effectively lost (it becomes incoherent) to the quasi-particles. In perturbation theory, this may be said to have happened because an effective $J \to \infty$ as $T \to 0$; this eliminates the local moment and an electron out of the problem. The asymptotic low temperature properties are obtained by adding to H^* the leading irrelevant operators:

$$H(\text{low } T) = H^* + \tilde{t}(\psi^\dagger_0 \psi_1 + \text{c.c.}) + \widetilde{U} n_{0\sigma} n_{0-\sigma} \,. \tag{12.17}$$

It is found that $\widetilde{U} = \tilde{t} \approx T^0_K \approx 3T_K$. The quantity $\tilde{t}$ then leads to a resonance of width T^0_K at the Fermi energy and $\widetilde{U}$ leads to quasi-particle interactions induced by the Kondo effect.

Nozières [10] observed that Eq. (12.17) is equivalent to scattering from a local potential with phase-shift:

$$\delta_\sigma(\epsilon) = \frac{\pi}{2} + 0.3\frac{\epsilon}{T_K} + \sum_\sigma \phi_{\sigma\sigma'} \delta n_{\sigma'} \,. \tag{12.18}$$

The Fermi-liquid term $\phi_{\sigma\sigma'}$ was directly related to T_K by assuming that the Kondo resonance moves rigidly with the chemical potential. This guarantees $\widetilde{U} = \tilde{t}$ in Eq. (12.17).

12.6 Problems in Applying Ideas from the Kondo Effect to Heavy Fermions

There are two principal questions which have been raised that we shall now discuss. The magnetic moments interact (via the conduction electrons) with each other through RKKY interaction. How then is one to talk about the Kondo effect of the individual moments? It used to be argued that if T_K is high compared to K_0 , the characteristic RKKY energy, the effect of the RKKY interactions may be neglected because below T_K the moments are locally compensated to singlets and such singlets do not interact with each other! This point of view is incorrect. The interaction between the moments modifies the Kondo effect in an interesting and important way which we shall soon discuss.

The other question arises from the idea, often discussed, of a "Kondo length" or the related idea of the "exhaustion principle" which confronts us if we believe that an electron in an energy scale within T_K of the Fermi energy is needed to compensate a moment. The concept of a Kondo length arises if we think of the Kondo effect as a bound state with energy T_K ; then the spatial extent of the bound state is the "Kondo length" v_F/T_K.

Actually there is no experiment nor any calculation [19] which shows the relevance of the concept of a Kondo length. I seriously doubt that if the ground state correlation function $\langle \vec{S} \cdot \vec{\sigma}(r) \rangle$ is calculated, any length scale besides k_F^{-1} will make its appearance. It is possible that for finite frequency (of the order of T_K) properties, such a length scale appears. The point may well be that the Kondo effect is not a bound state problem but a problem of resonance. The total phase-shift is simply fixed by Friedel's oscillations. It is also well to remember the compensation theorem [20], which states that no magnetic moment is ever induced in the conduction electrons.

On the other side, there is a considerable body of phenomenological argument that every moment is compensated in the heavy fermions. I think it is best to think that a given conduction electron state interacts with the local moment in each unit cell; the fact that the incoming electron state at a given unit cell arises from the interference due to similar scattering at the other cells merely changes the magnitudes of the parameters $V_\ell(\vec{k})$. This multiple interference of course gives rise to new bands in angular momentum channels of the local moments.

12.7 Theories Employing Kondo Resonances Plus Bloch's Theorem

Most of the approaches to the heavy fermion problem combine in one fashion or another the known solution to the single impurity Kondo problem with the requirement that the solution to a lattice problem obey Bloch symmetry. For an isolated impurity the Kondo problem is equivalent at low temperatures to the scattering of conduction electrons with the same local symmetry as that of the localized orbital with a phase shift given by Eq. (12.18). A local potential $V(r - R_i)$ can now be found to simulate the first two terms. If now the lattice problem is regarded as the scattering of conduction electrons by a periodic potential

$$V_{\text{lattice}}(r) = \sum_i V(r - R_i) ,$$

the solution is a hybridized band with width of order T_K as in Fig. 12.7. A slightly more complicated situation arises if orbital degeneracy *etc.* is considered.

The philosophy is implemented in one way or another in all of the following schemes:

1. Green's function decoupling [4].
2. $1/N$ expansions [21].
3. Self-consistent Green's function schemes using Kondo vertex [22].

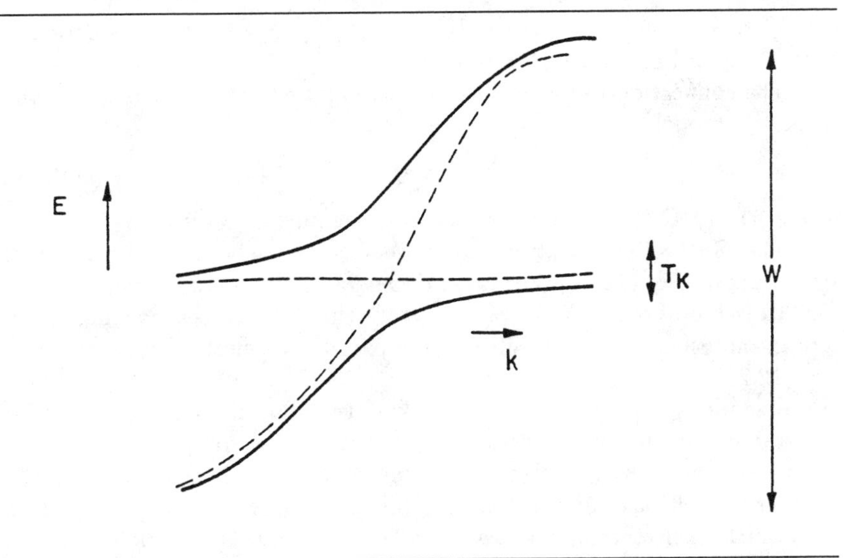

FIGURE 12.7
The band-structure in the pseudo one-electron solutions of the heavy fermion problem.

4. Gutzwiller variational schemes [23].
5. Most directly, phase shift from Kondo plus band structure calculations [24].

The Green's function decoupling schemes were the first to get results of the form discussed above in the mixed-valence context simply by writing down equations of motion and decoupling them in terms involving U which tend to block the motion of local electrons. These results are equivalent to dropping the term U in Eq. (12.2) by renormalizing the hybridization; *i.e.*, an effective Hamiltonian

$$H_{\text{eff}} = \sum_{k\sigma} \epsilon_k c^\dagger_{k\sigma} c_{k\sigma} + \sum_{i\sigma} \tilde{\epsilon}_0 \, a^\dagger_{i\sigma} a_{i\sigma} + \left(\sum_{k\sigma} \widetilde{V}_{k\sigma} \, c^\dagger_{k\sigma} a_{i\sigma} + \text{c.c.} \right), \quad (12.19)$$

with

$$\widetilde{V}_{k\sigma} = \langle 1 - n_{-i\sigma} \rangle^{\frac{1}{2}} V_k \,, \quad (12.20)$$

$$\tilde{\epsilon}_0 = \epsilon_0 + U \langle n_{-\sigma} \rangle \,. \quad (12.21)$$

The physics simply is that it takes two powers of V for a correlated particle to go from one site i to another site j. For U very large it can do so only if the site j is unoccupied. In mean field theory this hopping reduction is counted as in Eq. (12.20).

The connection to the Kondo problem arises from the fact that in the Anderson model

$$\widetilde{V}^2 \rho \approx T_K \ , \tag{12.22}$$

where $\rho \approx 1/W$ is the density of states of conduction electrons.

The Gutzwiller variational method has precisely the same effective Hamiltonian, Eq. (12.19). For large degeneracy, the blocking factor in Eq. (12.20) is modified as might be expected; this has the consequence of altering conditions for the stability of the pseudo one-electron type solution, Eq. (12.21).

One nice point illustrated by the variational solution is the reason for the stability of the simple pseudo one-electron type solution. As U is increased, the effective one-electron level moves up, Eq. (12.21), depleting the number of electrons in the localized orbitals and increasing the occupation of conduction electron states *above the Fermi energy*. This is the *incoherent part* of $\langle n_k \rangle$, which has a perfectly good discontinuity at the Fermi level consistent with Luttinger's theorem. Keeping U fixed, if one decreases ϵ_0 below the Fermi level, $\langle n_f \rangle$ increases as the cost of transferring electrons to the uncorrelated band increases. Ultimately $\langle n_f \rangle = 1$, so that $V = 0$ marking a transition to localized (magnetic) state of the correlated orbitals and a new Fermi surface for the conduction electrons.

The pseudo one-electron theories of the type discussed above have many virtues and a few serious shortcomings. They are based on the correct symmetry— that of a Fermi liquid— and therefore give various correct qualitative features of the experiments, like the Fermi surface; they have the correct scale of $d\Sigma/d\omega$, therefore the right order of magnitude of the specific heat and the susceptibility (although not the correct ratio between them). $d\Sigma/dk = 0$ in such theories— so they give the correct answers for the compressibility and the cancellations of renormalization in the transport properties as discussed earlier. The qualitative temperature dependence of the resistivity and its relationship to the specific heat is also explained.

The shortcomings have to do with the fact that they do not address the question of interactions between the Kondo resonances and the interplay of such interactions with the scattering of conduction electrons. They dispense with the local moments right away and with that the RKKY interaction between the moments. As we will see below the effect of such interactions cannot be treated in perturbation theory about the single impurity type solutions. The magnetic correlations and the excitation spectra produced by them are based on band-structures of the type shown in Fig. 12.7. The magnetic correlation are then of the Lindhard or RPA form:

$$\sum_k \frac{f_{k+q} - f_k}{\omega + \epsilon_{k+q} - \epsilon_k} g(k,q) ,$$

with ϵ_k taken from the "band structure" and $g(k,q)$ the appropriate matrix elements to take into account the variation of the orbital character in the bands with k. Since the excitation spectra and the quasi-particle interactions are incorrectly given by such theories, they predict incorrectly the instabilities— antiferromagnetic, superconducting *etc.*— of the heavy Fermi liquid.

12.8 Two Impurity Problem

In order to study the influence of the interactions among the local moments on their renormalization by the conduction electrons, we have solved the two Kondo impurity problem [25] by Wilson's method. Actually we take a somewhat simplified version of the two impurity problem— but one which, we believe, contains all the essential physics. Besides shedding light on the heavy fermion problem, the two impurity problem is very interesting in its own right. It belongs to the class of problems in which backward scattering is crucial and which are not soluble by known analytic methods. Earlier, we talked briefly of the path integral formulation of the single impurity Kondo problem. A similar formulation is possible for the two impurity problem. In this, (see Fig. 12.8) conduction electrons flip at times $t_{1A}, t_{2A}, \ldots$, corresponding to impurity spin at site A and at $t_{1B}, t_{2B}, \ldots$, corresponding to impurity at site B. The new and interesting physics arises because the sets of times $\{t_{iA}\}$ and $\{t_{iB}\}$ are self-consistently correlated. This implies that a phase relationship can develop between the two moments even while they dynamically respond to the conduction electrons. Indeed the numerical renormalization method solution discussed below shows such a phase relationship.

For two impurities it is best to work in a basis set of even and odd parity about the mid-plane between the two impurities. In terms of operators, c_{ke}, c_{ko} for even and odd states, and with the simplification of confining the magnitude of $\vec{k}$ to the Fermi surface, the two impurity problem is equivalent to

$$H = \sum_{k,k'} \Big\{ \left(\vec{S}_1 + \vec{S}_2\right) \cdot \left(J_e c^\dagger_{k'e} \vec{\sigma} c_{ke} + J_o c^\dagger_{k'o} \vec{\sigma} c_{ko}\right)$$

$$+ \left(\vec{S}_1 - \vec{S}_2\right) \cdot \left(iJ_m c^\dagger_{k'e} \vec{\sigma} c_{ko} + \text{h.c.}\right) \Big\} + H_e + H_o , \qquad (12.23)$$

where J_e and J_o are the coupling constants for the even and the odd chan-

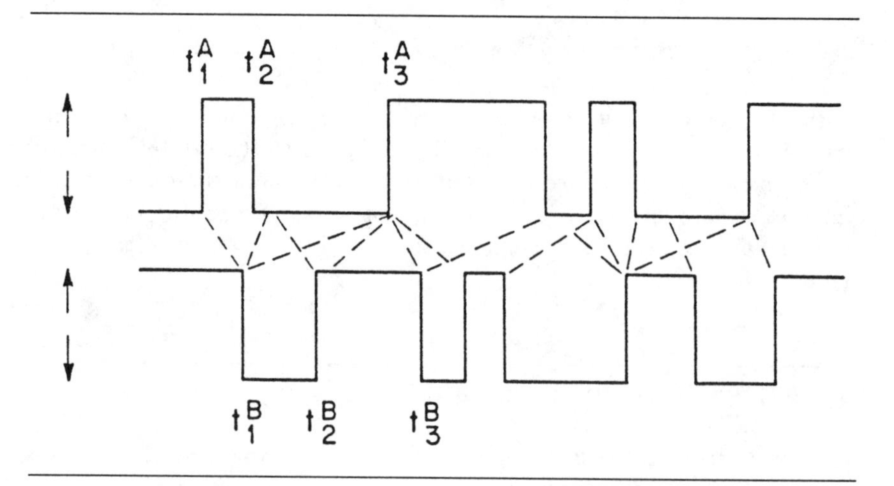

FIGURE 12.8
Path integral representation of the two Kondo impurity problem illustrating the correlations between the times of spin flips of conduction electrons at the two impurities.

nels, J_m mixes the two channels and scatters form the singlet to the triplet state of the two impurities, and H_e and H_o are the kinetic energy operators for the even- and odd-channel electrons. In this model the magnetic interaction between the two impurities is given by

$$I_0\,\vec{S}_1 \cdot \vec{S}_2 \quad \text{where} \quad I_0 = J_e^2 + J_o^2 - 2J_m^2 \,. \tag{12.24}$$

We found that the ground state of a pair of spin one-half impurities with nonzero separation is always a singlet. The principal physical results of this solution are the following.

a. For the initial interaction between the moments I_0, ferromagnetic (> 0) or antiferromagnetic (< 0), with $I_0 > -2.2T_K$, the low-temperature behavior is that of a correlated Kondo effect. The fixed-point Hamiltonian is that for two independent Kondo impurities, but the ground-state expectation value $\langle \vec{S}_1 \cdot \vec{S}_2 \rangle$ is not zero, as for two independent spins, even for $|I_0|/T_K \ll 1$. The quantitative behavior is given in Ref. [4]. The low-energy (or low-temperature) properties, determined by the leading irrelevant operators about the fixed point, depend on both I_0 and T_K as shown in Fig. 3 of Ref. 4. The universality in the properties characteristic of the single impurity Kondo problem is lost in the two impurity problem.

The phase shift of the conduction electrons in this regime is asymptotically $\frac{\pi}{2}$ in both the even and odd parity channels, and larger than $\frac{\pi}{2}$ in one and less than $\frac{\pi}{2}$ in the other, with the sum equal to π as naively expected for interacting resonances. At the fixed point there is then particle-hole symmetry in each channel separately. The fixed-point Hamiltonian contains no coupling between the odd and the even channels— the problem asymptotically becomes that of two pointlike orthogonal scatterers.

b. For larger antiferromagnetic interactions $I_0 < -2.2T_K$, no Kondo effect occurs. The asymptotic phase shift is zero. The two magnetic moments are not in a singlet state until $I_0 \ll -T_K$, but the total spin including the conduction electrons is zero. The Kondo correlations clearly persist in this regime.

c. The transition between the regimes described in items **a** and **b** is marked by a unstable fixed point (a critical point) at which the staggered susceptibility and the specific-heat coefficient γ diverge. The ground-state correlation $\langle \vec{S}_1 \cdot \vec{S}_2 \rangle$ approaches $-\frac{1}{4}$. The unstable fixed point is a new kind of fixed point at and around which a local description of the impurity and conduction electron degrees of freedom in terms of a local Fermi liquid breaks down.

The behavior in the very strongly ferromagnetic or antiferromagnetic regimes $|I_0| \gg T_K$ is straightforward. For strong antiferromagnetic I_0 the two moments lock to a singlet at a temperature well above T_K. This singlet is then decoupled from the conduction electrons and may be treated perturbatively. The surprise in the solution is how slowly the behavior approaches this limit as $|I_0|/T_K$ increases. For strong ferromagnetic coupling $I_0 \gg T_K$, the two impurities form an $S = 1$ state which undergoes a two-channel Kondo renormalization to ultimately a singlet state for the whole system, but with $\langle \vec{S}_1 \cdot \vec{S}_2 \rangle$ close to the triplet value of $\frac{1}{4}$.

Of the new results noted in items **a**, **b** and **c** above, especially interesting is the existence of the critical point at a finite value of a coupling constant, noted in **c**. A physical explanation of the observed behavior has recently been given [26].

Just as for the Kondo problem the asymptotic low temperature behavior for the two impurity problem is expressible by an effective Hamiltonian consisting of a fixed point Hamiltonian H^*, a one-electron Hamiltonian $H^{(1)}$ and an interaction Hamiltonian H^{int},

$$H_{\text{eff}}(12) = H^* + H^{(1)} + H^{\text{int}} , \tag{12.25}$$

where H^* describes the ground state. In the ground state we lose an electron both in the even parity and the odd parity channel. This is in itself a remarkable result which is quite contrary to what one gets from considering the problem as a "sum" of two one impurity Kondo problems. In this case the scattering from the left impurity interferes with the scattering from the right impurity such that an even parity combination of states moves down in energy with respect to the odd parity combination of states. Integrated up to the chemical potential, the even parity resonance would then have more than one electron and the odd parity channel less than one electron. The sum in the two channels is fixed by Friedel sum rule to be two. Our result that there is one electron in each channel is therefore very interesting. This means the two impurity problem cannot be represented by local potentials at the two sites.

$H^{(1)}$ describes the hopping of quasi-particles in the even and odd channels and H^{int} the interactions among the quasi-particles. These interactions are the intrachannel and interchannel repulsions and the interchannel exchange. The nature of the interchannel exchange of quasi-particles reflects the nature of the interaction among the moments, *i.e.*, I_0. It is antiferromagnetic for I_0 antiferromagnetic and ferromagnetic for I_0 ferromagnetic. The parameters in $H^{(1)}$ and H^{int} reflect the intial values J_e, J_o as well as I_0.

The two impurity problem has been solved with the simplifying assumption that the exchange matrices in the even, odd and the mixing channel are momentum independent. For the single impurity problem, it has been shown [27] that any momentum dependence can be removed by going to a new basis set at the expense of introducing a potential scattering term. The total asymptotic phase shift then is, remarkably, still $\frac{\pi}{2}$ from the spin channel plus the phase shift due to potential scattering plus that due to loss of particle-hole symmetry. I expect that the same is true for the two impurity problem, with the net result that the even and the odd parity resonances, having different effective potential scattering, are displaced with respect to each other. This point remains to be proved. If proved, it gives the proper genesis of the heavy fermion bands in the lattice problem.

12.9 The Lattice Problem

We have obtained the effective quasi-particle Hamiltonian for the problem of two magnetic impurities in a conduction electron sea. It is expressed in terms of even and odd states about the two impurities. It can in turn be re-expressed in terms of a tight binding basis set for quasi-particles in the lattice with the two magnetic sites i and j singled out. Let us denote this by $H_{ij}(1,\ldots,N)$. I believe that the essential low temperature physics of the heavy fermion lattice problem at low temperatures is given by

$$H_{\mathrm{eff}} = \sum_{i<j} H_{ij} \,. \tag{12.26}$$

The argument is: Suppose we solved the three impurity problem and obtained an effective low temperature Hamiltonian H_{ijk}. This can be decomposed into

$$H_{ijk} = \overline{H}_{ij} + \overline{H}_{jk} + \overline{H}_{ki} + \overline{H}_{ijk} , \tag{12.27}$$

where the first three terms are the parts of H_{ijk} which can be expressed in terms of pairwise interactions and the last cannot. The parts $\overline{H}_{ij}$ have the same symmetries as H_{ij} and simply renormalize the coefficients in the latter. Therefore they merely provide numerical improvement and do not add anything new. The last part consists of interactions with six (irreducible) legs. In Fermi-liquid theory they are never considered since they provide corrections to thermodynamics and transport properties that are higher powers of T/E_F than the four-legged terms in H_{ij}.

H_{eff} of Eq. (12.26) therefore contains all the essential physics of the heavy fermion problem at low temperature. All Ward identities and correlation functions and instabilities can be discussed in terms of it, but not the numerical values in any given experiment. H_{eff} thus provides a Fermi-liquid theory for heavy fermions much like the Fermi-liquid theory for ^{3}He in which only interrelationships among various normal state properties and indications of magnetic or superconducting instabilities are provided. To compare with experiments on actual materials, the analog of H_{eff} including the effects of orbital degeneracy, crystal fields and spin-orbit coupling must be derived or guessed.

12.10 Correspondence with Experiments

The theory predicts magnetic correlations among the quasi-particles at low temperatures. This is also one of the most important recent experimental discoveries [28]. The magnetic correlations have the character of itinerant electrons yet they are characteristic of interaction between quasi-particles at near neighbor sites. This is predicted by H_{eff} and unlike the correlations of the pseudo one-electron theories. In such theories, the correlations arise essentially from a Lindhard function calculated from a one-electron band structure.

The theory, of course, predicts everything that can be predicted from the pseudo one-electron approach and shows the subtleties of the problem beyond it. The density of states of the quasi-particles is specified by a modified Kondo temperature due to the magnetic correlations. The band structure of the quasi-particles is to be obtained from a non-local potential. One of the consequences of the underlying $\langle \vec{S}_i \cdot \vec{S}_j \rangle$ correlations is that it is not the individual moments that are compensated; those separated by $|i-j|$

such that I_0 is significant are collectively compensated. I believe that the non-locality of the quasi-particle and such collective compensation are the essential ingredients for an understanding of the remarkable impurity effects observed. A (non-magnetic) impurity causes potential scattering over the region of the non-locality. The result is to change the scattering cross-section from the unitary limit πk_F^2 to $\sigma_{\text{pot}} \approx \pi a^2$, where a is the range of non-locality which will typically be on the scale of one or two lattice constants. To this we must add the spin dependent cross-section of similar magnitude. This can be separated out by examining the magnetic field dependence of the impurity resistivity.

The collective compensation has the effect that even a non-magnetic impurity uncovers or deconfines the local moments around it. We believe it is for this reason that spin glass or antiferromagnetic behavior is discovered with non-magnetic impurities.

H_{eff} leads for large enough initial I_0 to antiferromagnetic instabilities among the heavy quasi-particles. The magnetic structures are again based on simple near-neighbor interactions which is remarkable for itinerant fermions. This is consistent with experiments [28].

The antiferromagnetic correlations of the quasi-particles and the enhancement of the large momentum transfer susceptibility has important consequences in the particle-particle channel as well. One of the big puzzles in heavy fermion superconductivity has been the fact that various experimental results could only be understood [29] on the basis of an anisotropic superconducting state in which the gap function has a line, or lines, of zeroes at the Fermi surface. On the one hand such states are disallowed [30] in the spin triplet manifold. On the other hand, the heavy fermion phenomenon is dominated by spin fluctuations and there is strong prejudice from our experience in liquid ^{3}He that spin fluctuations [31] promote triplet superconductivity. This puzzle has now been resolved. Antiferromagnetic spin fluctuations promote neither conventional singlet superconductivity nor triplet superconductivity. They do promote anisotropic singlet states all of which have lines of zeroes of the gap function on the Fermi surface. Impressive confirmation of anisotropic superconducting states has now been obtained by a set of experiments which see multiple superconducting antiferromagnetic transitions. These have been understood [32] in a phenomenological theory including the effects of vector superconducting and antiferromagnetic order parameters.

In conclusion, it seems to us that the basic principles for a qualitative understanding of all the phenomena observed in heavy fermions are now in place. For a comprehensive understanding of experimental results, phenomenological Hamiltonians of the form given by Eq. (12.26) including the effects of orbital degeneracy, crystal fields and spin-orbit coupling need to be studied.

Acknowledgements

The point of view expressed here is the result of collaborations with B. A. Jones, E. Abrahams, K. Miyake and S. Schmitt-Rink and discussions with G. Aeppli, P. W. Anderson, B. Batlogg, D. Bishop, Z. Fisk, H. R. Ott and F. Steglich.

Bibliography

[1] C.M. Varma, in *Fermi Surface Effects*, ed. J. Kondo and A. Yoshimori (Springer Verlag, Heidelberg, 1987).

[2] C.M. Varma, in *Windsurfing the Fermi Sea*, Vol. I, ed. T.T.S. Kuo and J. Speth (North Holland, Amsterdam, 1986).

[3] For a beautiful review of Fermi-liquid theory, see *The Theory of Quantum Liquids*, D. Pines and P. Nozières (W.A. Benjamin, NY, 1966).

[4] C.M. Varma and Y. Yafet, Phys. Rev. **B13** (1976) 2950.

[5] For a review of experimental results on heavy fermions, see for example, G. Stewart, Rev. Mod. Phys. **56** (1984) 755.

[6] For a review, see C.M. Varma, Rev. Mod. Phys. **48** (1976) 219.

[7] J.R. Schrieffer and P. Wolff, Phys. Rev. **149** (1966) 491.

[8] K. Yamada, K. Yosida and K. Hanzawa, Prog. Theor. Phys. **75** (1984) 450.

[9] C.M. Varma, Phys. Rev. Lett. **55** (1985) 2723.

[10] P. Nozières, J. Low Temp. Phys. **17** (1974) 31.

[11] C. Pethick and D. Pines, priv. comm.

[12] P.W. Anderson, Comments in Solid State Physics **5** (1973) 73.

[13] K. Wilson, Rev. Mod. Phys. **47** (1975) 773.

[14] N. Andrei, K. Furuya and J.H. Lowenstein, Rev. Mod. Phys. **55** (1985) 331.

[15] A.M. Tsvelick and P.B. Wiegmann, Adv. Phys. **32** (1983) 453.

[16] K. Yamada and K. Yosida, Prog. Theor. Phys. Suppl. **46** (1970) 244.

[17] P.W. Anderson, Phys. Rev. **124** (1961) 41.

[18] V. Zlatic and B. Horvatic, Phys. Rev. **B28** (1983) 6904.

[19] See for instance, H. Ishii, Prog. Theor. Phys. **55** (1976) 1373.

[20] The Anderson-Clogston compensation theorem is discussed by H. Shiba, Prog. Theor. Phys. **54** (1975) 967.

[21] P. Coleman, Phys. Rev. **B29** (1984) 3035; A. Auerbach and K. Levin, Phys. Rev. Lett. **57** (1986) 877; A. Millis and P.A. Lee, Phys. Rev. **B35** (1987) 3394.

[22] K. Miyake, T. Matsuura, H. Jichu and Y. Nagaoka, Prog. Theor. Phys. **72** (1984) 1063; H. Jichu, T. Matsuura, and Y. Kuroda, *ibid* **72** (1984) 366; H. Fukuyama, *Theory of Heavy Fermions*, ed. T. Kasuya and T. Saso (Springer Verlag, NY, 1985).

[23] C.M. Varma, in *Moment Formation in Solids*, ed. W.J. Buyers (Plenum, NY, 1984); C.M. Varma, W. Weber and L.J. Randall, Phys. Rev. **B33** (1986) 1015; T.M. Rice and K. Ueda, Phys. Rev. Lett. **55** (1985) 995; B.H. Brandow, Phys. Rev. **B33** (1986) 215; P. Fazekas, preprint; H. Shiba, preprint.

[24] H. Razahfimandimby, P. Fulde and J. Keller, Z. Phys. **B54** (1984) 111.

[25] B.A. Jones and C.M. Varma, Phys. Rev. Lett. **58** (1987) 843; B.A. Jones, C.M. Varma and J.W. Wilkins, Phys. Rev. Lett. **61** (1988) 125.

[26] B.A. Jones and C.M. Varma, Phys. Rev. **B40** (1989) 324.

[27] D.M. Cragg and P. Lloyd, J. Phys. **C12** (1979) 3301.

[28] G. Aeppli *et al.*, Phys. Rev. **B32** (1985) 7579; Phys. Rev. Lett. **58** (1987) 808.

[29] S. Schmitt-Rink, K. Miyake and C.M. Varma, Phys. Rev. Lett. **57** (1985) 2575; P. Hirschfelder, D. Volhardt and P. Wolfle, Solid State Comm. **59** (1986) 111.

[30] G.E. Volovik and L.P. Gorkov, JETP Lett. **39** (1984) 550; E.I. Blount, Phys. Rev. **B32** (1985) 2935.

[31] K. Miyake, S. Schmitt-Rink and C.M. Varma, Phys. Rev. **34** (1986) 6554; D.J. Scalapino, E. Loh, Jr. and J.E. Hirsch, *ibid*, **B34** (1986) 3480.

[32] The experimental references and the theory may be found in E.I. Blount, C.M. Varma and G. Aeppli, Phys. Rev. Lett. (1989) to be pub.

13

Electroweak Interactions With Nuclei

J. D. Walecka
Continuous Electron Beam Accelerator Facility
Newport News, VA 23606

13.1 Introduction

Let me begin at the beginning. Why do we do nuclear physics? Why is nuclear physics interesting? The atomic nucleus is a unique form of matter. It consists of many baryons in close proximity. All the forces of nature are present in the nucleus— strong, electromagnetic, weak, and gravitation (if we include neutron stars, which are nothing more than enormous nuclei held together by the gravitational attraction). Nuclei provide unique microscopic laboratories in which to test the structure of the fundamental interactions. In addition, the nuclear many-body problem is of intrinsic intellectual interest. Furthermore, most of the mass and energy in the visible universe comes from nuclei and nuclear reactions. If the goal of physics is to understand nature, then surely we must understand the nucleus. Finally, in sum, nuclear physics is the study of the *structure of matter*.

Why do we study electroweak interactions with nuclei [1–3]? The last decade and a half has seen the development of a unified theory of the electroweak interactions [4–6]. This is surely one of the great intellectual achievements of our era. The basis for this unification lies in a local gauge theory built on the symmetry structure $SU(2)_W \otimes U(1)_W$. It is essential to continue to put this theory to rigorous tests in all possible domains. In order

TABLE 13.1
β-decay rates for ^{6}He [9].

$\omega_\beta = 0.877 \pm 0.023$ sec^{-1}	Theory
$= 0.864 \pm 0.003$ sec^{-1}	Experiment

to do this, we need intense sources of electroweak probes (e^-, ν_ℓ, μ^-). Once the nature of the fundamental interaction is understood, the electroweak interaction provides a powerful tool for studying nuclear structure. We have a clean probe, and we know what we measure.

The current picture of the strong interactions is based on quarks and gluons as the underlying set of degrees of freedom. The theory of the strong interactions binding quarks into the observed hadrons, baryons and mesons, is quantum chromodynamics (QCD). It is a Yang-Mills non-abelian gauge theory based on a threefold local color symmetry— $SU(3)_c$ [7]. The symmetry structure of the "Standard Model" of the strong and electroweak interactions is thus $SU(3)_c \otimes SU(2)_W \otimes U(1)_W$. We will discuss how electroweak interaction with nuclei can be used to probe this assumed structure of the fundamental interactions. We will see how nuclei can be used to study the strong interaction, confinement aspects of QCD.

Let me start the discussion by defining the "traditional approach" to the nuclear many-body problem as developed in detail, for example in [8]. The underlying set of degrees of freedom is assumed to consist of structureless nucleons, the proton p and the neutron n. One starts from static two-body potentials fit to two-body scattering data. The non-relativistic, many-particle Schrödinger equation is then solved in some approximation, providing energy levels and wavefunctions for the system. One then constructs electroweak currents from the properties of free nucleons, and these are used to probe the system. Here a unified analysis of electromagnetic and weak interactions with nuclei has proved rewarding [3]. In selected cases, it is possible to achieve an accuracy of 5-10% using this approach. As an example, consider the processes illustrated in Fig. 13.1. An analysis of electromagnetic processes in ^{6}Li allows one to completely determine the nuclear transition densities within the p-shell, and an absolute calculation of the weak rates and cross sections with this system is then possible [9]. The β-decay rate is given in Table 13.1 and the prediction for the charge-changing neutrino cross sections is shown in Fig. 13.2.

The traditional approach to nuclear physics has had a great many successes; however, it is inadequate for a more detailed understanding of

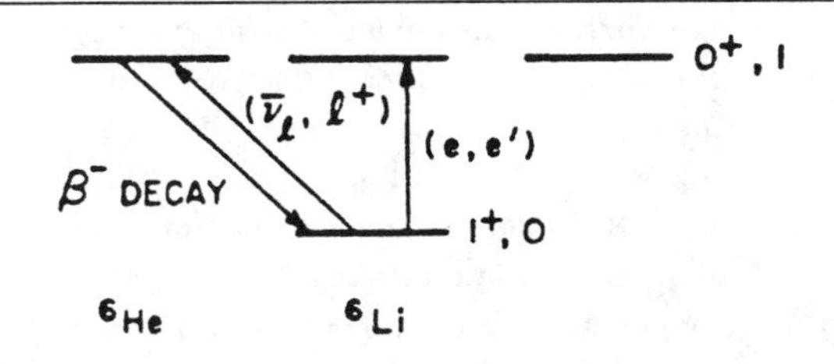

FIGURE 13.1
Electroweak processes for the A=6 nuclei.

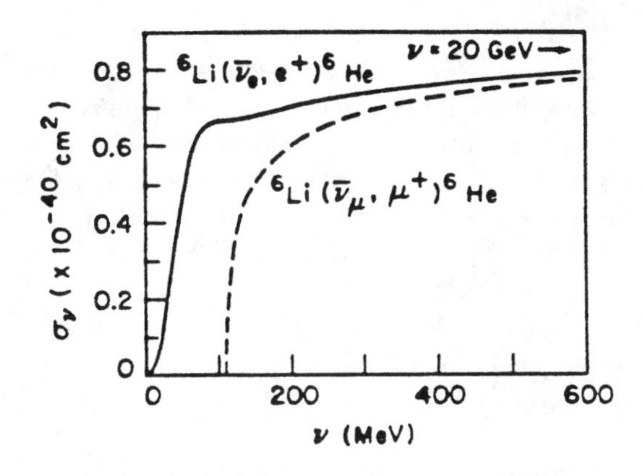

FIGURE 13.2
Antineutrino cross sections on ^{6}Li *vs.* neutrino energy [9]. The high energy limit is indicated by the arrow on the right.

nuclear structure. It is an oversimplification to picture the nucleus as a collection of free nucleons interacting through static potentials obtained from two-body scattering data. A more appropriate set of degrees of freedom for the nucleus consists of the *hadrons*, the strongly interacting baryons and mesons. One can give many reasons for this. For example, the long-range part of the Paris potential, probably the most accurate two-nucleon potential currently available, is derived from the exchange of mesons, predominantly $\pi(J^\pi = 0^-, T = 1)$, $\sigma(0^+, 0)$, $\omega(1^-, 0)$ and $\rho(1^-, 1)$. We have *experimental proof* that the long-range part of this interaction is governed by meson exchange. In addition, one of the most important recent advances in nuclear physics is the unambiguous identification through electron scattering (e, e') of exchange currents in nuclei. These are additional currents present in the many-baryon system which arise from the flow of charged mesons between baryons.

Furthermore, an important goal of nuclear physics is to study the properties of nuclear matter under *extreme conditions*. This is crucial in astrophysics, for example, where one studies the properties of condensed stellar objects, and supernovae. It is also central to high-energy heavy-ion reactions, where high density and high temperature properties of nuclear matter are studied under laboratory conditions.

In any extended description of nuclear structure, it is important to incorporate *general principles* of physics, in particular, quantum mechanics, special relativity, and microscopic causality. The only consistent theoretical framework we have for dealing with such a relativistic, interacting many-particle system is *relativistic quantum field theory based on a local lagrangian density*. I like to refer to such a field theory formulated in terms of the hadronic degrees of freedom as *quantum hadrodynamics* (QHD).

The theory of QCD is simple at short distances, where the theory is essentially free (asymptotic freedom). It is a complicated, strong-coupling theory at large distances (confinement) [7]. An important thesis of this lecture is that the appropriate set of degrees of freedom depends on the distance scale *at which we probe the nucleus*.

13.2 Electron Scattering

Why does electron scattering (e, e') provide such a powerful tool for studying nuclear structure [1]? First, the interaction is known; it is given by quantum electrodynamics (QED), the most accurate physical theory we have. Second, electrons provide a clean probe; we know what we measure. The electrons interact with the local electromagnetic charge and current densities in the target. Third, the interaction is relatively weak, of the order of the fine structure constant α, and one can make measurements without greatly disturbing the structure of the target.

In electron scattering one measures a macroscopic diffraction pattern in the laboratory, and this pattern is essentially the Fourier transform of the static and transition charge and current densities. Upon inversion of the Fourier transform, one has a determination of the detailed microscopic spatial distribution of the charge and current densities themselves. This is a unique source of information about nuclear structure.

Electron scattering also provides a versatile probe. There is kinematic flexibility (discussed below). Furthermore, in addition to the Coulomb interaction with the charge density and the transverse-photon-exchange interaction with the convection current density, there is an interaction with

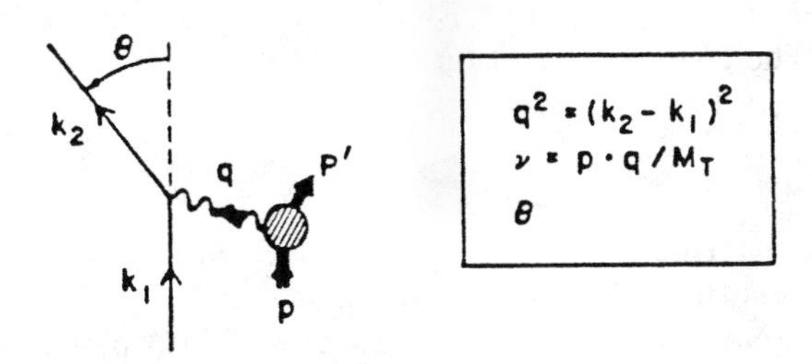

FIGURE 13.3
Kinematic situation for target response in semileptonic processes.

the magnetization current density in the nucleus resulting from the intrinsic spin magnetization of the nucleons.

Let us first concentrate on the lepton variables in the process (e, e') as illustrated in Fig. 13.3. There are three free variables (k_1, k_2, θ), the initial and final wave numbers of the electrons and the scattering angle, or equivalently (ν, q^2, θ), where $\nu = q \cdot p/M_T = \epsilon_1 - \epsilon_2$ is the electron energy loss in the laboratory system, q^2 is the four-momentum transfer, and Θ is the scattering angle. The first two combinations in the latter set are Lorentz invariants.

The cross section for electron scattering can be written in the form $d\sigma \propto \eta_{\mu\nu} W_{\mu\nu}$. Here $\eta_{\mu\nu}$ is a known tensor constructed from the lepton variables. The target response is completely characterized in terms of the response tensor defined by

$$W_{\mu\nu} = (2\pi)^3 V \overline{\sum_i} \sum_f \langle i|J_\nu(0)|f\rangle\langle f|J_\mu(0)|i\rangle(E)\delta^{(4)}(p' - p + q) \,, \quad (13.1)$$

where $J_\mu(x) = (\vec{J}(x), i\rho(x))$ is the electromagnetic current density in the target. The average over initial nuclear states, and sum over final states, simplifies this tensor, since it can only be constructed from the two independent four vectors p_μ and q_μ remaining in the problem. The coefficients in this expansion must be functions of the two Lorentz scalars q^2 and $q \cdot p$. In this fashion one is led directly to the basic electron scattering cross section, a result which I hope is familiar to all of you.

$$\left(\frac{d^2\sigma}{d\Omega_2 d\epsilon_2}\right)^{\text{ERL}}_{ee'} = \frac{\alpha^2 \cos^2 \frac{\theta}{2}}{4\epsilon_1^2 \sin^4 \frac{\theta}{2}} \frac{1}{M_T} \left[W_2(q^2, q \cdot p) + 2W_1(q^2, q \cdot p) \tan^2 \frac{\Theta}{2}\right] . \quad (13.2)$$

The extreme relativistic limit (ERL) shall mean that the lepton mass is neglected. The Mott cross section

$$\sigma_M = \frac{\alpha^2 \cos^2 \frac{\theta}{2}}{4\epsilon_1^2 \sin^4 \frac{\theta}{2}} \tag{13.3}$$

is that for scattering a relativistic Dirac electron from a point charge. The two response surfaces W_1 and W_2 can be separated by keeping their arguments fixed and making a straight-line Rosenbluth plot against $\tan^2 \theta/2$, or by working at $\theta = 180°$ where only W_1 contributes. The assumptions that go into the derivation of this result are general: one-photon exchange, Lorentz invariance and current conservation.

For transitions to a discrete state of the target, one can perform the integration over final electron energies

$$\int \frac{d\epsilon_2}{M_T} W_i(q^2, q \cdot p) = \omega_i(q^2) r \; , \tag{13.4}$$

where the recoil factor is defined by

$$r^{-1} = 1 + \frac{2\epsilon_1 \sin^2 \frac{\theta}{2}}{M_T} \; . \tag{13.5}$$

Let us assume further that the nuclear transition densities are localized in space and that the initial and final nuclear states are eigenstates of angular momentum. A multipole decomposition of the electromagnetic interaction can be carried out with the result that [1]

$$\omega_1 = \frac{2\pi}{2J_i + 1} \sum_{J=1}^{\infty} \left(|\langle J_f || \widehat{T}_J^{\text{mag}}(q) || J_i \rangle|^2 + |\langle J_f || \widehat{T}_J^{\text{el}}(q) || J_i \rangle|^2 \right) \frac{E'}{M_T} \; , \tag{13.6}$$

$$\omega_2 - \frac{q^2}{\vec{q}^{\,2}} \omega_1 = \frac{q^4}{\vec{q}^{\,4}} \frac{4\pi}{2J_i + 1} \sum_{J=0}^{\infty} |\langle J_f || \widehat{M}_J(q) || J_i \rangle|^2 \frac{E'}{M_T} \; . \tag{13.7}$$

The multipole operators are defined by (here $q \equiv |\vec{q}|$)

$$\widehat{M}_{JM}(q) = \int j_J(qx) Y_{JM}(\Omega_x) \hat{\rho}(\vec{x}) d\vec{x} \; , \tag{13.8}$$

$$\widehat{T}_{JM}^{\text{el}}(q) = \frac{1}{q} \int \nabla \times \left[j_J(qx) \vec{Y}_{JJ1}^M(\Omega_x) \right] \cdot \hat{\vec{J}}(\vec{x}) d\vec{x} \; , \tag{13.9}$$

$$\widehat{T}_{JM}^{\text{mag}}(q) = \int \left[j_J(qx) \vec{Y}_{JJ1}^M(\Omega_x) \right] \cdot \hat{\vec{J}}(\vec{x}) d\vec{x} \; . \tag{13.10}$$

They are irreducible tensor operators in the nuclear Hilbert space. This decomposition permits one to extract the M-dependence and angular mo-

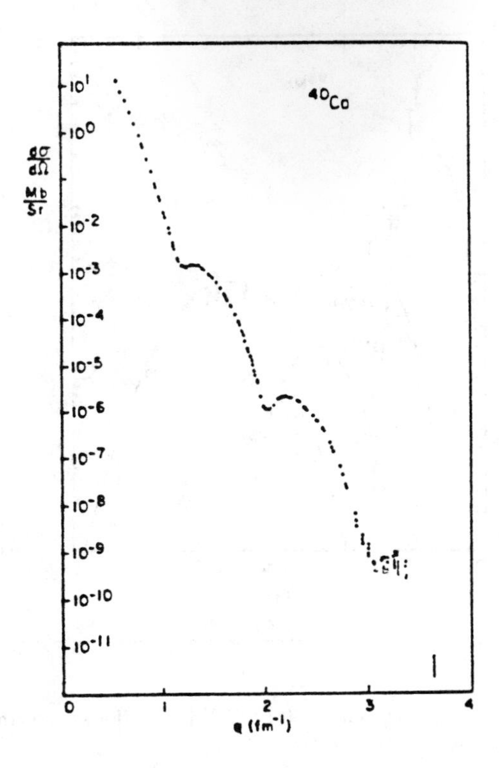

FIGURE 13.4
Elastic (e, e) cross section for ^{40}Ca *vs.* momentum transfer [10]. The scattering here is from the charge distribution.

mentum selection rules for the nuclear matrix elements through the use of the Wigner-Eckart theorem. The longitudinal multipoles

$$\hat{L}_{JM}(q) = \frac{i}{q} \int \left[\nabla \left(j_J(qx) Y_{JM}(\Omega_x) \right) \right] \cdot \vec{J}(\vec{x}) d\vec{x} \tag{13.11}$$

have been eliminated here through the use of current conservation.

We briefly discuss some selected examples of electron scattering results. Fig. 13.4 shows the measured elastic monopole charge scattering cross section for ^{40}Ca [10]. Note the logarithmic scale on the left hand side of this figure! This is a state-of-the-art example of a measurement of the macroscopic diffraction pattern referred to above. The high q data comes from Saclay.

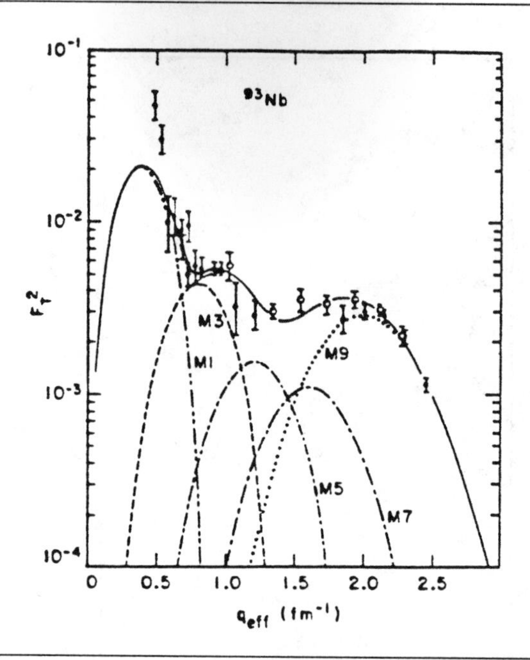

FIGURE 13.5
Elastic transverse (e, e) form factor for ^{93}Nb [11]. The scattering here is from the magnetization distribution.

Fig. 13.5 shows elastic magnetic scattering from ^{93}Nb [11] whose shell-model configuration is $(1g_{9/2})$. Note how all the magnetic multipoles show up as the momentum transfer is increased (parity and time reversal limit one to odd magnetic multipoles in elastic scattering). A measurement of the elastic magnetic cross section at all q allows one to determine the spatial distribution of the ground-state magnetization, and hence of the last valence proton in this nucleus. One can actually see that there is a little current loop in the ground state of this nucleus! The high q data is from Bates.

Fig. 13.6 shows an inelastic spectrum from ^{24}Mg taken at large angle and large momentum transfer [12]. Note how the state at 15.0 MeV dominates the spectrum. It can be identified as a stretched, high-spin magnetic excitation, in this case predominantly to the $[(1d_{5/2})^{-1}(1f_{7/2})]_{6^-1}$ configuration (The large isovector magnetic moment of the nucleon implies isovector excitations dominate under these conditions). Fig. 13.7 shows the inelastic form factor for this state, the area under the peak, as a function

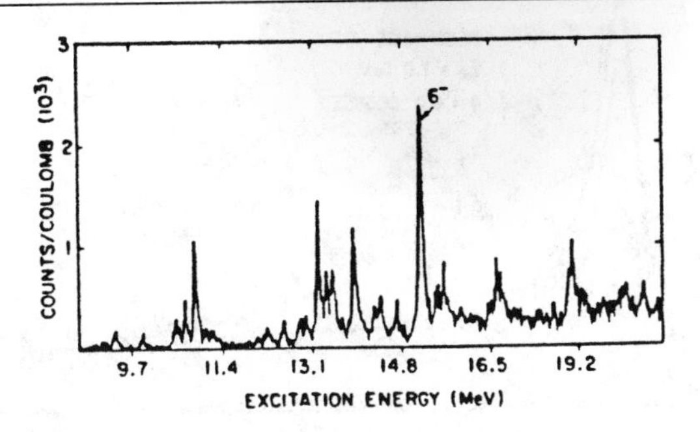

FIGURE 13.6
^{24}Mg (e, e') data taken at $\theta = 160^\circ$ and $q = 2.13$ fm^{-1}. From Ref. [12].

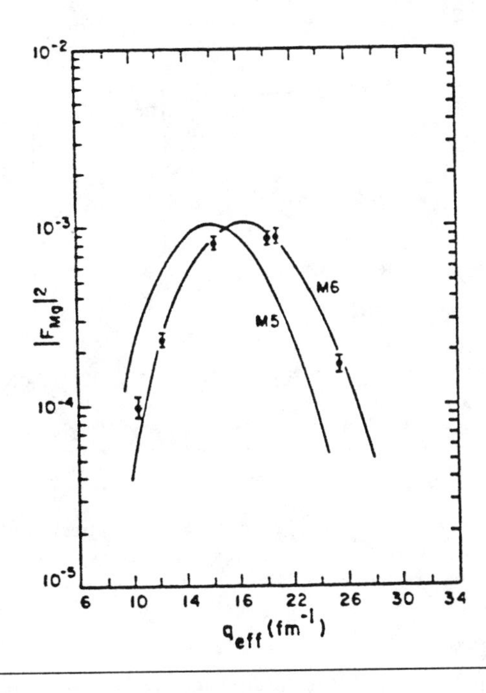

FIGURE 13.7
Transverse form factor *vs.* q for the state at 15.0 MeV in Fig. 13.6 [12].

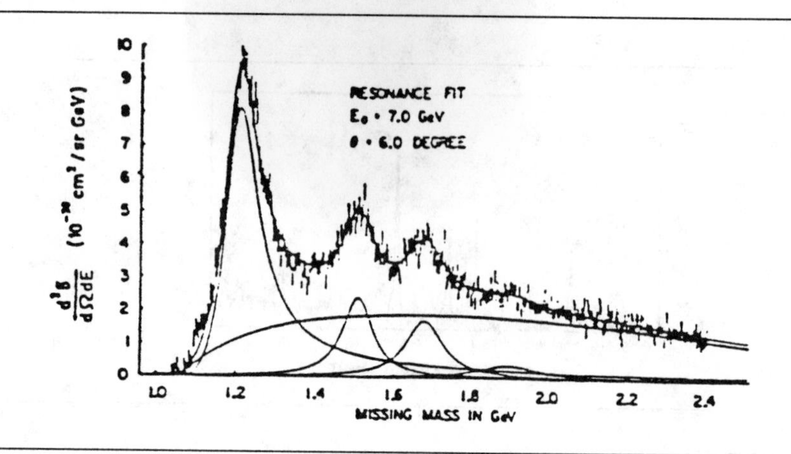

FIGURE 13.8
The SLAC experimental inelastic spectrum at $\epsilon_1 = 7$ GeV, $\theta = 6^\circ$, resolved into Breit-Wigner resonances [13].

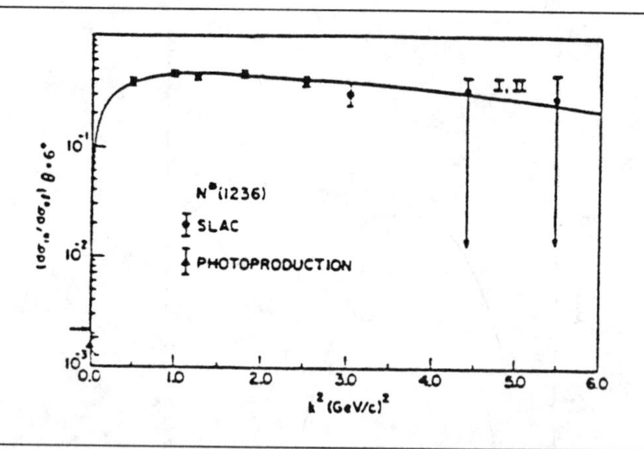

FIGURE 13.9
$d\sigma_{in}/d\sigma_{el}$ at 6° for the $3/2^+, 3/2(1236)$ resonance. Experimental points are from SLAC Group A and the resonance analysis of Breidenbach [14]. The theoretical curves are from Ref. [15].

of momentum transfer. It exhibits a characteristic M6 multipolarity. The data was taken at Bates.

Fig. 13.8 shows the proton resonance spectrum observed in inelastic electron scattering at SLAC at $E_o = 7.0$ GeV and $\theta_e = 6.0^\circ$ [13,14]. Note the elastic peak has been suppressed in this figure. Fig. 13.9 shows the ratio

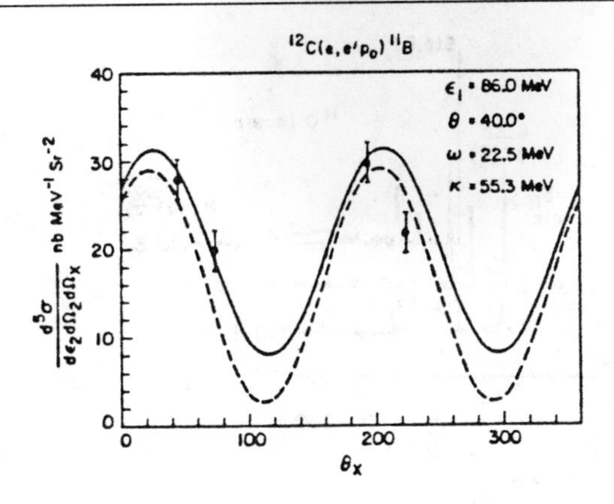

FIGURE 13.10
^{12}C$(e, e'p_0)$ cross section at $\omega = 22.5$ MeV and $q = 55.3$ MeV. The data and theory are from [16,17].

of the inelastic to elastic cross section for electroexcitation of the $\Delta(1236)$. The theoretical curve is from Ref [15].

Coincidence electron scattering $(e, e'X)$ where the angular distributions of one or more emitted hadrons are measured with respect to the momentum transfer $\vec{q}$ will form a major future thrust of the field of electromagnetic interactions with nuclei. An extensive discussion of the general phenomenological analysis of such processes is developed in Ref [1]. Fig. 13.10 shows coincidence data for $(e, e'p)$ on the giant dipole resonance in ^{12}C [16,17]. A characteristic dipole pattern is seen for the protons. Such experiments at higher momentum transfer promise to exhibit other collective oscillations.

Fig. 13.11 shows a mapping of the longitudinal response surface for ^{16}O as a function of energy loss and momentum transfer in the reaction $(e, e'p)$ [18]. Measurement of the final energies determines the excitation energy of the residual nucleus, and the form factor then measures the Fourier transform of the resulting hole state. Such experiments provide our most direct studies of the single particle structure of nuclei. The limitation in all the existing coincidence experiments is duty factor and kinematic flexibility. This is a prime motivation for the new generation of electron accelerators and stretcher rings for nuclear physics.

As one example of the need for relativistic models of nuclear structure, we discuss an extension of the traditional analysis to allow for additional *two-body* currents arising from the exchange of charged mesons between

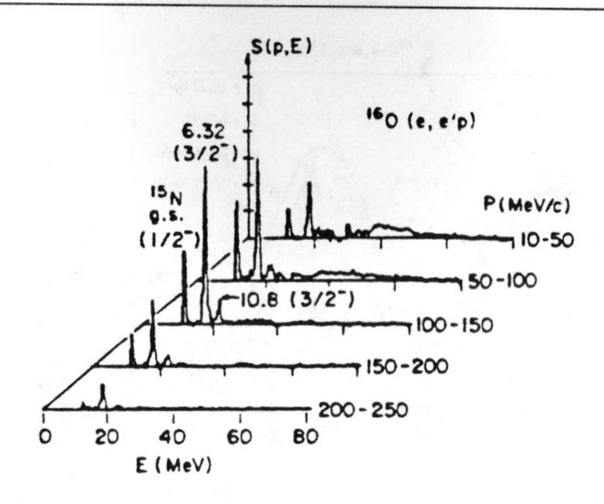

FIGURE 13.11
Longitudinal response surface for the $^{16}\mathrm{O}(e, e'p)$ reaction *vs.* energy and momentum of residual hole state [18].

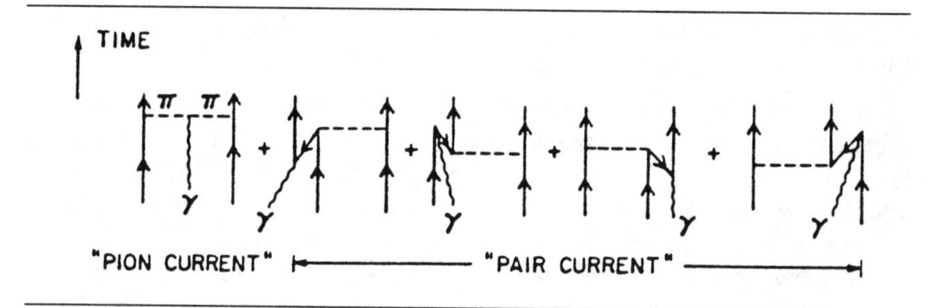

FIGURE 13.12
Time-ordered Feynman diagrams retained in the one-pion-exchange current calculation in Ref. [19].

nucleons. Although many exchange current calculations exist, we focus on those of Dubach, Koch, and Donnelly (DDK) for concreteness [19]. These authors keep the static limit (leading terms in $1/M$) of the time-ordered Feynman diagrams shown in Fig. 13.12. Each of these processes clearly represents an additional contribution to the current. If the two-nucleon potential is of the form $V = V_{\text{neutral}} + V_{\text{OPEP}}$ where the first term arises from neutral meson exchange and the second is the one-pion-exchange potential, then the electromagnetic current of DDK is conserved. Furthermore,

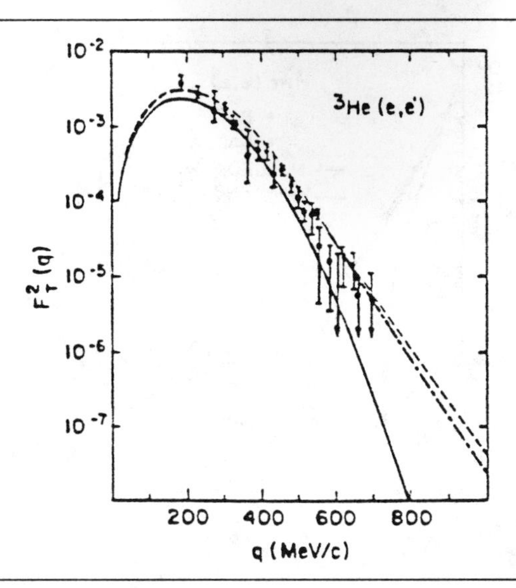

FIGURE 13.13
Elastic transverse form factor for ^{3}He(e,e) with (dashed) and without (solid) one-pion-exchange currents [19].

the threshold pion-electroproduction parts of the graphs satisfy the Kroll-Ruderman Theorem. The approach of DDK thus yields the correct form of the longest-range part of the exchange current. In this approach, the charge density operator is unmodified, and only the current density $\vec{J}(\vec{x})$ receives exchange current contributions, which are pure isovectors.

Assume that ^{3}He can be described by a $(1s_{1/2})^{-1}$ harmonic oscillator shell-model configuration. The magnetic moment calculated with the inclusion of exchange currents $\mu = -2.078$ nm is now closer to the experimental value $\mu = -2.127$ nm than is the Schmidt value $\mu = -1.913$ nm, indicating we are in the right ballpark. The effect on the elastic magnetic electron scattering form factor at intermediate q is shown in Fig. 13.13. Here the oscillator parameter has been determined from elastic charge scattering. This figure illustrates *the marginal role of exchange currents in the traditional nuclear physics domain.*

Fig. 13.14 taken from Ref. [20] illustrates the state-of-the-art with elastic magnetic electron scattering from ^{3}He. The dashed line shows the result obtained from the best three-body calculation done in the traditional approach (the three-body wave function is obtained by solving the Faddeev equations with potentials fit to the two-body data). The best three-body

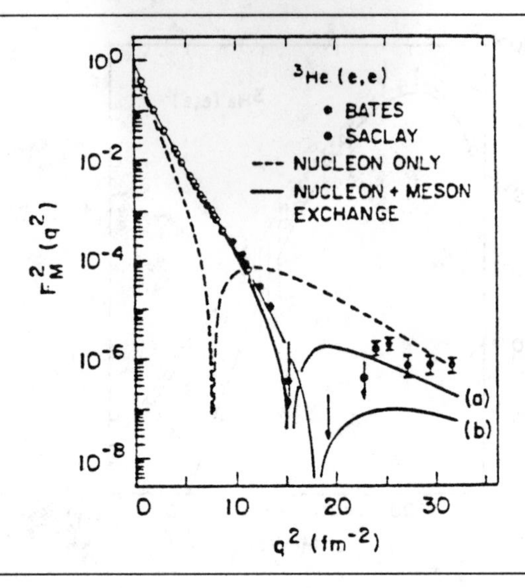

FIGURE 13.14
Elastic magnetic form factor for ^{3}He(e,e) out to high q^2 [20]. Two exchange current theories are shown; (a) from Ref. [22] and (b) from Ref. [21].

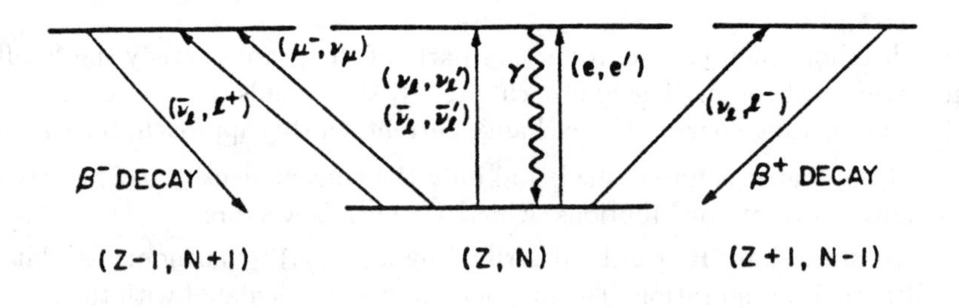

FIGURE 13.15
Semileptonic electroweak processes with nuclei.

calculation in the traditional picture clearly fails at high q^2. Also shown in Fig. 13.14 are two meson-exchange-current calculations [21,22], which include the pion exchange current discussed above, as well as other contributions. The difference between these two curves is a good measure of the present theoretical uncertainty. Note that the exchange current contribution is now a *dominant effect* at large q^2. This application *illustrates the need for QHD.*

These results also graphically illustrate that the appropriate set of degrees of freedom depends on the distance scale at which we probe the system.

13.3 Semileptonic Weak Interactions

Let us extend this discussion to the semileptonic weak interactions (Fig. 13.15). At energies small compared to the masses of the W and Z weak intermediate bosons ($\sim$ 100 GeV), semileptonic interactions in the Standard Model can be described by the following effective contact lagrangian densities

$$\mathcal{L}_{\text{eff}}^{(\pm)} = \frac{iG}{\sqrt{2}}\Big\{ [\bar{e}\gamma_\lambda(1+\gamma_5)\nu_e + (e \rightleftharpoons \mu)]\mathcal{J}_\lambda^{(+)} + [\bar{\nu}_e\gamma_\lambda(1+\gamma_5)e + (e \rightleftharpoons \mu)]\mathcal{J}_\lambda^{(-)}\Big\} , \tag{13.12}$$

$$\mathcal{L}_{\text{eff}}^{(\nu)} = \frac{iG}{\sqrt{2}}[\bar{\nu}_e\gamma_\lambda(1+\lambda_5)\nu_e + (e \rightleftharpoons \mu)]\mathcal{J}_\lambda^{(0)} , \tag{13.13}$$

$$\mathcal{L}_{\text{eff}}^{(e)} = \frac{-iG}{\sqrt{2}}[\bar{e}\gamma_\lambda(1+\gamma_5)e - 4\sin^2\Theta_W \bar{e}\gamma_\lambda e + (e \rightleftharpoons \mu)]\mathcal{J}_\lambda^{(0)} . \tag{13.14}$$

The first expression describes charge-changing neutrino processes, the second describes neutrino scattering, and the last expression describes the weak-neutral-current interaction of charged leptons. The Fermi constant determined from μ-decay is[1]

$$G = \frac{1.02 \times 10^{-5}}{M^2} . \tag{13.15}$$

The hadronic current densities entering into these lagrangians have the following general structure in the Standard Model:

$$\mathcal{J}_\mu = J_\mu + J_{\mu 5} , \qquad \text{V-A} \tag{13.16}$$

$$\mathcal{J}_\mu^{(\pm)} = \mathcal{J}_\mu^{V_1} \pm i\mathcal{J}_\mu^{V_2} , \qquad \text{Isovectors} \tag{13.17}$$

$$J_\mu^\gamma = J_\mu^S + J_\mu^{V_3} , \qquad \text{E. M. Current} \tag{13.18}$$

$$J_\mu^{(\pm)} = J_\mu^{V_1} \pm iJ_\mu^{V_2} , \qquad \text{CVC} \tag{13.19}$$

$$\mathcal{J}_\mu^{(0)} = \mathcal{J}_\mu^{V_3} - 2\sin^2\Theta_W J_\mu^\gamma . \qquad \text{WSG Standard Model} \tag{13.20}$$

[1] Here we assume $G^{(\pm)} = G\cos\Theta_c \cong G$ (see later)

The first relation says that the weak hadronic current has both Lorentz vector and axial-vector parts. The second exhibits the isovector nature of the weak charge-raising and -lowering hadronic currents. The third exhibits the isospin structure of the electromagnetic current, and the fourth is the statement of the conserved vector current theory (CVC) [23]. It states that the vector part of the charge-raising and -lowering currents forms an isovector triplet with the isovector part of the electromagnetic current. In consequence, the matrix elements of these operators can be related through the Wigner-Eckart theorem applied to isospin. The physical content of this relation is that the electromagnetic and weak interactions couple to different isospin components of the same current! One can make a strong argument that CVC provided the original insight into the unification of the electroweak interactions. The final relation is the form of the weak neutral current in the Standard Model. The third isovector component of the charge-changing current in Eq. (13.17) is mixed with the electromagnetic current of Eq. (13.18) through $\sin^2 \Theta_W$ where Θ_W is the Weinberg angle.

Consider next the topic of parity violation in electron scattering. The contribution of the weak neutral current interaction of the electron in Eq. (13.14) is completely masked by electromagnetic interactions, unless one looks at an effect that necessarily involves the weak interaction. Parity violation, which is forbidden to all orders in the electromagnetic interaction, is such an effect. Consider the difference in cross section for the scattering of right-and left-handed longitudinally polarized electrons

(13.21)

which can only arise from parity violation. We calculate the difference in cross section arising from the interference of one-photon and Z^0 exchange (Eq. (13.14))

γ + Z^0

(13.22)

An analysis in terms of response tensors leads to the following general result [1]:

$$\left[\frac{d\sigma_\uparrow - d\sigma_\downarrow}{d\sigma_\uparrow + d\sigma_\downarrow}\right] \cdot \left[W_2^\gamma \cos^2\frac{\Theta}{2} + 2W_1^\gamma \sin^2\frac{\Theta}{2}\right] = \frac{Gq^2}{4\pi\alpha\sqrt{2}}$$
$$\times\left\{b\left[W_2^{\rm int}\cos^2\frac{\Theta}{2} + 2W_1^{\rm int}\sin^2\frac{\Theta}{2}\right]\right.$$
$$\left. -a\left[\left(\frac{2W_8^{\rm int}}{M_T}\right)\sin\frac{\Theta}{2}\left(q^2\cos^2\frac{\Theta}{2} + \vec{q}^{\,2}\sin^2\frac{\Theta}{2}\right)^{\frac{1}{2}}\right]\right\}. \quad (13.23)$$

This expression assumes Lorentz invariance and a V-A coupling; in the Standard Model the lepton couplings have the following form:

$$a = -(1 - 4\sin^2\Theta_W) \cong 0\,,$$
$$b = -1\,. \quad (13.24)$$

We also assume that the hadronic states have definite parity. (Parity admixtures in the nuclear states give a very different q^2 dependence, and except perhaps at very low q^2, the above interference term dominates [24]). The response surfaces in this expression arise from the following combination of currents:

$$W_{1,2}^{\rm int} \quad \text{from} \quad (J^{(0)}J^\gamma + J^\gamma J^{(0)})\,,$$
$$W_8^{\rm int} \quad \text{from} \quad (J_5^{(0)}J^\gamma + J^\gamma J_5^{(0)})\,. \quad (13.25)$$

Note that Eq. (13.23) holds at all q^2 and $q \cdot p$.

A multipole analysis of all the semileptonic processes can again be carried out just as in electron scattering. For example, the ERL cross section for neutrino (antineutrino) induced semileptonic weak processes takes the form [3]

$$\left(\frac{d\sigma}{d\Omega}\right)^{\rm ERL}_{\nu \atop \bar\nu} = \frac{rG^2\epsilon_2^2\cos^2\frac{\Theta}{2}}{2\pi^2}\frac{4\pi}{2J_i+1}\left\{\sum_{J=0}^{\infty}|\langle J_f\|\widehat{\mathcal{M}}_J(q) - \frac{q_0}{q}\hat{\mathcal{L}}_J(q)\|J_i\rangle|^2\right.$$
$$+\left(\frac{q^2}{2\vec{q}^{\,2}} + \tan^2\frac{\Theta}{2}\right)\sum_{J=1}^{\infty}\left(|\langle J_f\|\hat{\mathcal{T}}_J^{\rm mag}(q)\|J_i\rangle|^2 + |\langle J_f\|\hat{\mathcal{T}}_J^{\rm el}(q)\|J_i\rangle|^2\right)$$
$$\left.\mp\tan\frac{\Theta}{2}\left(\frac{q^2}{\vec{q}^{\,2}} + \tan^2\frac{\Theta}{2}\right)^{\frac{1}{2}}\sum_{J=1}^{\infty}\left(2{\rm Re}\langle J_f\|\hat{\mathcal{T}}_J^{\rm mag}(q)\|J_i\rangle\langle J_f\|\hat{\mathcal{T}}_J^{\rm el}(q)\|J_i\rangle^*\right)\right\}.$$
$$(13.26)$$

The multipole operators appearing in this expression are computed with the weak current $\mathcal{J}_\mu(x)$; hence each is a sum of two terms with opposite parities.

Let us use the previous analysis to derive some relations that are *independent of the details of nuclear structure*. They serve as tests of the Standard Model in the strong interaction, confinement regime appropriate to nuclear physics. The basic idea is to use the nucleus as a filter to isolate various pieces of the current. In particular, we shall use the isospin selection rules.

Consider $T = 0 \rightarrow T = 1$ transitions. In this case only the isovector parts of the currents in Eqs. (13.16-13.20) can contribute, and the effective currents are given by

$$\mathcal{J}_\mu^{(0)} \doteq J_\mu^{V_3}\left(1 - 2\sin^2\Theta_W\right) + J_{\mu 5}^{V_3}\,, \tag{13.27}$$

$$J_\mu^\gamma \doteq J_\mu^{V_3}\,. \tag{13.28}$$

With the aid of the general analysis in terms of response surfaces one can then derive relations between semileptonic electroweak cross sections [2,25]. (Such relations were anticipated by Weinberg [4].) For example,

$$\frac{d\sigma_{\nu_l\nu_l'} - d\sigma_{\bar\nu_l\bar\nu_l'}}{\left[d\sigma_{\nu_l l^-} - d\sigma_{\bar\nu_l l^+}\right]^{\mathrm{ERL}}} = \tfrac{1}{2}(1 - 2\sin^2\Theta_W)\,. \tag{13.29}$$

The right hand side of this ratio of neutrino cross sections must be a constant depending only on the Weinberg angle. A better test of the unification of the electroweak interactions, which also involves the electron scattering cross section, is the relation

$$\tfrac{1}{2}\left[d\sigma_{\nu_l l^-} + d\sigma_{\bar\nu_l l^+}\right]^{\mathrm{ERL}} - \left[d\sigma_{\nu_l\nu_l'} + d\sigma_{\bar\nu_l\bar\nu_l'}\right] = \frac{G^2q^4}{4\pi^2\alpha^2}(\sin 2\Theta_W)^2 d\sigma_{ee'}^{\mathrm{ERL}}\,. \tag{13.30}$$

Consider next $T = 0 \rightarrow T = 0$ transitions. In this case there can only be isoscalar contributions, and the effective currents satisfy the relation

$$\mathcal{J}_\mu^{(0)} \doteq -2\sin^2\Theta_W J_\mu^\gamma\,. \tag{13.31}$$

This leads to a direct proportionality between the cross sections for neutrino and electron scattering

$$d\sigma_{\substack{\nu_l\nu_l' \\ \bar\nu_l\bar\nu_l'}} = \sin^4\Theta_W \frac{G^2q^4}{2\pi^2\alpha^2} d\sigma_{ee'}^{\mathrm{ERL}}\,. \tag{13.32}$$

Now these are truly *remarkable results*! They are independent of nuclear structure, and they hold for all q^2 (and $q \cdot p$). Thus they hold at wavelengths where only the gross features of nuclear structure are important, down to wavelengths where the baryon/meson degrees of freedom are critical, and on down to distances where the quark/gluon degrees of freedom manifest themselves. They are based on a minimal set of assumptions: Lorentz invariance, the structure of the currents in Eqs. (13.16-13.20), isotopic spin invariance in the nuclear system, and the one-photon exchange analysis of (e, e').

Consider Eq. (13.32). It states that one should be able to lay the cross sections (appropriately scaled) on top of each other. They should be identical! And this holds at all q^2, and hence through all of the diffraction structure (seen, for example, in Fig. 13.4)! No matter how complicated an exchange current contribution one comes up with in electron scattering, for example, exactly the same contribution must be present in neutrino scattering through the weak neutral current. This, to me, provides *a true test of the unification of the electroweak interactions.*

Let us turn our attention to an example of parity violation. Consider elastic scattering from a 0^+ nucleus. In this case there can only be a charge monopole form factor, and the parity violation asymmetry $\mathcal{A} \equiv (d\sigma_\uparrow - d\sigma_\downarrow)/(d\sigma_\uparrow + d\sigma_\downarrow)$ in Eq. (13.23) takes the form [1]

$$\mathcal{A} = \frac{Gq^2}{2\pi\alpha\sqrt{2}} b \left[\mathrm{Re} \frac{F^{(0)}(q^2)}{F^{(\gamma)}(q^2)} \right] . \tag{13.33}$$

This relation assumes only Lorentz invariance and a V-A structure of the weak neutral current (WNC).[2] Now assume further that the ground state has isospin T=0 and use the result in Eq. (13.31). The electron scattering and WNC form factors must then be *identical*,

$$F^{(0)}(q^2) = -2\sin^2\Theta_W F^{(\gamma)}(q^2) . \tag{13.34}$$

This is again a truly *remarkable result*! All the comments made in the discussion above again apply. Eq. (13.34) is a relation which must hold between the charge and WNC form factors at all q^2. Substitution of Eq. (13.34) into Eq. (13.33) leads to

$$\mathcal{A} = \frac{Gq^2}{\pi\alpha\sqrt{2}} \sin^2\Theta_W . \tag{13.35}$$

This result is originally due to Feinberg [26]. An experiment to measure $\mathcal{A}$ for ^{12}C is underway at Bates. To me, the most important aspect to test is the linear dependence on q^2 in Eq. (13.35); we really want to test Eq. (13.34) in detail at all q^2. This again provides a true test of the unification of the electroweak interactions. Note that the parity-violation experiment, as hard as it is, is easier than a test of Eq. (13.32) which involves the cross section for neutrino scattering.

[2]There is no contribution from parity admixtures in the ground state in this case with one photon exchange.

TABLE 13.2
Quark quantum numbers.

Field/Particle	T	T_3	Q	B	S	C	$Y = B + S + C$
u	$\frac{1}{2}$	$\frac{1}{2}$	$\frac{2}{3}$	$\frac{1}{3}$	0	0	$\frac{1}{3}$
d	$\frac{1}{2}$	$-\frac{1}{2}$	$-\frac{1}{3}$	$\frac{1}{3}$	0	0	$\frac{1}{3}$
s	0	0	$-\frac{1}{3}$	$\frac{1}{3}$	-1	0	$-\frac{2}{3}$
c	0	0	$\frac{2}{3}$	$\frac{1}{3}$	0	1	$\frac{4}{3}$

13.4 The Standard Model and QCD

In order to understand the simplicity of the above relations, we must examine the underlying structure of hadrons. There is now convincing evidence that hadrons are composed of a simpler substructure, quarks, the lightest four of which are assigned the quantum numbers shown in Table 13.2. Since the existence of quarks is *inferred*, it is useful to review the basic arguments for their existence.

• If one assumes that the baryons are composed of three quarks (qqq) and the mesons of a quark-antiquark pair $(\bar{q}q)$, and if the quarks are assigned the quantum numbers in Table 13.2, then one obtains a concise description of the observed supermultiplets of hadrons, and predictions for new ones.

• If one assumes that the electroweak currents are constructed from point-like, Dirac quark fields with the quantum numbers in Table 13.2, then a marvelously simple, accurate, and predictive description of these currents is obtained. The electromagnetic current, for example, is simply given by

$$J_\mu^\gamma = i\left[\tfrac{2}{3}\left(\bar{u}\gamma_\mu u + \bar{c}\gamma_\mu c\right) - \tfrac{1}{3}\left(\bar{d}\gamma_\mu d + \bar{s}\gamma_\mu s\right)\right] \, . \tag{13.36}$$

The charge-raising weak current is given by

$$\begin{aligned}\mathcal{J}_\mu^{(+)} = {} & i\bar{u}\gamma_\mu(1+\gamma_5)(d\cos\Theta_c + s\sin\Theta_c) \\ & + i\bar{c}\gamma_\mu(1+\gamma_5)(-d\sin\Theta_c + s\cos\Theta_c) \, .\end{aligned} \tag{13.37}$$

The one complexity here is the occurrence of the Cabbibo angle Θ_c, dictated by the empirically observed strangeness-changing weak interactions, which slightly mixes the quark fields appearing in this expression. It disappears from the WNC in the Standard Model, which is given by

$$\mathcal{J}_{\mu}^{(0)} = \frac{i}{2}\big[\bar{u}\gamma_{\mu}(1+\gamma_5)u + \bar{c}\gamma_{\mu}(1+\gamma_5)c - \bar{d}\gamma_{\mu}(1+\gamma_5)d - \bar{s}\gamma_{\mu}(1+\gamma_5)s\big] - 2\sin^2\Theta_W J_{\mu}^{\gamma} \ . \qquad (13.38)$$

Indeed, it was the argument of Glashow *et al.* [6] that there should be no strangeness-changing WNC that led them to predict the existence of the c quark— a prediction later fully confirmed by experiment.

We shall define the *nuclear domain* by truncating the Hilbert space so that it contains only u and d quarks. We can still have any number of pairs of these quarks present (and hence any number of mesons) so that the states can still be very complicated. The quark field can then be written in the form of an isodoublet

$$\psi \doteq \begin{pmatrix} u \\ d \end{pmatrix} \qquad ;\text{nuclear domain} \ . \qquad (13.39)$$

We further redefine the charge-changing Fermi constant by[3]

$$G^{(\pm)} \equiv G\cos\Theta_c \ . \qquad (13.40)$$

The currents in Eqs. (13.36–13.38) can then be recast in the form

$$J_{\mu}^{\gamma} = i\big[\bar{\psi}\gamma_{\mu}\tfrac{1}{2}\tau_3\psi + \tfrac{1}{6}\bar{\psi}\gamma_{\mu}\psi\big] \ , \qquad (13.41)$$

$$\mathcal{J}_{\mu}^{(\pm)} = i\bar{\psi}\gamma_{\mu}(1+\gamma_5)\tau_{\pm}\psi \ , \qquad (13.42)$$

$$\mathcal{J}_{\mu}^{(0)} = i\bar{\psi}\gamma_{\mu}(1+\gamma_5)\tfrac{1}{2}\tau_3\psi - 2\,\sin^2\Theta_W J_{\mu}^{\gamma} \ . \qquad (13.43)$$

These relations imply the previously assumed structure of the currents in Eqs. (13.16-13.20)!

We note that there is now a correction term to the WNC due to the $(\bar{s}s)$ and $(\bar{c}c)$ pairs in the nucleus

$$\delta\mathcal{J}_{\mu}^{(0)} \equiv \mathcal{J}_{\mu}^{S} = \frac{i}{2}\big[\bar{c}\gamma_{\mu}(1+\gamma_5)c - \bar{s}\gamma_{\mu}(1+\gamma_5)s\big] \ . \qquad (13.44)$$

The contribution of this term, which takes us outside of the "nuclear domain," is expected to be small, but it is very difficult to estimate quantitatively.

• Dynamic evidence for the existence of a pointlike substructure in the nucleon is obtained from deep inelastic electron scattering (e, e'). The situation is illustrated qualitatively in Fig. 13.16, where we sketch the two-dimensional response surface νW_2 against q^2 and $1/x \equiv 2M\nu/q^2$. For elastic scattering[4] or inelastic scattering to discrete levels, there is a form factor

[3]The factor of $\cos\Theta_c$ can easily be incorporated in the previous Eqs. (13.29) and (13.30).

[4]For elastic scattering from the nucleon $x = 1$.

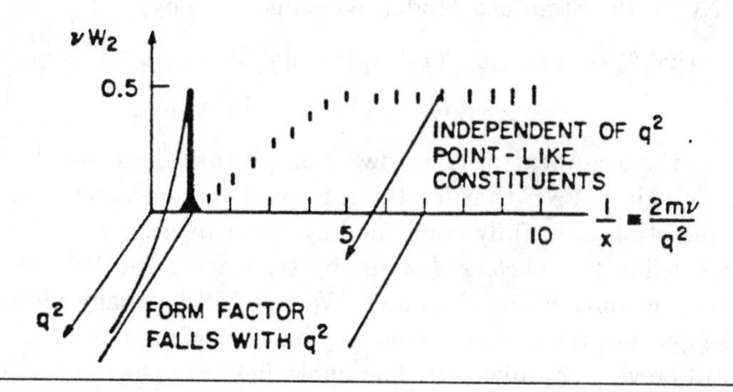

FIGURE 13.16
Qualitative sketch of SLAC results on deep inelastic electron scattering (e, e') from the nucleon.

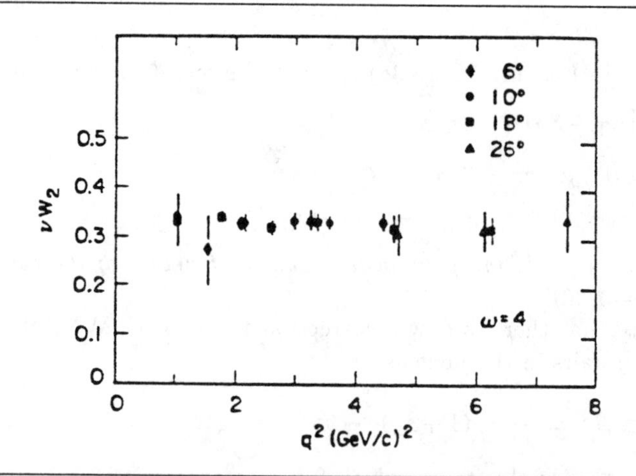

FIGURE 13.17
Experimental demonstration that the response surface νW_2 for the proton is independent of q^2 at fixed $\omega = 1/x = 2m\nu/q^2$ in deep inelastic scattering. Taken from [27].

which falls off with q^2. We have seen many examples of this in the nuclear case. At large values of

$$\omega \equiv \frac{1}{x} = \frac{2M\nu}{q^2}\,, \tag{13.45}$$

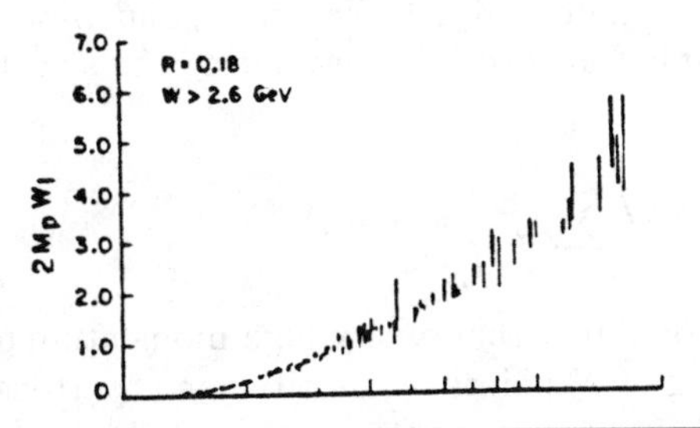

FIGURE 13.18
Response surface $2mW_1$ for the proton as a function of the Bjorken scaling variable $\omega = 1/x$ [27].

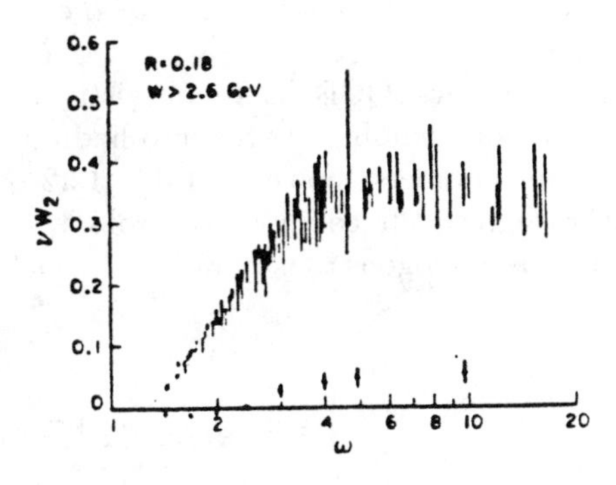

FIGURE 13.19
Same as Fig. 13.18 for νW_2 [27]. (The modification of this function in ^{56}Fe observed by the EMC at CERN is shown at the bottom of this figure.)

(*i.e.*, in the deep inelastic region) the cross section is observed to be *independent of* q^2. This is graphically illustrated by the SLAC data shown in Fig. 13.17 [27]. The absence of a form factor here indicates that one is scattering from pointlike objects. The statement of *Bjorken scaling* is that the two-dimensional response surface becomes a function of the single variable in this region

$$\nu W_2(\nu, q^2) \longrightarrow F_2(x) \; . \tag{13.46}$$

The structure functions for the proton determined by the SLAC experiments are shown in Figs. 13.18 and 13.19 [27].

The quark-parton model (due to Feynman, and Bjorken and Paschos [28]) provides a simple framework for evaluating the structure function in the scaling region

$$F_2(x) = \sum_N P(N) \left\langle \sum_i Q_i^2 \right\rangle_N x f_N(x) \ . \tag{13.47}$$

Here $P(N)$ is the probability that a very high momentum proton will have a structure consisting of N pointlike constituents ("partons"). The second factor is the sum of the squares of the charges of the partons in this configuration. The quantity $f_N(x)$ is the probability that a parton will carry a fraction x of the longitudinal momentum of the nucleon in this frame.

Also sketched in Fig. 13.19, for orientation, is the magnitude of the EMC effect (see *e.g.*, [29]). It represents the difference in cross section per nucleon for $^{56}\text{Fe}(\mu, \mu')$ and $^{2}\text{H}(\mu, \mu')$. *The EMC effect is a clear and unambiguous demonstration of the modification of the quark structure of nucleons in the nucleus.*

Quantum chromodynamics (QCD) is a relativistic quantum field theory of the strong interactions binding quarks into hadrons [7]. In addition to the "flavor" quantum numbers shown in Table 13.2 quarks are given an additional intrinsic degree of freedom called "color", which takes three values $i = R, G, B$. This is analogous to isospin for the nucleons. The quark field then becomes

$$\psi = \begin{pmatrix} u \\ d \\ s \\ c \end{pmatrix} \longrightarrow \begin{pmatrix} u_R & u_G & u_B \\ d_R & d_G & d_B \\ s_R & s_G & s_B \\ c_R & c_G & c_B \end{pmatrix} \equiv (\psi_R, \psi_G, \psi_B) \equiv \psi_i \ , \tag{13.48}$$

where $i = R,\ G,\ B$. We introduce the three-component fields

$$\underline{\psi} \equiv \begin{pmatrix} \psi_R \\ \psi_G \\ \psi_B \end{pmatrix} \ . \tag{13.49}$$

This is actually a very compact notation; each field has many flavors, and each flavor is a 4-component Dirac field. The *electroweak currents are assumed to be independent of color.* They are written for example, as

$$\underline{\bar{\psi}}\gamma_\mu \underline{\psi} = \bar{\psi}_R \gamma_\mu \psi_R + \bar{\psi}_G \gamma_\mu \psi_G + \bar{\psi}_B \gamma_\mu \psi_B \ . \tag{13.50}$$

Thus each color field has identical electroweak couplings as indicated in Fig. 13.20.

QCD is a local gauge theory built on color. It is invariant under local unitary transformations on the three-component field in Eq. (13.49), and hence the theory possesses the local symmetry $SU(3)_c$. It is a non-

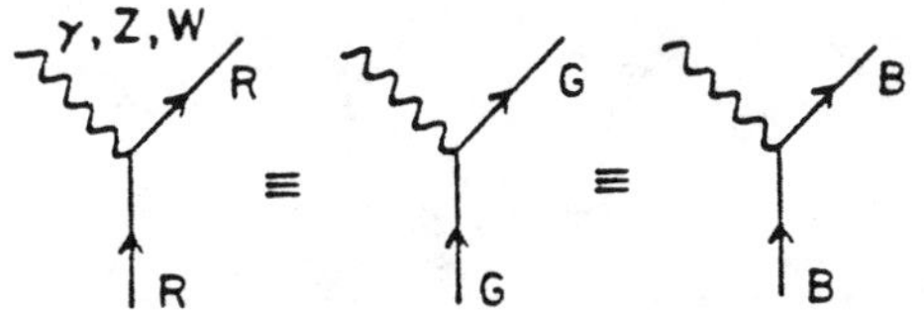

FIGURE 13.20
Weak and electromagnetic quark couplings are independent of color.

abelian Yang-Mills theory. To construct such a theory one first introduces 8 massless gauge boson fields, the *gluon* fields, $A_\mu^a(x)$ with $a = 1, 2, \ldots, 8$; there is one for each generator. (These are the analogues of the photon field $A_\mu(x)$ in the abelian theory QED). The use of the covariant derivative (the analogue of $(p - eA)$ in QED) then leads to the QCD lagrangian

$$\mathcal{L}_{\text{QCD}} = -\tfrac{1}{4} F_{\mu\nu}^a F_{\mu\nu}^a - \bar{\underline{\psi}} \gamma_\mu \left(\frac{\partial}{\partial x_\mu} - \frac{i}{2} g \underline{\lambda}^a A_\mu^a(x) \right) \underline{\psi} \,. \tag{13.51}$$

The SU(3) matrices $\underline{\lambda}^a$ appearing in the covariant derivative satisfy the algebra of the generators

$$[\tfrac{1}{2}\underline{\lambda}^a, \tfrac{1}{2}\underline{\lambda}^b] = i\, f^{abc} \tfrac{1}{2}\underline{\lambda}^c \,. \tag{13.52}$$

The field tensor appearing in Eq. (13.51) must be more complicated than the Maxwell tensor of QED to maintain local gauge invariance. It is given by

$$F_{\mu\nu}^a = \frac{\partial}{\partial x_\mu} A_\nu^a - \frac{\partial}{\partial x_\nu} A_\mu^a + g f^{abc} A_\mu^b A_\nu^c \,. \tag{13.53}$$

The cubic and quartic cross terms in the square of this quantity imply that the QCD lagrangian is intrinsically non-linear in the gluon fields. (Note that there are induced non-linearities in QED coming from electron loops; here, however, the lagrangian itself is intrinsically non-linear.)

Since $\mathcal{L}_{\text{QCD}}$ is written in terms of the $\underline{\psi}$ in Eq. (13.49), the strong color interactions are independent of flavor as illustrated in Fig. 13.21 (just multiply out the matrix product using the definition in Eq. (13.48)). Equation (13.51) has been written for massless quarks, but a mass term of the form

$$\delta\mathcal{L}_{\text{mass}} = -\, \bar{\underline{\psi}}\, \underline{M}\, \underline{\psi} \,, \tag{13.54}$$

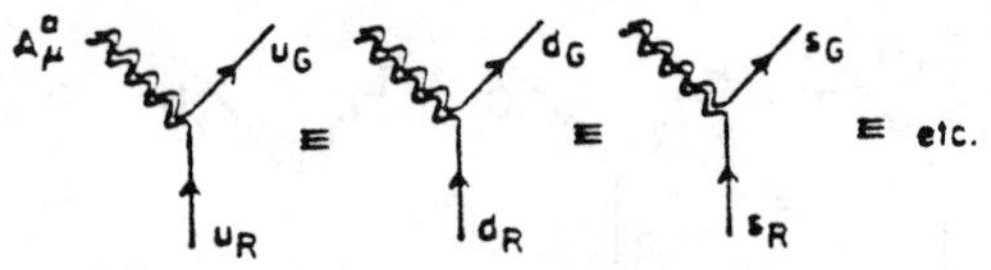

FIGURE 13.21
Strong color interactions of the quarks are independent of flavor.

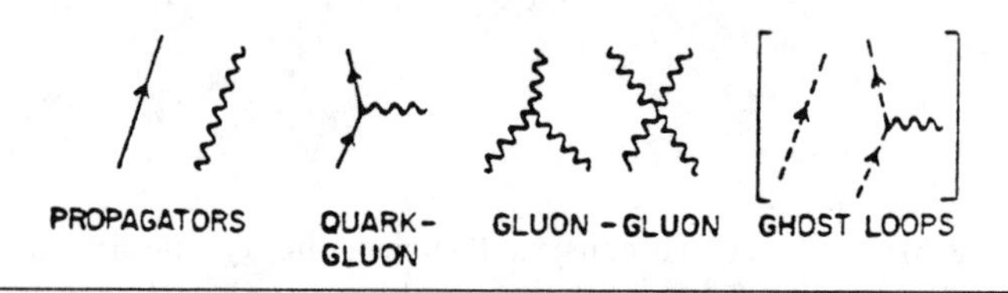

FIGURE 13.22
Components of Feynman rules for QCD.

where $\underline{M} = \begin{pmatrix} \underline{M} & & \\ & \underline{M} & \\ & & \underline{M} \end{pmatrix}$ is the identity matrix with respect to color which evidently retains local $SU(3)_c$ invariance.

The Feynman rules for QCD may now be derived, and they have the components illustrated in Fig. 13.22.[5] The ghost loops in the gluon amplitudes are a technical complexity; they are there to ensure that one generates the correct Lorentz invariant, gauge invariant, unitary S-matrix in this non-abelian gauge theory.

QCD has two remarkable properties [7]:

- **Asymptotic Freedom** At large momenta (small distances) the theory is essentially free. Due to the non-linear gluon couplings the charge is antishielded, and the renormalized coupling constant measured at low momentum transfers is *larger* than the effective coupling constant seen at high momenta, or short distances. This is in contrast to the situation in QED where the pairs arising from vacuum polarization shield the bare charge. When the effective coupling is small, one can do perturbation theory.

[5]They are given in detail, for example, in [30].

FIGURE 13.23
Model of the nucleon.

□ **Confinement** Quarks and color are confined to the interior of hadrons. We never see the underlying degrees of freedom in QCD as asymptotic free states in the laboratory! There is strong evidence from lattice gauge theory calculations, where QCD is solved as on a finite space-time lattice, that confinement is indeed a dynamic property of QCD. Confinement is a manifestation of the strong-coupling aspect of QCD.

It is important to note that *in the Standard Model of the strong, electromagnetic, and weak interactions based on* $SU(3)_c \otimes SU(2)_W \otimes U(1)_W$, *the gluons have no weak or electromagnetic interactions.*

Let us combine these results to get a simple expression for the parity violation asymmetry parameter measured in deep inelastic electron scattering from the deuteron $^2\text{H}(e,e')$. We use the model of the nucleon shown in Fig. 13.23— we work in the nuclear domain and assume quark configurations $p = (uud)$ and $n = (udd)$; the states may have any number of additional gluons. The general expression for $\mathcal{A}$ in Eq. (13.23) may now be employed, with $\Theta \to 0$ as in the SLAC experiments. For simplicity, it will be assumed that $\sin^2\Theta_W = \frac{1}{4}$ so that $a = 0$. The quark-parton model in Eq. (13.47) then provides a simple expression for the required ratio of response functions

$$\left(\frac{\nu W_2^{int}}{\nu W_2^{\gamma}}\right)_{^2H} = \frac{\left\langle 2\sum_i Q_i^{\gamma} Q_i^{(0)}\right\rangle^n_{3-\text{quark}} + (n \rightleftharpoons p)}{\left\langle \sum_i (Q_i^{\gamma})^2\right\rangle^n_{3-\text{quark}} + (n \rightleftharpoons p)} = \frac{4}{5} \,. \tag{13.55}$$

The n and p response surfaces are added incoherently since one is adding cross sections. Everything else cancels in this ratio, and only the indicated ratios of charges remain. They are immediately read off from Eq. (13.38) and Table 13.2; the result in Eq. (13.55) is just $\frac{4}{5}$. The asymmetry is then immediately given by the simple expression [1]

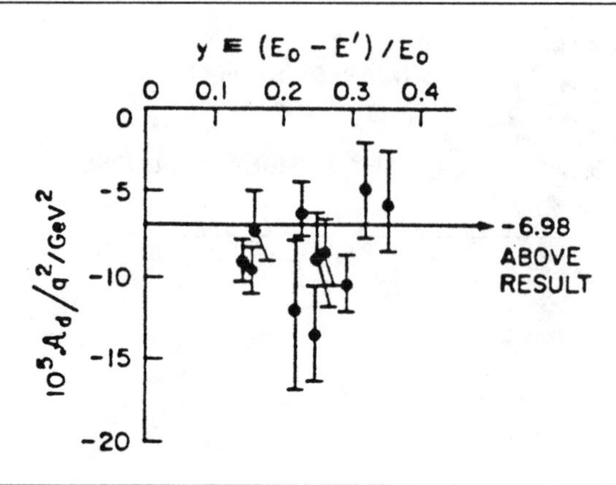

FIGURE 13.24
Result in Eq. (13.56) compared with SLAC data for parity-violation asymmetry in deep inelastic (e, e') from ^{2}H [31].

$$\mathcal{A}_d = \frac{-Gq^2}{4\pi\alpha\sqrt{2}} \cdot \frac{4}{5} \, . \tag{13.56}$$

This result is compared with the SLAC data in Fig. 13.24 [31]. The agreement is all that one could ask.

13.5 Picture of the Nucleus and Outlook

On the basis of this discussion, we arrive at the picture of the nucleus illustrated in Fig. 13.25. We proceed to make some observations concerning this figure.

- The structure of confinement in a many-baryon system is an unsolved problem. QCD is simple at short distances, or high momenta. QCD is a complicated, strong-coupling theory at large distances— at the boundary of the confinement region. If QCD is to be a correct theory it must reproduce *nuclear physics*; in particular it must reproduce meson exchanges, baryon dynamics in nuclei, and meson dynamics in nuclei.
- The electroweak interactions couple to the quarks. The electroweak interactions see through the hadronic structure to the interior quark structure. The gluons providing the confinement are *absolutely neutral* to the electroweak interactions. The electroweak interactions are *colorblind*.

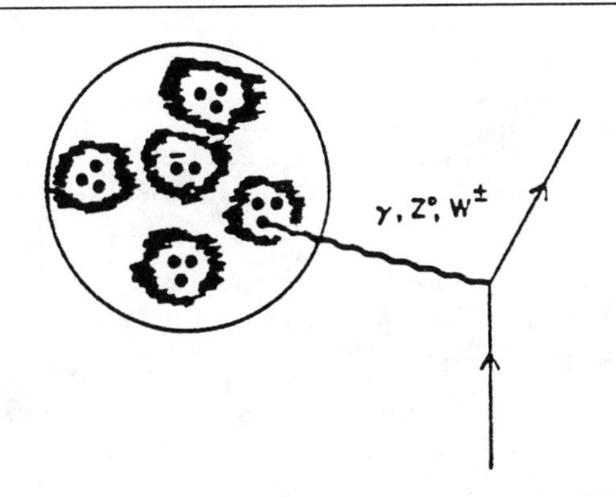

FIGURE 13.25
Picture of the nucleus in the Standard Model.

- We can now understand the simplicity of the previous relations, derived within the framework of the Standard Model, which are independant of the details of nuclear structure (Eqs. (13.29), (13.30), (13.32) and (13.35)). No matter how complicated the nuclear structure in terms of baryons (qqq) or mesons (($\bar{q}q$) pairs) the electroweak interactions *see through* the structure to the interior quark structure of the hadrons. The electroweak currents are simple in terms of quarks. The various semileptonic electroweak processes then just probe different components of these basic quark currents.

Let me spend the last five minutes talking about the Continuous Electron Beam Accelerator Facility (CEBAF). I have tried to crystallize my own thoughts on CEBAF.

CEBAF will provide the most precise accessible probe of matter. The interaction is known, one knows what is being measured. It is a unique time for nuclear physics. What we are really discussing is a tool and a capability for the next generation of nuclear scientists.

To me, CEBAF's scientific goal is to study the structure of the nuclear many-body system, its quark substructure, and the strong and electroweak interactions governing the behavior of this fundamental form of matter.

We have appointed a Program Advisory Committee (PAC) to give us guidance on the scientific program, (I give you the initial membership in Table 13.3). The way we proceeded in the first meeting was to establish a set of Steering Committees coming from the national users community. These were the natural outgrowth of the workshops and summer study programs that had been held over the past several years. The Steering

TABLE 13.3
CEBAF program advisory committee.

Function	Advise CEBAF Directorate on scientific directions and relative scientific priorities for the experimental program	
Membership	J. Schiffer (Argonne), Chairman	R. McKeown (Caltech)
	P. Barnes (Carnegie-Mellon)	E. Moniz (MIT)
	W. Bertozzi (MIT)	I. Sick (Basel)
	T. W. Donnelly (MIT)	H. Thiessen (LANL)
	R. Eisenstein (Illinois)	C. Williamson (MIT)—*ex officio*
	J. Friar (LANL)	S. Wojcicki (Stanford)
	S. Kowalski (MIT)	
First Meeting	February 13-15, 1987	Report completed

TABLE 13.4
Subject areas for CEBAF steering committees

$(e, e'$ nucleon)
$(e, e'K)$ and $(e, e'\pi)$
$(e, e'$ two nucleon)
$(e, e'$ multihadron)
$p(\vec{e}, e)p$ – parity
few nucleon systems
$(e, e'N^*)$
$(e, e'X)$ deep inelastic transition to x-scaling

Committees were centered on the possible experimental programs that can be carried out at CEBAF. These are listed in Table 13.4. CEBAF provides a continuous beam. It allows one to do coincidence experiments where one can look at the hadron produced in coincidence with the scattered electron, and study the process of hadronization. Listed are simply the particles that can be detected in coincidence: $(e, e'$ nucleon); $(e, e'K^+)$ and $(e, e'\pi)$, the first of which gives a tagged hyperon, and, in fact, with good resolution one can do precise hypernuclear spectroscopy; $(e, e'$ two nucleon) where hopefully, one can get at the short-range correlation of two nucleons, or at multiquark clustering, in the nuclear system. There is the parity-violation experiment. We are going to have a polarized beam. We are going to do that experiment. The PAC said our top priority should be the few-nucleon systems— both to look at the wave functions for the few-nucleon systems and also to try and extract the form factors of the nucleons. We will study the properties of the isobars through $N(e,e')N^*$. And we will hopefully push the kinematic regime in all these experiments out towards asymptopia, or at least asymptopia as seen at SLAC in the deep-inelastic scattering experiments.

The PAC has met a total of three times to provide advice and feedback on the letters of intent and advice on the scientific program and required equipment. A formal call for the first round of proposals for experiments at CEBAF has been issued, and they are due on October 31, 1989.

Our own institutional goal at CEBAF is to build a world-class user-friendly laboratory for nuclear physics research and graduate education, centered around a high-intensity 4 GeV CW electron accelerator.

I would like to close with two conclusions that help to define the future in electronuclear physics:

- The electroweak interactions see the quark structure of nuclei.
- Nuclear physics is the study of the strong-interaction aspects of QCD.

Bibliography

[1] J.D. Walecka, "Electron Scattering", Lectures given at Argonne National Laboratory, Argonne, Illinois, ANL-83-50 (1983); Continuous Electron Beam Accelerator Facility, Newport News, Virginia (1987).

[2] J.D. Walecka, "Electroweak Interactions with Nuclei", in *Proc. Conf. on Intersections Between Particle and Nuclear Physics*, ed. R. Mischke (Steamboat Springs, Colorado, May, 1984) A.I.P. Conf. Proc. No. **123** (A.I.P., N.Y.) p. 1.

[3] J.D. Walecka, "Semi-Leptonic Weak Interactions in Nuclei", in *Proc. Weak-Interactions-1977*, ed. D.B. Lichtenberg, A.I.P. Conf. Proc. No. **37** (A.I.P., N.Y.) p. 125.

[4] S. Weinberg, Phys. Rev. Lett. **19** (1967) 1264; Phys. Rev. **D5** (1972) 1412.

[5] A. Salam and J.C. Ward, Phys. Lett. **13** (1964) 168.

[6] S.L. Glashow *et al.*, Phys. Rev. **D2** (1970) 1285.

[7] F. Wilczek, Ann. Rev. of Nucl. and Part. Sci. **32** (1982) 177.

[8] A.L. Fetter and J.D. Walecka, *Quantum Theory of Many-Particle Systems* (McGraw-Hill, New York, 1971).

[9] T.W. Donnelly *et al.*, Phys. Lett. **B44** (1973) 330.

[10] I. Sick *et al.*, *Lect. Notes in Physics* **108** (Springer, Berlin, 1979).

[11] R.C. York and G.A. Peterson, Phys. Rev. **C19** (1979) 574.

[12] H. Zarek *et al.*, Phys. Rev. Lett. **38** (1977) 750.

[13] E. Bloom *et al.*, reported by W. K. Panofsky, in *Int. Conf. on High-Energy Physics*, Vienna 1968 (CERN, Geneva, 1968) p. 23.

[14] M. Breidenbach, Ph.D. Thesis, Massachusetts Institute of Technology, 1971 (unpublished).

[15] P.L. Pritchett *et al.*, Phys. Rev. **184** (1969) 1825.

[16] J. Calarco *et al.*, *Giant Multipole Resonances* (Harwood, 1980) p. 438.

[17] W. Kleppinger *et al.*, Ann. Phys. **146** (1983) 349.

[18] J. Mougey, Nucl. Phys. **A335** (1980) 35.

[19] J. Dubach *et al.*, Nucl. Phys. **A271** (1976) 279.

[20] J.M. Cavedon *et al.*, Phys. Rev. Lett. **49** (1982) 986.

[21] D. Riska, Nucl. Phys. **A350** (1980) 227.

[22] E. Hadjimichael *et al.*, Phys. Rev. **C27** (1983) 831.

[23] R.P. Feynman and M. Gell-Mann, Phys. Rev. **109** (1958) 193.

[24] B.D. Serot, Nucl. Phys. **A322** (1979) 408.

[25] J.D. Walecka, "Neutrino Interactions with Nuclei", in *Proc. Second LAMPF II Workshop*, ed. H.A. Thiessen *et al.*, LA-9572-C, Vol. II (Los Alamos National Laboratory, Los Alamos, New Mexico, 1982) p. 560.

[26] G. Feinberg, Phys. Rev. **D12** (1975) 3575.

[27] J.I. Friedman and H.W. Kendall, Ann. Rev. Nucl. Sci. **22** (1972) 203.

[28] J.D. Bjorken and E.A. Paschos, Phys. Rev. **185** (1969) 1975.

[29] R.L. Jaffe, Phys. Rev. Lett. **50** (1983) 228.

[30] B.D. Serot and J.D. Walecka, "The Relativistic Nuclear Many–Body Problem", *Advances in Nuclear Physics*, ed. J.W. Negele and E. Vogt, Vol. **16** (Plenum Press, New York, 1986).

[31] C.Y. Prescott *et al.*, Phys. Lett. **B77** (1978) 347;*ibid.* **B84** (1979) 524.

The Addison-Wesley **Advanced Book Program** would like to offer you the opportunity to learn about our new physics and scientific computing titles in advance. To be placed on our mailing list and receive pre-publication notices and special offers, just **fill out this card completely** and return to us, postage paid. Thank you.

Title and Author of this book: ____________________ **Date purchased:** __________

Name ____________________
Title ____________________
School/Company ____________________
Department ____________________
Street Address ____________________
City ____________________ State __________ Zip __________
Telephone/s(   ) ____________________ (   ) ____________________

Where did you buy/obtain this book?

☐ Bookstore
☐ Campus Bookstore
☐ Other ____________________
☐ Mail Order
☐ Toll Free # to Publisher
☐ School (Required for Class)
☐ Professional Meeting
☐ Publisher's Representative

What professional scientific and engineering associations are you an active member of?

☐ AAPT (Amer Assoc of Physics Teachers)
☐ AIP (Amer Institute of Physics)
☐ Other ____________________
☐ APS (Amer Physical Society)
☐ Sigma Pi Sigma
☐ SPS (Society of Physics Students)
☐ AAAS (Amer Assoc for the Advancement of Science)

Check your areas of interest.

(10) ✓ **Physics**

11 ☐ Quantum Mechanics
12 ☐ Particle/Astro Physics
13 ☐ Condensed Matter
14 ☐ Mathematical Physics
15 ☐ Nuclear Physics
16 ☐ Electron/Atomic Physics
17 ☐ Plasma/Fusion Physics
29 ☐ Other ____________________
18 ☐ Materials Science
19 ☐ Biological Physics
20 ☐ High Polymer Physics
21 ☐ Chemical Physics
22 ☐ Fluid Dynamics
23 ☐ History of Physics
24 ☐ Statistical Physics
25 ☐ Geophysics
26 ☐ Medical Physics
27 ☐ Optics
28 ☐ Vacuum Physics

Are you more interested in: ☐ **theory** ☐ **experimentation?**

Are you currently writing, or planning to write a textbook, research monograph, reference work, or create software in any of the above areas?

☐ Yes ☐ No
Area: ____________________

(If Yes) **Are you interested in discussing your project with us?**

☐ Yes ☐ No

Physics

fold and staple

No Postage
Necessary
if Mailed in the
United States

BUSINESS REPLY MAIL
FIRST CLASS PERMIT NO. 828 REDWOOD CITY, CA 94065

Postage will be paid by Addressee:

ADDISON-WESLEY
PUBLISHING COMPANY, INC.®
Advanced Book Program
390 Bridge Parkway, Suite 202
Redwood City, CA 94065-1522